Discrimination Law Issues for the Safety Professional

Occupational Safety and Health Guide Series

Series Editor

Thomas D. Schneid
Eastern Kentucky University
Richmond, Kentucky

Forthcoming Titles

The Comprehensive Handbook of School Safety
E. Scott Dunlap

Workplace Safety and Health: Assessing Current Practices and Promoting Change in the Profession
Thomas D. Schneid

Discrimination Law Issues for the Safety Professional

THOMAS D. SCHNEID

CRC Press
Taylor & Francis Group
Boca Raton   London   New York

CRC Press is an imprint of the
Taylor & Francis Group, an **informa** business

My thanks to my wife, Jani, and my daughters, Shelby, Madison, and Kasi, for permitting me the time to research and write this text.

Special thanks to my parents, Bob and Rosella, for instilling the importance of education in all of their children.

And thank you to Jane for her assistance in researching and writing this text.

Contents

Foreword

Discrimination. It is a very thought-provoking word when we hear it mentioned.

When Tom asked me to write a foreword for this very important text, I was honored to help play a part in the education of today's safety practitioners. Dr. Schneid and I have worked together since the late 1980s. We have authored several texts together, and this text on *Discrimination Law Issues for the Safety Professional* is a "must have" in today's challenging and changing times.

As you go through each chapter in the text, take the time to think carefully about how the effects of discrimination may have played a positive or negative role in your *own* professional life. Look at the case "Ricci v DeStafano" and look at Supreme Court Justice Anton Scalia's concurrence where he writes (in part) …

> Title VII's disparate-impact provisions "place a racial thumb on the scales, often requiring employers to evaluate the racial outcomes of their policies, and to make decisions based on (because of) those racial outcomes." "That type of racial decision making is, as the Court explains, discriminatory."

The key to operating in a safe *and* legal environment requires thought and a careful understanding of the laws, interpretations and policies that affect *everything* we do in safety and health.

Each of the chapters in Dr. Schneid's latest text has an important bearing on your work in the safety and health field.

The history is critically important, along with some far-reaching legislation including (but *not* limited to) Title VII, ADA, equal pay, and other chapters in the appendices.

Discrimination case law changes regularly, and you should always avail yourself of competent legal counsel in any situation. Do not think that what you may have learned years ago has not been altered by case law over the years.

An interesting study of each of the cases cited in the text will help to shape your learning in this field as well.

This text may be used in one of the many courses at Eastern Kentucky University taught by Dr. Schneid and his colleagues. It is also sure to be used as a reference on many university and corporate bookshelves across the nation.

Read, learn, and understand that the knowledge is in where to find accurate information, and always keep current with rules and regulations.

Take the time to make sure this text has a prominent place on your bookshelf, read it, highlight the pages, and ensure that it is a useful and often-used reference text.

Be safe yourself, and remember that your safety career crosses many disciplines. Learn every day as you take your journey in this profession.

Michael J. Fagel, PhD., CEM
MPPA Program
Northwestern University

Preface

Safety professionals work in an environment where they communicate, direct, and interact with a large number of employees and others on a daily basis. The safety professional's actions and inactions, decisions, guidance, gestures, and, most importantly, words directly reflect on themselves as well as their companies. Often without even realizing the repercussions of their actions, a safety professional may have discriminated against an individual based upon their race, age, sex, color, creed, national origin, disability, or other protected criteria under the law.

Safety professionals are not expected to be experts in the law, however, it is important that safety professionals possess a general understanding of the various laws and regulations offering protection to employees and individuals against discrimination in the workplace, as well as the "red flags" where safety professionals should proceed with caution. Safety professionals should also understand that they are agents of their company and their actions or inactions in potentially discriminating against an employee or individual directly impacts their companies. In the event of a claim of discrimination, safety professionals should be aware of the administrative processes provided under the various laws, as well as their possible role in the administrative process, including the litigation process.

In acquiring general knowledge and taking a proactive approach, prudent safety professionals can identify situations in which potential discrimination against an employee or individual may surface, and be able to take immediate action to address the potential discriminatory situation. Absent this basic knowledge of the laws and regulations prohibiting discrimination, the safety professional may base his or her decision making in a vacuum without awareness of applicable safety laws, regulations, and functions, inadvertently creating a potentially discriminatory situation.

As with any laws, the laws prohibiting discrimination in the workplace are constantly being clarified by the courts and recently being expanded by Congress. As noted above, safety professionals are not expected to be experts in these laws and should always contact their human resource office or legal counsel with any and all questions of issues. However, safety professionals should possess a level of knowledge within these areas to be able to identify the issues and situations in order that they can seek additional guidance before finalizing their decision. Knowing of the laws and regulations is the first step in assuring compliance and eliminating discrimination from the workplace.

The Author

 Thomas D. Schneid is a tenured professor in the Department of Safety, Security and Emergency Management (formerly Loss Prevention and Safety) at Eastern Kentucky University and serves as the graduate program director for the online and on-campus Master of Science degree program in Safety, Security and Emergency Management.

Tom has worked in the safety and human resources fields for over 30 years at various levels, including corporate safety director and industrial relations director. Tom has represented numerous corporations in OSHA and labor-related litigations throughout the United States. He has earned a B.S. in education, M.S., and CAS in safety as well as his Juris Doctor (J.D. in law) from West Virginia University and LL.M. (Graduate Law) from the University of San Diego. Tom is a member of the bar for the U.S. Supreme Court, 6th Circuit Court of Appeals and a number of federal districts as well as the Kentucky and West Virginia Bar.

Tom has authored and/or coauthored numerous texts, including *Corporate Safety Compliance: Law, OSHA and Ethics* (2008); *Americans With Disabilities Act: A Compliance Guide* (1994); *ADA: A Manager's Guide* (1993); *Legal Liabilities for Safety and Loss Prevention Professionals* (2010); *Fire and Emergency Law Casebook* (1996); *Creative Safety Solutions* (1998); *Occupational Health Guide to Violence in the Workplace* (1999); *Legal Liabilities in Emergency Management* (2001); and *Fire Law* (1995). Tom has also coauthored several texts including *Food Safety Law* (1997), *Legal Liabilities for Safety and Loss Prevention Professionals* (1997), *Physical Hazards in the Workplace* (2001), and *Disaster Management and Preparedness* (2000), as well as over 100 articles on safety and legal topics.

Introduction

Diversity is the fabric of which America was made, and it will continue to be an essential part of the American future. The "melting pot" that is America creates a unique combination of many nationalities, races, religions, ages, and genders, as well as many skills and abilities, which have been interwoven, through chance or design, to create that quilt work which is America. As inscribed on the Statue of Liberty,

> Give me your tired, your poor,
> Your huddled masses yearning to breathe free,
> The wretched refuse of your teeming shore.
> Send these, the homeless, tempest-tossed to me.
> I lift my lamp beside the golden door.

Individuals and families, whether coming through Ellis Island early in the century or arriving on a commercial aircraft at your local airport yesterday, have come to America for a wide variety of reasons; however, virtually all possessed the desire to live free and the opportunity to work and create a better life for themselves and their families. This rich and ongoing mixture of individuals, with their unique talents, skills, and abilities, creates an exceptional pool of prospective employees for the public and private sectors within the American workplace and permits us to compete successfully in the global market. Without the talents and ambition of our forefathers as well as the skills and abilities of our current individuals, America would not be the global economic power that was achieved and is being maintained today.

Most individuals simply want a level playing field through which to showcase their talents. America has historically been the place of new beginnings, a place where individuals are provided an opportunity and a place where individuals are judged on their abilities and not on their lineage. Over the decades, education is often found to be the great equalizer. However, America does have a storied history of discrimination and harassment of individuals and groups within the workplace. In order to rectify the past and current situations involving discrimination and harassment in the workplace and to maintain that level playing field today, Congress and the courts have created a myriad of laws and regulations, addressing a broad spectrum of protected classes of qualified individuals, developed to create an environment void of any discrimination or harassment.

Although there are many texts on the market addressing discrimination in the workplace, it is the author's hope that this text will educate individuals and employers to *prevent* acts and omissions in the workplace that can result in discrimination. It is the author's opinion that most individuals and employers do not purposefully try to discriminate but simply do not realize the unlawful results of their unintended actions. And with the multitude of laws and regulations addressing the prohibition of discrimination in the workplace, often legal actions result from individuals and employers simply not being knowledgeable in the requirements of the law. It is vitally important that individuals and employers ask employees how they can assist or help and *listen* intently to the responses before reacting. It is the author's opinion that most cases involving discrimination in the workplace could be addressed and rectified before the initiation of legal action or agency involvement if individuals and employers simply knew what the laws require and listen to their employees. As Elbert Hubbard noted, "Don't make excuses—make good." Knowing the laws and regulations is the first step in preventing, avoiding, or eliminating discrimination from your workplace. Hopefully this text will be your guide to taking this first step!

1 History and Overview

Laws and institutions are constantly tending to gravitate. Like clocks, they must be occasionally cleansed, and wound up, and set to true time.

Henry Ward Beecher

The purpose of law is to prevent the strong always having their way.

Ovid

LEARNING OBJECTIVES

1. Acquire an understanding of the historical underpinnings of antidiscrimination laws.
2. Develop an understanding of the fabric of laws prohibiting discrimination in the workplace.
3. Acquire an understanding of the safety professional's role and responsibilities in creating and maintaining diversity in the workplace.

HYPOTHETICAL SITUATION

You are the safety professional is a large manufacturing facility. During your weekly safety inspections, an employee asks to talk with you privately. After moving away from the production area, the employee tells you that he is being called names with racial references by his supervisor. The employee states that when he asked the supervisor to stop calling him these names, the supervisor transferred him to the hardest job in the department. The employee expresses to you that he needs this job and doesn't want to cause any problems. He simply wants the supervisor to stop the name calling and to get moved back to his job. What should the safety professional do?

OVERVIEW

Throughout history, there has been harassment and discrimination in societies throughout the world, often specifically related to the workplace, based on a variety of factors including, but not limited to, ethnicity, religion, skin color, nationality, gender, pregnancy, disability, and numerous other characteristics or beliefs. One of the unique strengths of the United States since the beginning has been its consistent ability to welcome diversity in society, and specifically in the workplace, and build upon the strengths of the individual rather than selectively eliminating individuals and groups from the workforce based on their beliefs, physical makeup, or heritage. However, despite the best efforts of American society, the government, and most employers, discrimination

1

still exists. Over time, a patchwork of laws and regulations have been enacted to address and redress current and past discriminations, primarily in the workplace, and provide a broad spectrum of protections against discrimination to the American worker.

Safety professionals should be aware that although the United States has taken the lead in the world arena to prohibit discrimination through the enactment of laws and regulations for the workplace and in our society, this has been a daunting and ongoing task. Discrimination, especially discrimination based on race, came to our shores along with the *Mayflower*. The issues involving racial discrimination are woven into the history of our country. In the 1860s, the Civil War was fought primarily over the issue of racial discrimination. Violence in many forms resulting from racial discrimination issues erupted through the United States during the nineteenth and twentieth centuries. Although there were laws and regulations prohibiting discrimination, many states, primarily in the South, continued to maintain the discriminatory practices of the past. The first comprehensive laws prohibiting discrimination in our workplaces, schools, and in our society were not enacted until the 1960s.

Safety professionals should be aware that, as a member of the management team, your company or organizational entity possesses a duty and responsibility to protect employees against discrimination in their workplace. Safety professionals are often on the front lines in identifying potential issues and situations where potential discrimination may exist in the workplace. Although most safety professionals do not possess direct responsibilities within the area of antidiscrimination policy enforcement and most safety professionals are not "experts" as to the laws and regulations addressing discrimination in the workplace, it is important that safety professionals acquire a knowledge of the various laws and regulations as well as a working knowledge of the processes and procedures to follow when an issue involving potential discrimination in the workplace arises.

As with the Occupational Safety and Health Act (hereinafter referred to as the "OSH Act") enacted on April 28, 1971[1], many of the federal laws prohibiting discrimination in the workplace had their origins in the turbulent decade of the 1960s. However, since the beginning of this century, there has been a slow progression in the development in laws, executive orders, and regulations through which the federal government attempted to regulate and thus prohibit discrimination in the American workplace, which culminated with the relatively recent enactment of a patchwork of various laws specifically enacted to address the relevant issue involving a specific discriminatory act, issue, or action.

Safety professionals should be aware that each of the protected classes, and thus each of the forms of discrimination, has been regulated and the laws enacted at different times in our history, incorporating different philosophies, and often utilizing different methods to regulate and prohibit discrimination in the workplace. This patchwork of laws, executive orders and regulations, as well as court decisions, creates the quilt of protections against discrimination for workers in today's workplace. Additionally, safety professionals should be aware that this patchwork of regulatory protection is a "work in progress" and is constantly moving and adjusting to address current issues as well as changes in our society and workplace.

Although we will focus on the laws and regulations enacted at the federal level, safety professionals should be aware that most states also have enacted antidiscrimination laws and regulations that can provide far greater protection or prohibitions against discrimination than the laws and regulations at the federal level. Additionally, safety professionals should be aware that the Constitution of the United States also addresses discrimination in the Fifth, Thirteenth, and Fourteenth Amendments. The federal level patchwork of antidiscriminations laws and regulations that have developed over time include the following:

- Title VII of the Civil Rights Act of 1964 (commonly referred to as "Title VII or 7")[2]
- The Americans with Disabilities Act (commonly referred to as the "ADA")[3]
- The Civil Rights Act of 1866[4]
- The Civil Rights Act of 1871[5]
- The Civil Rights Act of 1991[6]
- The Age Discrimination in Employment Act (commonly referred to as 'ADEA")[7]
- The Equal Pay Act (commonly referred to as the "EPA")[8]
- The Genetic Information Nondiscrimination Act (commonly referred to as "GINA")[9]
- Lilly Ledbetter Fair Pay Act of 2009[10]
- Vocational Rehabilitation Act of 1973[11]
- Family and Medical Leave Act of 1993[12]
- Immigration Reform and Control Act of 1986[13]
- Glass Ceiling Act of 1991[14]

As can be seen, this patchwork of laws, regulations, and executive orders prohibits discrimination, primarily focused on the workplace, based on race, color, national origin, religion, sex, age, physical or mental disability, and ethnicity while also providing protections for genetic information, wages, and the ability to be hired and promoted in the workplace can impact the safety function. Although safety professionals seldom possess direct responsibility for compliance with these laws and regulations, the safety function often interacts, either directly or indirectly, with each and every one of these laws and regulations at some point throughout the course of the safety working activities. The safety professional does not function in a vacuum and, thus, through interaction with management, employees, and the daily safety functions, safety professionals will often be confronted with issues that look and sound safety related but primarily are focused on one or more of the above laws or regulations.

Safety professionals should be aware that these antidiscrimination laws were often specifically focused on the workplace and are designed to keep the employer from making employment-related decisions that could disadvantage employees based on their race, sex, age, color, disability, or any of the other protected classes. Safety professionals should be aware that most issues of discrimination involve hiring, promotion, training, termination, demotion, layoffs, or other terms and conditions involved in the workplace, including safety issues.

Safety professionals should be aware that there is a difference between discrimination and harassment. The laws and regulations prohibiting discrimination in the workplace are primarily focused on policies and decisions. Harassment, on the other hand, is unwelcome conduct that creates a hostile work environment. Harassment is only illegal if it is against one or more of the employees who are in one or more of the protected classes created by one of the federal or state laws. To this end, safety professionals should be aware that not all types of harassment are automatically illegal. Harassment can be illegal if: (1) a management team member harasses an employer and the harassment results in "tangible employment action" (such as a job transfer, demotion, or termination); or the more common (2) Harassment resulting in a hostile work environment.

A hostile work environment is a form of harassment where the employee may not have suffered a "tangible employment action" but the employee is subject to unwelcome conduct which: (1) is based on the employee's protected classification (such as age or sex); (2) is severe or pervasive that it alters the terms and conditions of the employee's employment situation; and (3) creates a hostile or abusive environment for the employee. Safety professionals should be aware that conduct by other employees or management may be considered "unwelcome" if the employee doesn't invite the comments, actions, etc., and/or if the employee doesn't want the actions, comments, etc., to happen. Some examples of unwelcome conduct includes unwelcome touching, racial slurs, cartoons of an offensive nature, jokes, horseplay, insults, and gestures.

Safety professionals should be aware that there are three (3) predominate theories under which discrimination can take form. The first theory is that of Disparate Treatment. Under the Disparate Treatment theory, the discrimination takes the form of intentionally treating individuals differently due to their race, sex, or other protected classification or characteristic. For example, the disparate treatment theory would apply where an employer denies a job to an individual because the individual is of a specific race or the safety professional only permits the male members of the safety team to travel to the safety conference.

Under the Disparate Impact theory, safety professionals should be aware that conduct that appears fair on its face but can detrimentally affect a number of individuals in a protected class can also constitute discrimination. However, safety professionals should be aware that if the employer has a job-related or business reason for this different treatment, this conduct may be permissible. For example, the safety professional is incurring a number of back injuries in the dock area where employees are lifting constantly. The company, in an effort to reduce back injuries, requires all employees to be able to lift X lbs. This conduct may disproportionally affect female employees. If the company possesses a valid business reason, such as reduction of back injuries, this may be appropriate; however, even with a valid business reason, this practice may constitute discrimination if an alternative method, such as lifting assist equipment, exists that would not disproportionally affect the female employees.

The third theory of discrimination that safety professionals should be aware is that of Failure to Accommodate. This theory is especially pertinent for safety professionals under the Americans with Disabilities Act. The Failure to Accommodate theory comes into play for safety professionals when their company is required to accommodate a qualified individual with a disability, a religious practice, or other protected activity and the company fails to accommodate the individual's disability.

Safety professionals should be aware that discrimination is not always blatantly obvious, such as failing to promote an employee due to their race or sex. In fact, the type of discrimination where there is "smoking gun" direct evidence of discrimination is very rare. In most cases, evidence of discrimination is circumstantial or "piecemeal," suggesting that the logical conclusion for the action or inaction by the employer must be the result of discrimination. Safety professionals should be aware that an employee does not automatically win a discrimination case simply by filing a claim of discrimination. The employee has the burden of proof for several elements in order to be successful with a claim. In general, an employee filing a claim of discrimination must prove that (1) he or she is qualified for the job; (2) he or she is a member of a protected class; (3) the employee suffered an adverse action; and (4) there is a reason to suspect the company or management team member possessed an improper motive.

Safety professionals should be aware that the primary governmental agency on the federal level tasked with enforcing Title VII, ADA, and the related laws and regulations is the Equal Employment Opportunity Commission (hereinafter referred to as the EEOC). An overview of the responsibilities, authorities, and locations of the Equal Employment Opportunity Commission as identified on the EEOC website follows next.

Safety professionals should be aware that there are a number of different laws and regulations prohibiting discrimination in the workplace, and each and every one of these laws and regulations impact, whether directly or indirectly, the safety function. Although safety professionals do not need to be discrimination experts, it is imperative that safety professionals possess a working knowledge of these laws and regulations in order to avoid any type of potential discrimination in their work activities,

REVIEW (FROM THE EEOC WEBSITE)

The U.S. Equal Employment Opportunity Commission (EEOC) is responsible for enforcing federal laws that make it illegal to discriminate against a job applicant or an employee because of the person's race, color, religion, sex (including pregnancy), national origin, age (40 or older), disability, or genetic information. It is also illegal to discriminate against a person because the person complained about discrimination, filed a charge of discrimination, or participated in an employment discrimination investigation or lawsuit.

Most employers with at least 15 employees are covered by EEOC laws (20 employees in age discrimination cases). Most labor unions and employment agencies are also covered.

The laws apply to all types of work situations, including hiring, firing, promotions, harassment, training, wages, and benefits.

AUTHORITY AND ROLE

The EEOC has the authority to investigate charges of discrimination against employers who are covered by the law. Our role in an investigation is to fairly and accurately assess the allegations in the charge and then make a finding. If we find that discrimination has occurred, we will try to settle the charge. If we aren't successful, we have the authority to file a lawsuit to protect the rights of individuals and the interests of the public. We do not, however, file lawsuits in all cases where we find discrimination.

We also work to prevent discrimination before it occurs through outreach, education, and technical assistance programs.

The EEOC provides leadership and guidance to federal agencies on all aspects of the federal government's equal employment opportunity program. EEOC assures federal agency and department compliance with EEOC regulations, provides technical assistance to federal agencies concerning EEO complaint adjudication, monitors and evaluates federal agencies' affirmative employment programs, develops and distributes federal sector educational materials and conducts training for stakeholders, provides guidance and assistance to our Administrative Judges who conduct hearings on EEO complaints, and adjudicates appeals from administrative decisions made by federal agencies on EEO complaints.

LOCATION

We carry out our work through our headquarters offices in Washington, D.C. and through 53 field offices serving every part of the nation.[15]

actions, or decisions. Through knowledge and diligence, safety professionals can avoid even a hint of any type of discrimination in their safety actions or inactions that may detract from the important goals and objective of creating and maintaining a safety and healthful workplace for all employees.

CHAPTER QUESTIONS

1. The American with Disabilities Act provides protection to:
   a. Individuals of specific religions
   b. Individuals with disabilities
   c. Individuals of specific races
   d. None of the above
2. Title VII provides protection against discrimination based on:
   a. Race
   b. Sex
   c. Color
   d. All of the above
3. Most antidiscrimination laws require employers to possess at least
   a. 1 employee
   b. 10 employees
   c. 15 employees
   d. 50 employees
4. The primary federal agency charged with enforcing the antidiscrimination laws is
   a. The U.S. Department of Labor
   b. The Equal Employment Opportunity Commission

    c. The Occupational Safety and Health Commission

    d. None of the above

5. Please review the hypothetical situation provided at the beginning of this chapter. Please provide a brief response as to what you would do in this situation.

Answers: 1—b; 2—d; 3—c; 4—b; 5—individual response.

Supreme Court of the United States

Frank RICCI et al., Petitioners, v. John DeSTEFANO et al.*

Nos. 07-1428, 08-328.

Argued April 22, 2009.

Decided June 29, 2009.

Background: White firefighters and one Hispanic firefighter sued city and city officials, alleging that city violated Title VII by refusing to certify results of promotional examination, based on city's belief that its use of results could have disparate impact on minority firefighters. The United States District Court for the District of Connecticut, Janet Bond Arterton, J., 554 F.Supp.2d 142, entered summary judgment for city and city officials. Firefighters appealed. The United States Court of Appeals for the Second Circuit, 530 F.3d 87, affirmed. Certiorari was granted.

    Holdings: The Supreme Court, Justice Kennedy, held that:

(1) City's refusal to certify results was violation of Title VII's disparate-treatment prohibition absent some valid defense;

(2) before employer can engage in intentional discrimination for asserted purpose of avoiding unintentional disparate impact, employer must have strong basis in evidence to believe it will be subject to disparate-impact liability if it fails to take race-conscious action;

(3) city officials lacked strong basis in evidence to believe that examinations were not job-related and consistent with business necessity; and

(4) city officials lacked strong basis in evidence to believe there existed equally valid, less-discriminatory alternative to use of examinations that served City's needs but that city refused to adopt.

Reversed and remanded. Justice Scalia filed concurring opinion. Justice Alito filed concurring opinion joined by Justices Scalia and Thomas. Justice Ginsburg filed dissenting opinion joined by Justices Stevens, Souter, and Breyer.

---

* From Westlaw and modified for the purposes of this text.

2661 SYLLABUS[FN]

[FN*] The syllabus constitutes no part of the opinion of the Court but has been prepared by the Reporter of Decisions for the convenience of the reader. See *United States v. Detroit Timber & Lumber Co.,* 200 U.S. 321, 337, 26 S.Ct. 282, 50 L.Ed. 499.

New Haven, Conn. (City), uses objective examinations to identify those firefighters best qualified for promotion. When the results of such an exam to fill vacant lieutenant and captain positions showed that white candidates had outperformed minority candidates, a rancorous public debate ensued. Confronted with arguments both for and against certifying the test results—and threats of a lawsuit either way—the City threw out the results based on the statistical racial disparity. Petitioners, white and Hispanic firefighters who passed the exams but were denied a chance at promotions by the City's refusal to certify the test results, sued the City and respondent officials, alleging that discarding the test results discriminated against them based on their race in violation of, *inter alia,* Title VII of the Civil Rights Act of 1964. The defendants responded that had they certified the test results, they could have faced Title VII liability for adopting a practice having a disparate impact on minority firefighters. The District Court granted summary judgment for the defendants, and the Second Circuit affirmed.

Held: The City's action in discarding the tests violated Title VII. Pp. 2672–2682.

(a) Title VII prohibits intentional acts of employment discrimination based on race, color, religion, sex, and national origin, 42 U.S.C. § 2000e-2(a)(1) (disparate treatment), as well as policies or practices that are not intended to discriminate but in fact have a disproportionately adverse effect on minorities, § 2000e-2(k)(1)(A)(i) (disparate impact). Once a plaintiff has established a prima facie case of disparate impact, the employer may defend by demonstrating that its policy or practice is "job related for the position in question and consistent with business necessity." *Ibid.* If the employer meets that burden, the plaintiff may still succeed by showing that the employer refuses to adopt an available alternative practice that has less disparate impact and serves the employer's legitimate needs. §§ 2000e-2(k)(1)(A)(ii) and (C). Pp. 2672–2673.

(b) Under Title VII, before an employer can engage in intentional discrimination for the asserted purpose of avoiding or remedying an unintentional, disparate impact, the employer must have a strong basis in evidence to believe it will be subject to disparate-impact liability if it fails to take the race-conscious, discriminatory action. The Court's analysis begins with the premise that the City's actions would violate Title VII's disparate-treatment prohibition absent some valid defense. All the evidence demonstrates that the City rejected the test results because the higher *2662 scoring candidates were white. Without some other justification, this express, race-based decision making is prohibited. The question, therefore, is whether the purpose of avoiding disparate-impact liability excuses what otherwise would be prohibited disparate-treatment discrimination. The Court has considered cases similar to the present litigation, but in the context of the Fourteenth Amendment's Equal Protection Clause. Such cases can provide helpful guidance in this statutory context. See *Watson v. Fort Worth Bank & Trust,* 487 U.S. 977, 993, 108 S.Ct. 2777, 101 L.Ed.2d 827. In those cases, the Court held that certain government actions to remedy past racial discrimination—actions that are themselves based on race—are constitutional only where there is a "strong basis in evidence" that the remedial actions were necessary. *Richmond v. J.A. Croson Co.,* 488 U.S. 469, 500, 109 S.Ct. 706, 102 L.Ed.2d 854; see also *Wygant v. Jackson Bd. of Ed.,* 476 U.S. 267, 277, 106 S.Ct. 1842, 90 L.Ed.2d 260. In announcing the strong-basis-in-evidence standard, the *Wygant* plurality recognized the tension between eliminating segregation and discrimination on the one hand and doing away with all governmentally imposed discrimination based on race on the other. 476 U.S., at 277, 106 S.Ct. 1842. It reasoned that "[e]videntiary support for the conclusion that remedial action is warranted becomes crucial when the remedial program is challenged in court by nonminority employees." *Ibid.* The same interests are at work in the interplay between Title VII's disparate-treatment and disparate-impact provisions. Applying the strong-basis-in-evidence standard to Title VII gives effect to both provisions, allowing violations of one in the name of compliance with the other only in certain, narrow circumstances. It also allows the disparate-impact prohibition to work in a manner that is consistent with other Title VII provisions, including the prohibition on adjusting employment-related test scores based on race, see § 2000e-2(*l*), and the section that expressly protects bona fide promotional exams, see § 2000e-2(h). Thus, the Court adopts the strong-basis-in-evidence standard as a matter of statutory construction in order to resolve any conflict between Title VII's disparate-treatment and disparate-impact provisions. Pp. 2673–2677.

(c) The City's race-based rejection of the test results cannot satisfy the strong-basis-in-evidence standard. Pp. 2677–2681.

(i) The racial adverse impact in this litigation was significant, and petitioners do not dispute that the City was faced with a prima facie case of disparate-impact liability. The problem for respondents is that such a prima facie case—essentially, a threshold showing of a significant statistical disparity, *Connecticut v. Teal,* 457 U.S. 440, 446, 102 S.Ct. 2525, 73 L.Ed.2d 130, and nothing more—is far from a strong basis in evidence that the City would have been liable under Title VII had it certified the test results. That is because the City could be liable for disparate-impact discrimination only if the exams at issue were not job related and consistent with business necessity, or if there existed an equally valid, less discriminatory alternative that served the City's needs but that the City refused to adopt. §§ 2000e-2(k)(1)(A), (C). Based on the record the parties developed through discovery, there is no substantial basis in evidence that the test was deficient in either respect. Pp. 2677–2678.

(ii)   The City's assertions that the exams at issue were not job related and consistent with business necessity are blatantly contradicted by the record, which demonstrates the detailed steps taken to develop and administer the tests and the *2663 painstaking analyses of the questions asked to assure their relevance to the captain and lieutenant positions. The testimony also shows that complaints that certain examination questions were contradictory or did not specifically apply to firefighting practices in the City were fully addressed, and that the City turned a blind eye to evidence supporting the exams' validity. Pp. 2678–2679.

(iii) Respondents also lack a strong basis in evidence showing an equally valid, less discriminatory testing alternative that the City, by certifying the test results, would necessarily have refused to adopt. Respondents' three arguments to the contrary all fail. First, respondents refer to testimony that a different composite-score calculation would have allowed the City to consider black candidates for then-open positions, but they have produced no evidence to show that the candidate weighting actually used was indeed arbitrary, or that the different weighting would be an equally valid way to determine whether candidates are qualified for promotions. Second, respondents argue that the City could have adopted a different interpretation of its charter provision limiting promotions to the highest-scoring applicants, and that the interpretation would have produced less discriminatory results; but respondents' approach would have violated Title VII's prohibition of race-based adjustment of test results, § 2000e-2(l). Third, testimony asserting that the use of an assessment center to evaluate candidates' behavior in typical job tasks would have had less adverse impact than written exams does not aid respondents, as it is contradicted by other statements in the record indicating that the City could not have used assessment centers for the exams at issue. Especially when it is noted that the strong-basis-in-evidence standard applies to this case, respondents cannot create a genuine issue of fact based on a few stray (and contradictory) statements in the record. Pp. 2679–2681.

(iv)   Fear of litigation alone cannot justify the City's reliance on race to the detriment of individuals who passed the examinations and qualified for promotions. Discarding the test results was impermissible under Title VII, and summary judgment is appropriate for petitioners on their disparate-treatment claim. If, after it certifies the test results, the City faces a disparate-impact suit, then in light of today's holding the City can avoid disparate-impact liability based on the strong basis in evidence that, had it not certified the results, it would have been subject to disparate-treatment liability. P. 2681.

530 F.3d 87, reversed and remanded.

KENNEDY, J., delivered the opinion of the Court, in which ROBERTS, C.J., and SCALIA, THOMAS, and ALITO, JJ., joined. SCALIA, J., filed a concurring opinion. ALITO, J., filed a concurring opinion, in which SCALIA and THOMAS, JJ., joined. GINSBURG, J., filed a dissenting opinion, in which STEVENS, SOUTER, and BREYER, JJ., joined.

JUSTICE KENNEDY DELIVERED THE OPINION OF THE COURT.

In the fire department of New Haven, Connecticut—as in emergency-service agencies throughout the nation—firefighters prize their promotion to and within the officer ranks. An agency's officers command respect within the department and in the whole community; and, of course, added responsibilities command increased salary and benefits. Aware of the intense competition for promotions, New Haven, like many cities, relies on objective examinations to identify the best-qualified candidates.

In 2003, 118 New Haven firefighters took examinations to qualify for promotion to the rank of lieutenant or captain. Promotion examinations in New Haven (or City) were infrequent, so the stakes were high. The results would determine which firefighters would be considered for promotions during the next two years, and the order in which they would be considered. Many firefighters studied for months, at considerable personal and financial cost.

When the examination results showed that white candidates had outperformed minority candidates, the mayor and other local politicians opened a public debate that turned rancorous. Some firefighters argued the tests should be discarded because the results showed the tests to be discriminatory. They threatened a discrimination lawsuit if the City made promotions based on the tests. Other firefighters said the exams were neutral and fair. And they, in turn, threatened a discrimination lawsuit if the City, relying on the statistical racial disparity, ignored the test results and denied promotions to the candidates who had performed well. In the end the City took the side of those who protested the test results. It threw out the examinations.

Certain white and Hispanic firefighters who likely would have been promoted based on their good test performance sued the City and some of its officials. Theirs is the suit now before us. The suit alleges that, by discarding the test results, the City and the named officials discriminated against the plaintiffs based on their race, in violation of both Title VII of the Civil Rights Act of 1964, 78 Stat. 253, as amended, 42 U.S.C. § 2000e et seq. , and the Equal Protection Clause of the Fourteenth Amendment. The City and the officials defended their actions, arguing that if they had certified the results, they could have faced liability under Title VII for adopting a practice that had a disparate impact on the minority firefighters. The District Court granted summary judgment for the defendants, and the Court of Appeals affirmed.

We conclude that race-based action like the City's in this case is impermissible under Title VII unless the employer can demonstrate a strong basis in evidence that, had it not taken the action, it would have been liable under the disparate-impact statute. The respondents, we further determine, cannot meet that threshold standard. As a result, the City's action in discarding the tests was a violation of Title VII. In light of our ruling under the statutes, we need not reach the question *2665 whether respondents' actions may have violated the Equal Protection Clause.

I

This litigation comes to us after the parties' cross-motions for summary judgment, so we set out the facts in some detail. As the District Court noted, although "the parties strenuously dispute the relevance and legal import of, and inferences to be drawn from, many aspects of this case, the underlying facts are largely undisputed." 554 F.Supp.2d 142, 145 (Conn.2006).

A

When the City of New Haven undertook to fill vacant lieutenant and captain positions in its fire department (Department), the promotion and hiring process was governed by the city charter, in addition to federal and state law. The charter establishes a merit system. That system requires the City to fill vacancies in the classified civil-service ranks with the most qualified individuals, as determined by job-related examinations. After each examination, the New Haven Civil Service Board (CSB) certifies a ranked list of applicants who passed the test. Under the charter's "rule of three," the relevant hiring authority must fill each vacancy by choosing one candidate from the top three scorers on the list. Certified promotional lists remain valid for two years.

The City's contract with the New Haven firefighters' union specifies additional requirements for the promotion process. Under the contract, applicants for lieutenant and captain positions were to be screened using written and oral examinations, with the written exam accounting for 60% and the oral exam 40% of an applicant's total score. To sit for the examinations, candidates for lieutenant needed 30 months' experience in the department, a high-school diploma, and certain vocational training courses. Candidates for captain needed one year's service as a lieutenant in the department, a high-school diploma, and certain vocational training courses.

After reviewing bids from various consultants, the City hired Industrial/ Organizational Solutions (IOS), Inc. to develop and administer the examinations, at a cost to the City of $100,000. IOS is an Illinois company that specializes in designing entry-level and promotional examinations for fire and police departments. In order to fit the examinations to the New Haven Department, IOS began the test-design process by performing job analyses to identify the tasks, knowledge, skills, and abilities that are essential for the lieutenant and captain positions. IOS representatives interviewed incumbent captains and lieutenants and their supervisors. They rode with and observed other on-duty officers. Using information from those interviews and ride-alongs, IOS wrote job-analysis questionnaires and administered them to most of the incumbent battalion chiefs, captains, and lieutenants in the Department. At every stage of the job analyses,

IOS, by deliberate choice, oversampled minority firefighters to ensure that the results—which IOS would use to develop the examinations—would not unintentionally favor white candidates.

With the job-analysis information in hand, IOS developed the written examinations to measure the candidates' job-related knowledge. For each test, IOS compiled a list of training manuals, department procedures, and other materials to use as sources for the test questions. IOS presented the proposed sources to the New Haven fire chief and assistant fire chief for their approval. Then, using the approved sources, IOS drafted a multiple-choice test for each position. Each *2666 test had 100 questions, as required by CSB rules, and was written below a 10th-grade reading level. After IOS prepared the tests, the City opened a 3-month study period. It gave candidates a list that identified the source material for the questions, including the specific chapters from which the questions were taken.

IOS developed the oral examinations as well. These concentrated on job skills and abilities. Using the job-analysis information, IOS wrote hypothetical situations to test incident-command skills, firefighting tactics, interpersonal skills, leadership, and management ability, among other things. Candidates would be presented with these hypotheticals and asked to respond before a panel of three assessors.

IOS assembled a pool of 30 assessors who were superior in rank to the positions being tested. At the City's insistence (because of controversy surrounding previous examinations), all the assessors came from outside Connecticut. IOS submitted the assessors' resumes to City officials for approval. They were battalion chiefs, assistant chiefs, and chiefs from departments of similar sizes to New Haven's throughout the country. Sixty-six percent of the panelists were minorities, and each of the nine three-member assessment panels contained two minority members. IOS trained the panelists for several hours on the day before it administered the examinations, teaching them how to score the candidates' responses consistently using checklists of desired criteria.

Candidates took the examinations in November and December 2003. Seventy-seven candidates completed the lieutenant examination: 43 whites, 19 blacks, and 15 Hispanics. Of those, 34 candidates passed: 25 whites, 6 blacks, and 3 Hispanics. 554 F.Supp.2d, at 145. Eight lieutenant positions were vacant at the time of the examination. As the rule of three operated, this meant that the top 10 candidates were eligible for an immediate promotion to lieutenant. All 10 were white. *Ibid.* Subsequent vacancies would have allowed at least 3 black candidates to be considered for promotion to lieutenant.

Forty-one candidates completed the captain examination: 25 whites, 8 blacks, and 8 Hispanics. Of those, 22 candidates passed: 16 whites, 3 blacks, and 3 Hispanics. *Ibid.* Seven captain positions were vacant at the time of the examination. Under the rule of three, 9 candidates were eligible for an immediate promotion to captain: 7 whites and 2 Hispanics. *Ibid.*

B

The City's contract with IOS contemplated that, after the examinations, IOS would prepare a technical report that described the examination processes and methodologies and analyzed the results. But in January 2004, rather than requesting the technical report, City officials, including the City's counsel, Thomas Ude, convened a meeting with IOS Vice President Chad Legel. (Legel was the leader of the IOS team that developed and administered the tests.) Based on the test results, the City officials expressed concern that the tests had discriminated against minority candidates. Legel defended the examinations' validity, stating that any numerical disparity between white and minority candidates was likely due to various external factors and was in line with results of the Department's previous promotional examinations.

Several days after the meeting, Ude sent a letter to the CSB purporting to outline its duties with respect to the examination results. Ude stated that under federal law, "a statistical demonstration of disparate impact," standing alone, "constitutes a sufficiently serious claim of racial discrimination to serve as a predicate for *2667 employer-initiated, voluntar[y] remedies-even ... race-conscious remedies." App. to Pet. for Cert. in No. 07-1428, p. 443a; see also 554 F.Supp.2d, at 145 (issue of disparate impact "appears to have been raised by ... Ude").

1

The CSB first met to consider certifying the results on January 22, 2004. Tina Burgett, director of the City's Department of Human Resources, opened the meeting by telling the CSB that "there is a significant disparate impact on these two exams." App. to Pet. for Cert. in No. 07-1428, at 466a. She distributed lists showing the candidates' races and scores (written, oral, and composite) but not their names. Ude also described the test results as reflecting "a very significant disparate impact," *id.,* at 477a, and he outlined possible grounds for the CSB's refusing to certify the results.

Although they did not know whether they had passed or failed, some firefighter-candidates spoke at the first CSB meeting in favor of certifying the test results. Michael Blatchley stated that "[e]very one" of the questions on the written examination "came from the [study] material. ... [I]f you read the materials and you studied the material, you would have done well on the test." App. in No. 06-4996-cv (CA2), pp. A772-A773 (hereinafter CA2 App.). Frank Ricci stated that the test questions were based on the department's own rules and procedures and on "nationally recognized" materials that represented the "accepted standard[s]" for firefighting. *Id.,* at A785-A786. Ricci stated that he had "several learning disabilities," including dyslexia; that he had spent more than $1,000 to purchase the materials and pay his neighbor to read them on tape so he could "give it [his] best shot"; and that he had studied "8 to 13 hours a day to prepare" for the test. *Id.,* at A786, A789. "I don't even know if I made it," Ricci told the CSB, "[b]ut the people who passed should be promoted. When your life's on the line, second best may not be good enough." *Id.,* at A787-A788.

Other firefighters spoke against certifying the test results. They described the test questions as outdated or not relevant to firefighting practices in New Haven. Gary Tinney stated that source materials "came out of New York.... Their makeup of their city and everything is totally different than ours." *Id.,* at A774-

A775; see also *id.*, at A779, A780-A781. And they criticized the test materials, a full set of which cost about $500, for being too expensive and too long.

2

At a second CSB meeting, on February 5, the president of the New Haven firefighters' union asked the CSB to perform a validation study to determine whether the tests were job related. Petitioners' counsel in this action argued that the CSB should certify the results. A representative of the International Association of Black Professional Firefighters, Donald Day from neighboring Bridgeport, Connecticut, "beseech[ed]" the CSB "to throw away that test," which he described as "inherently unfair" because of the racial distribution of the results. *Id.*, at A830-A831. Another Bridgeport-based representative of the association, Ronald Mackey, stated that a validation study was necessary. He suggested that the City could "adjust" the test results to "meet the criteria of having a certain amount of minorities get elevated to the rank of Lieutenant and Captain." *Id.*, at A838. At the end of this meeting, the CSB members agreed to ask IOS to send a representative to explain how it had developed and administered the examinations. They also discussed *2668, asking a panel of experts to review the examinations and advise the CSB whether to certify the results.

3

At a third meeting, on February 11, Legel addressed the CSB on behalf of IOS. Legel stated that IOS had previously prepared entry-level firefighter examinations for the City but not a promotional examination. He explained that IOS had developed examinations for departments in communities with demographics similar to New Haven's, including Orange County, Florida; Lansing, Michigan; and San Jose, California.

Legel explained the exam-development process to the CSB. He began by describing the job analyses IOS performed of the captain and lieutenant positions: the interviews, ride-alongs, and questionnaires IOS designed to "generate a list of tasks, knowledge, skills and abilities that are considered essential to performance" of the jobs. *Id.*, at A931-A932. He outlined how IOS prepared the written and oral examinations, based on the job-analysis results, to test most heavily those qualities that the results indicated were "critica[l]" or "essentia[l]." *Id.*, at A931. And he noted that IOS took the material for each test question directly from the approved source materials. Legel told the CSB that third-party reviewers had scrutinized the examinations to ensure that the written test was drawn from the source material and that the oral test accurately tested real-world situations that captains and lieutenants would face. Legel confirmed that IOS had selected oral-examination panelists so that each three-member assessment panel included one white, one black, and one Hispanic member.

Near the end of his remarks, Legel "implor[ed] anyone that had ... concerns to review the content of the exam. In my professional opinion, it's facially neutral. There's nothing in those examinations ... that should cause somebody to think that one group would perform differently than another group." *Id.*, at A961.

4

At the next meeting, on March 11, the CSB heard from three witnesses it had selected to "tell us a little bit about their views of the testing, the process, [and] the methodology." *Id.,* at A1020. The first, Christopher Hornick, spoke to the CSB by telephone. Hornick is an industrial/organizational psychologist from Texas who operates a consulting business that "direct[ly]" competes with IOS. *Id.,* at A1029. Hornick, who had not "stud[ied]" the test at length or in detail" and had not "seen the job analysis data," told the CSB that the scores indicated a "relatively high adverse impact." *Id.,* at A1028, A1030, A1043. He stated that "[n]ormally, whites outperform ethnic minorities on the majority of standardized testing procedures," but that he was "a little surprised" by the disparity in the candidates' scores—although "[s]ome of it is fairly typical of what we've seen in other areas of the countr[y] and other tests." *Id.,* at A1028-A1029. Hornick stated that the "adverse impact on the written exam was somewhat higher but generally in the range that we've seen professionally." *Id.,* at A1030-A1031.

When asked to explain the New Haven test results, Hornick opined in the telephone conversation that the collective-bargaining agreement's requirement of using written and oral examinations with a 60/40 composite score might account for the statistical disparity. He also stated that "[b]y not having anyone from within the [D]epartment review" the tests before they were administered—a limitation the City had imposed to protect the security of the exam questions— "you inevitably get 2669 things in there" that are based on the source materials but are not relevant to New Haven. *Id.,* at A1034-A1035. Hornick suggested that testing candidates at an "assessment center" rather than using written and oral examinations "might serve [the City's] needs better." *Id.,* at A1039-A1040. Hornick stated that assessment centers, where candidates face real-world situations and respond just as they would in the field, allow candidates "to demonstrate how they would address a particular problem as opposed to just verbally saying it or identifying the correct option on a written test." *Ibid.*

Hornick made clear that he was "not suggesting that [IOS] somehow created a test that had adverse impacts that it should not have had." *Id.,* at A1038. He described the IOS examinations as "reasonably good test[s]." *Id.,* at A1041. He stated that the CSB's best option might be to "certify the list as it exists" and work to change the process for future tests, including by "[r]ewriting the Civil Service Rules." *Ibid.* Hornick concluded his telephonic remarks by telling the CSB that "for the future," his company "certainly would like to help you if we can." *Id.,* at A1046.

The second witness was Vincent Lewis, a fire program specialist for the Department of Homeland Security and a retired fire captain from Michigan. Lewis, who is black, had looked "extensively" at the lieutenant exam and "a little less extensively" at the captain exam. He stated that the candidates "should know that material." *Id.,* at A1048, A1052. In Lewis's view, the "questions were relevant for both exams," and the New Haven candidates had an advantage because the study materials identified the particular book chapters from which the questions were taken. In other departments, by contrast, "you had to know basically the ... entire book." *Id.,* at A1053. Lewis concluded that any disparate impact likely was due to a pattern that "usually whites outperform some of the minorities on testing," or that "more whites ... take the exam." *Id.,* at A1054.

The final witness was Janet Helms, a professor at Boston College whose "primary area of expertise" is "not with firefighters per se" but in "race and culture as they influence performance on tests and other assessment procedures." *Id.,* at A1060. Helms expressly declined the CSB's offer to review the examinations. At the outset, she noted that "regardless of what kind of written test we give in this country ... we can just about predict how many people will pass who are members of under-represented groups. And your data are not that inconsistent with what predictions would say were the case." *Id.,* at A1061. Helms nevertheless offered several "ideas about what might be possible factors" to explain statistical differences in the results. *Id.,* at A1062. She concluded that because 67% of the respondents to the job-analysis questionnaires were white, the test questions might have favored white candidates, because "most of the literature on firefighters shows that the different groups perform the job differently." *Id.,* at A1063. Helms closed by stating that no matter what test the City had administered, it would have revealed "a disparity between blacks and whites, Hispanics and whites," particularly on a written test. *Id.,* at A1072.

5

At the final CSB meeting, on March 18, Ude (the City's counsel) argued against certifying the examination results. Discussing the City's obligations under federal law, Ude advised the CSB that a finding of adverse impact "is the beginning, not the end, of a review of testing procedures" to determine whether they violated the *2670 disparate-impact provision of Title VII. Ude focused the CSB on determining "whether there are other ways to test for ... those positions that are equally valid with less adverse impact." *Id.,* at A1101. Ude described Hornick as having said that the written examination "had one of the most severe adverse impacts that he had seen" and that "there are much better alternatives to identifying [firefighting] skills." *Ibid.* Ude offered his "opinion that promotions ... as a result of these tests would not be consistent with federal law, would not be consistent with the purposes of our Civil Service Rules or our Charter[,] nor is it in the best interests of the firefighters ... who took the exams." *Id.,* at A1103-A1104. He stated that previous Department exams "have not had this kind of result," and that previous results had not been "challenged as having adverse impact, whereas we are assured that these will be." *Id.,* at A1107, A1108.

CSB Chairman Segaloff asked Ude several questions about the Title VII disparate-impact standard.

"CHAIRPERSON SEGALOFF: [M]y understanding is the group ... that is making to throw the exam out has the burden of showing that there is out there an exam that is reasonably probable or likely to have less of an adverse impact. It's not our burden to show that there's an exam out there that can be better. We've got an exam. We've got a result ....

"MR. UDE: Mr. Chair, I point out that Dr. Hornick said that. He said that there are other tests out there that would have less adverse impact and that [would] be more valid.

"CHAIRPERSON SEGALOFF: You think that's enough for us to throw this test upside-down ... because Dr. Hornick said it?

"MR. UDE: I think that by itself would be sufficient. Yes. I also would point out that ... it is the employer's burden to justify the use of the examination." *Id.,* at A1108-A1109.

Karen DuBois-Walton, the City's chief administrative officer, spoke on behalf of Mayor John DeStefano and argued against certifying the results. DuBois-Walton stated that the results, when considered under the rule of three and applied to then-existing captain and lieutenant vacancies, created a situation in which black and Hispanic candidates were disproportionately excluded from opportunity. DuBois-Walton also relied on Hornick's testimony, asserting that Hornick "made it extremely clear that ... there are more appropriate ways to assess one's ability to serve" as a captain or lieutenant. *Id.,* at A1120.

Burgett (the human resources director) asked the CSB to discard the examination results. She, too, relied on Hornick's statement to show the existence of alternative testing methods, describing Hornick as having "started to point out that alternative testing does exist" and as having "begun to suggest that there are some different ways of doing written examinations." *Id.,* at A1125, A1128.

Other witnesses addressed the CSB. They included the president of the New Haven firefighters' union, who supported certification. He reminded the CSB that Hornick "also concluded that the tests were reasonable and fair and under the current structure to certify them." *Id.,* at A1137. Firefighter Frank Ricci again argued for certification; he stated that although "assessment centers in some cases show less adverse impact," *id.,* at A1140, they were not available alternatives for the current round of promotions. It would take several years, Ricci explained, for the Department to develop an assessment-center protocol and the accompanying training *2671 materials. *Id.,* at A1141. Lieutenant Matthew Marcarelli, who had taken the captain's exam, spoke in favor of certification.

At the close of witness testimony, the CSB voted on a motion to certify the examinations. With one member recused, the CSB deadlocked 2 to 2, resulting in a decision not to certify the results. Explaining his vote to certify the results, Chairman Segaloff stated that "nobody convinced me that we can feel comfortable that, in fact, there's some likelihood that there's going to be an exam designed that's going to be less discriminatory." *Id.,* at A1159-A1160.

C

The CSB's decision not to certify the examination results led to this lawsuit. The plaintiffs—who are the petitioners here—are 17 white firefighters and 1 Hispanic firefighter, who passed the examinations but were denied a chance at promotions when the CSB refused to certify the test results. They include the named plaintiff, Frank Ricci, who addressed the CSB at multiple meetings.

Petitioners sued the City, Mayor DeStefano, DuBois-Walton, Ude, Burgett, and the two CSB members who voted against certification. Petitioners also named as a defendant Boise Kimber, a New Haven resident who voiced strong opposition to certifying the results. Those individuals are respondents in this Court. Petitioners filed suit under Rev. Stat. §§ 1979 and 1980, 42 U.S.C. §§ 1983 and 1985, alleging that respondents, by arguing or voting against certifying the results, violated and conspired to violate the Equal Protection Clause of the Fourteenth Amendment. Petitioners also filed timely charges of discrimination with the Equal Employment Opportunity Commission (EEOC); upon the EEOC's issuing right-to-sue letters, petitioners amended their complaint to assert that the City violated the disparate-treatment prohibition contained in Title VII of the Civil Rights Act of 1964, as amended. See 42 U.S.C. §§ 2000e-2(a) .

The parties filed cross-motions for summary judgment. Respondents asserted they had a good-faith belief that they would have violated the disparate-impact prohibition in Title VII, § 2000e-2(k), had they certified the examination results. It follows, they maintained, that they cannot be held liable under Title VII's disparate-treatment provision for attempting to comply with Title VII's disparate-impact bar. Petitioners countered that respondents' good-faith belief was not a valid defense to allegations of disparate treatment and unconstitutional discrimination.

The District Court granted summary judgment for respondents. 554 F.Supp.2d 142. It described petitioners' argument as "boil[ing] down to the assertion that if [respondents] cannot prove that the disparities on the Lieutenant and Captain exams were due to a particular flaw inherent in those exams, then they should have certified the results because there was no other alternative in place." *Id.,* at 156. The District Court concluded that, "[n]otwithstanding the shortcomings in the evidence on existing, effective alternatives, it is not the case that [respondents] *must* certify a test where they cannot pinpoint its deficiency explaining its disparate impact ... simply because they have not yet formulated a better selection method." *Ibid.* It also ruled that respondents' "motivation to avoid making promotions based on a test with a racially disparate impact ... does not, as a matter of law, constitute discriminatory intent" under Title VII. *Id.,* at 160. The District Court rejected petitioners' equal protection claim on the theory that respondents had not acted because of "discriminatory animus" toward petitioners. *2672 *Id.,* at 162. It concluded that respondents' actions were not "based on race" because "all applicants took the same test, and the result was the same for all because the test results were discarded and nobody was promoted." *Id.,* at 161.

After full briefing and argument by the parties, the Court of Appeals affirmed in a one-paragraph, unpublished summary order; it later withdrew that order, issuing in its place a nearly identical, one-paragraph *per curiam* opinion adopting the District Court's reasoning. 530 F.3d 87 (C.A.2 2008). Three days later, the Court of Appeals voted 7 to 6 to deny rehearing en banc, over written dissents by Chief Judge Jacobs and Judge Cabranes. 530 F.3d 88.

This action presents two provisions of Title VII to be interpreted and reconciled, with few, if any, precedents in the courts of appeals discussing the issue.

Depending on the resolution of the statutory claim, a fundamental constitutional question could also arise. We found it prudent and appropriate to grant certiorari. 555 U.S. ----, 129 S.Ct. 894, 172 L.Ed.2d 768 (2009). We now reverse.

II

[1] ☑ Petitioners raise a statutory claim, under the disparate-treatment prohibition of Title VII, and a constitutional claim, under the Equal Protection Clause of the Fourteenth Amendment. A decision for petitioners on their statutory claim would provide the relief sought, so we consider it first. See *Atkins v. Parker,* 472 U.S. 115, 123, 105 S.Ct. 2520, 86 L.Ed.2d 81 (1985); *Escambia County v. McMillan,* 466 U.S. 48, 51, 104 S.Ct. 1577, 80 L.Ed.2d 36 (1984) *(per curiam)* ("[N]ormally the Court will not decide a constitutional question if there is some other ground upon which to dispose of the case").

A

[2] ☑ Title VII of the Civil Rights Act of 1964, 42 U.S.C. § 2000e *et seq. ,* as amended, prohibits employment discrimination on the basis of race, color, religion, sex, or national origin. Title VII prohibits both intentional discrimination (known as "disparate treatment") as well as, in some cases, practices that are not intended to discriminate but in fact have a disproportionately adverse effect on minorities (known as "disparate impact").

[3] ☑ [4] ☑ As enacted in 1964, Title VII's principal nondiscrimination provision held employers liable only for disparate treatment. That section retains its original wording today. It makes it unlawful for an employer "to fail or refuse to hire or to discharge any individual, or otherwise to discriminate against any individual with respect to his compensation, terms, conditions, or privileges of employment, because of such individual's race, color, religion, sex, or national origin." § 2000e-2(a)(1); see also 78 Stat. 255. Disparate-treatment cases present "the most easily understood type of discrimination," *Teamsters v. United States,* 431 U.S. 324, 335, n. 15, 97 S.Ct. 1843, 52 L.Ed.2d 396 (1977), and occur where an employer has "treated [a] particular person less favorably than others because of" a protected trait. *Watson v. Fort Worth Bank & Trust,* 487 U.S. 977, 985-986, 108 S.Ct. 2777, 101 L.Ed.2d 827 (1988). A disparate-treatment plaintiff must establish "that the defendant had a discriminatory intent or motive" for taking a job-related action. *Id.,* at 986, 108 S.Ct. 2777.

The Civil Rights Act of 1964 did not include an express prohibition on policies or practices that produce a disparate impact. But in *Griggs v. Duke Power Co.,* 401 U.S. 424, 91 S.Ct. 849, 28 L.Ed.2d 158 (1971), the Court interpreted the Act to prohibit, in some cases, employers' facially *2673 neutral practices that, in fact, are "discriminatory in operation." *Id.,* at 431, 91 S.Ct. 849. The *Griggs* Court stated that the "touchstone" for disparate-impact liability is the lack of "business necessity": "If an employment practice which operates to exclude [minorities] cannot be shown to be related to job performance, the practice is prohibited." *Ibid.;* see also *id.,* at 432, 91 S.Ct. 849 (employer's burden to demonstrate that practice has "a manifest relationship to the employment in question"); *Albemarle Paper Co. v. Moody,* 422 U.S. 405, 425, 95 S.Ct. 2362, 45 L.Ed.2d 280 (1975).

Under those precedents, if an employer met its burden by showing that its practice was job-related, the plaintiff was required to show a legitimate alternative that would have resulted in less discrimination. *Ibid.* (Allowing the complaining party to show "that other tests or selection devices, without a similarly undesirable racial effect, would also serve the employer's legitimate interest.")

Twenty years after *Griggs,* the Civil Rights Act of 1991, 105 Stat. 1071, was enacted. The Act included a provision codifying the prohibition on disparate-impact discrimination. That provision is now in force along with the disparate-treatment section already noted. Under the disparate-impact statute, a plaintiff establishes a prima facie violation by showing that an employer uses "a particular employment practice that causes a disparate impact on the basis of race, color, religion, sex, or national origin." 42 U.S.C. § 2000e-2(k)(1)(A)(i). An employer may defend against liability by demonstrating that the practice is "job related for the position in question and consistent with business necessity." *Ibid.* Even if the employer meets that burden, however, a plaintiff may still succeed by showing that the employer refuses to adopt an available alternative employment practice that has less disparate impact and serves the employer's legitimate needs. §§ 2000e-2(k)(1)(A)(ii) and (C).

B

Petitioners allege that when the CSB refused to certify the captain and lieutenant exam results based on the race of the successful candidates, it discriminated against them in violation of Title VII's disparate-treatment provision. The City counters that its decision was permissible because the tests "appear[ed] to violate Title VII's disparate-impact provisions." Brief for Respondents 12.

[5] ☑ Our analysis begins with this premise: The City's actions would violate the disparate-treatment prohibition of Title VII absent some valid defense. All the evidence demonstrates that the City chose not to certify the examination results because of the statistical disparity based on race: that is, how minority candidates had performed when compared to white candidates. As the District Court put it, the City rejected the test results because "too many whites and not enough minorities would be promoted were the lists to be certified." 554 F.Supp.2d, at 152; see also *ibid.* (respondents' "own arguments ... show that the City's reasons for advocating noncertification were related to the racial distribution of the results"). Without some other justification, this express, race-based decision making violates Title VII's command that employers cannot take adverse employment actions because of an individual's race. See § 2000e-2(a)(1).

The District Court did not adhere to this principle, however. It held that respondents' "motivation to avoid making promotions based on a test with a racially disparate impact ... does not, as a matter of law, constitute discriminatory intent." 554 F.Supp.2d, at 160. And the Government makes a similar argument in this *2674 Court. It contends that the "structure of Title VII belies any claim that an employer's intent to comply with Title VII's disparate-impact provisions constitutes prohibited discrimination on the basis of race." Brief for United States as *Amicus Curiae* 11. But both of those statements turn upon the City's objective-avoiding disparate-impact liability—while ignoring the City's conduct in the name of reaching that objective. Whatever the City's ultimate aim—however well intentioned or benevolent it might have seemed—the City made its employment decision because of race. The City rejected the test results solely because the higher-scoring candidates were white. The question is not whether that conduct was discriminatory but whether the City had a lawful justification for its race-based action.

[6] ☑ We consider, therefore, whether the purpose to avoid disparate-impact liability excuses what otherwise would be prohibited disparate-treatment discrimination. Courts often confront cases in which statutes and principles point in different directions. Our task is to provide guidance to employers and courts for situations when these two prohibitions could be in conflict absent a rule to reconcile them. In providing this guidance, our decision must be consistent with the important purpose of Title VII—that the workplace be an environment free of discrimination, where race is not a barrier to opportunity.

[7] ☑ [8] ☑ With these principles in mind, we turn to the parties' proposed means of reconciling the statutory provisions. Petitioners take a strict approach, arguing that under Title VII, it cannot be permissible for an employer to take race-based adverse employment actions in order to avoid disparate-impact liability-even if the employer knows its practice violates the disparate-impact provision. See Brief for Petitioners 43. Petitioners would have us hold that, under Title VII, avoiding unintentional discrimination cannot justify intentional discrimination. That assertion, however, ignores the fact that, by codifying the disparate-impact provision in 1991, Congress has expressly prohibited both types of discrimination. We must interpret the statute to give effect to both provisions where possible. See, for example, *United States v. Atlantic Research Corp.*, 551 U.S. 128, 137, 127 S.Ct. 2331, 168 L.Ed.2d 28 (2007) (rejecting an interpretation that would render a statutory provision "a dead letter"). We cannot accept petitioners' broad and inflexible formulation.

[9] ☑ [10] ☑ Petitioners next suggest that an employer in fact must be in violation of the disparate-impact provision before it can use compliance as a defense in a disparate-treatment suit. Again, this is overly simplistic and too restrictive of Title VII's purpose. The rule petitioners offer would run counter to what we have recognized as Congress's intent that "voluntary compliance" be "the preferred means of achieving the objectives of Title VII." *Firefighters v. Cleveland,* 478 U.S. 501, 515, 106 S.Ct. 3063, 92 L.Ed.2d 405 (1986); see also *Wygant v. Jackson Bd. of Ed.,* 476 U.S. 267, 290, 106 S.Ct. 1842, 90 L.Ed.2d 260 (1986) (O'Connor, J., concurring in part and concurring in judgment). Forbidding employers to act unless they know, with certainty, that a practice violates the disparate-impact provision would bring compliance efforts to a near standstill. Even in the limited situations when this restricted standard could be met, employers likely would hesitate before taking voluntary action for fear of later being proven wrong in the course of litigation and then held to account for disparate treatment.

[11] ☑ [12] ☑ At the opposite end of the spectrum, respondents and the Government assert that an employer's good-faith *2675 belief that its actions are necessary to comply with Title VII's disparate-impact provision should be enough to justify race-conscious conduct. But the original, foundational prohibition of Title VII bars employers from taking adverse action "because of ... race."

§ 2000e-2(a)(1). And when Congress codified the disparate-impact provision in 1991, it made no exception to disparate-treatment liability for actions taken in a good-faith effort to comply with the new, disparate-impact provision in subsection (k). Allowing employers to violate the disparate-treatment prohibition based on a mere good-faith fear of disparate-impact liability would encourage race-based action at the slightest hint of disparate impact. A minimal standard could cause employers to discard the results of lawful and beneficial promotional examinations even where there is little if any evidence of disparate-impact discrimination. That would amount to a de facto quota system, in which a "focus on statistics ... could put undue pressure on employers to adopt inappropriate prophylactic measures." *Watson,* 487 U.S., at 992, 108 S.Ct. 2777 (plurality opinion). Even worse, an employer could discard test results (or other employment practices) with the intent of obtaining the employer's preferred racial balance. That operational principle could not be justified, for Title VII is express in disclaiming any interpretation of its requirements as calling for outright racial balancing. § 2000e-2(j). The purpose of Title VII "is to promote hiring on the basis of job qualifications, rather than on the basis of race or color." *Griggs,* 401 U.S., at 434, 91 S.Ct. 849.

In searching for a standard that strikes a more appropriate balance, we note that this Court has considered cases similar to this one, albeit in the context of the Equal Protection Clause of the Fourteenth Amendment. The Court has held that certain government actions to remedy past racial discrimination-actions that are themselves based on race-are constitutional only where there is a " 'strong basis in evidence' " that the remedial actions were necessary. *Richmond v. J.A. Croson Co.,* 488 U.S. 469, 500, 109 S.Ct. 706, 102 L.Ed.2d 854 (1989) (quoting *Wygant, supra,* at 277, 106 S.Ct. 1842 (plurality opinion)). This suit does not call on us to consider whether the statutory constraints under Title VII must be parallel in all respects to those under the Constitution. That does not mean the constitutional authorities are irrelevant, however. Our cases discussing constitutional principles can provide helpful guidance in this statutory context. See *Watson, supra,* at 993, 108 S.Ct. 2777 (plurality opinion).

Writing for a plurality in *Wygant* and announcing the strong-basis-in-evidence standard, Justice Powell recognized the tension between eliminating segregation and discrimination on the one hand and doing away with all governmentally imposed discrimination based on race on the other. 476 U.S., at 277, 106 S.Ct. 1842. The plurality stated that those "related constitutional duties are not always harmonious," and that "reconciling them requires ... employers to act with extraordinary care." *Ibid.* The plurality required a strong basis in evidence because "[e]videntiary support for the conclusion that remedial action is warranted becomes crucial when the remedial program is challenged in court by nonminority employees." *Ibid.* The Court applied the same standard in *Croson,* observing that "an amorphous claim that there has been past discrimination ... cannot justify the use of an unyielding racial quota." 488 U.S., at 499, 109 S.Ct. 706.

[13] ☑ The same interests are at work in the interplay between the disparate-treatment and disparate-impact provisions of *2676 Title VII. Congress has imposed liability on employers for unintentional discrimination in order to rid

the workplace of "practices that are fair in form, but discriminatory in operation." *Griggs, supra,* at 431, 91 S.Ct. 849. But it has also prohibited employers from taking adverse employment actions "because of" race. § 2000e-2(a)(1). Applying the strong-basis-in-evidence standard to Title VII gives effect to both the disparate-treatment and disparate-impact provisions, allowing violations of one in the name of compliance with the other only in certain, narrow circumstances. The standard leaves ample room for employers' voluntary compliance efforts, which are essential to the statutory scheme and to Congress's efforts to eradicate workplace discrimination. See *Firefighters, supra,* at 515. And the standard appropriately constrains employers' discretion in making race-based decisions: It limits that discretion to cases in which there is a strong basis in evidence of disparate-impact liability, but it is not so restrictive that it allows employers to act only when there is a provable, actual violation.

Resolving the statutory conflict in this way allows the disparate-impact prohibition to work in a manner that is consistent with other provisions of Title VII, including the prohibition on adjusting employment-related test scores on the basis of race. See § 2000e-2(*l*). Examinations like those administered by the City create legitimate expectations on the part of those who took the tests. As is the case with any promotion exam, some of the firefighters here invested substantial time, money, and personal commitment in preparing for the tests. Employment tests can be an important part of a neutral selection system that safeguards against the very racial animosities Title VII was intended to prevent. Here, however, the firefighters saw their efforts invalidated by the City in sole reliance upon race-based statistics.

If an employer cannot rescore a test based on the candidates' race, § 2000e-2(*l*), then it follows *a fortiori* that it may not take the greater step of discarding the test altogether to achieve a more desirable racial distribution of promotion-eligible candidates—absent a strong basis in evidence that the test was deficient and that discarding the results is necessary to avoid violating the disparate-impact provision. Restricting an employer's ability to discard test results (and thereby discriminate against qualified candidates on the basis of their race) also is in keeping with Title VII's express protection of bona fide promotional examinations. See § 2000e-2(h) ("[N]or shall it be an unlawful employment practice for an employer to give and to act upon the results of any professionally developed ability test provided that such test, its administration or action upon the results is not designed, intended or used to discriminate because of race"); cf. *AT & T Corp. v. Hulteen,* 556 U.S. —, —, 129 S.Ct. 1962, 1970, 173 L.Ed.2d 898 (2009).

For the foregoing reasons, we adopt the strong-basis-in-evidence standard as a matter of statutory construction to resolve any conflict between the disparate-treatment and disparate-impact provisions of Title VII.

Our statutory holding does not address the constitutionality of the measures taken here in purported compliance with Title VII. We also do not hold that meeting the strong-basis-in-evidence standard would satisfy the Equal Protection Clause in a future case. As we explain below, because respondents have not met their burden under Title VII, we need not decide whether a legitimate fear of disparate impact is ever sufficient to justify discriminatory treatment under the Constitution.

*2677 [14] ☑ Nor do we question an employer's affirmative efforts to ensure that all groups have a fair opportunity to apply for promotions and to participate in the process by which promotions will be made. But once that process has been established and employers have made clear their selection criteria, they may not then invalidate the test results, thus upsetting an employee's legitimate expectation not to be judged on the basis of race. Doing so, absent a strong basis in evidence of an impermissible disparate impact, amounts to the sort of racial preference that Congress has disclaimed, § 2000e-2(j), and is antithetical to the notion of a workplace where individuals are guaranteed equal opportunity regardless of race.

[15] ☑ [16] ☑ Title VII does not prohibit an employer from considering, before administering a test or practice, how to design that test or practice in order to provide a fair opportunity for all individuals, regardless of their race. And when, during the test-design stage, an employer invites comments to ensure the test is fair, that process can provide a common ground for open discussions toward that end. We hold only that, under Title VII, before an employer can engage in intentional discrimination for the asserted purpose of avoiding or remedying an unintentional disparate impact, the employer must have a strong basis in evidence to believe it will be subject to disparate-impact liability if it fails to take the race-conscious, discriminatory action.

C

The City argues that, even under the strong-basis-in-evidence standard, its decision to discard the examination results was permissible under Title VII. That is incorrect. Even if respondents were motivated as a subjective matter by a desire to avoid committing disparate-impact discrimination, the record makes clear there is no support for the conclusion that respondents had an objective, strong basis in evidence to find the tests inadequate, with some consequent disparate-impact liability in violation of Title VII.

[17] ☑ [18] ☑ On this basis, we conclude that petitioners have met their obligation to demonstrate that there is "no genuine issue as to any material fact" and that they are "entitled to judgment as a matter of law." Fed. Rule Civ. Proc. 56(c). On a motion for summary judgment, "facts must be viewed in the light most favorable to the nonmoving party only if there is a 'genuine' dispute as to those facts." *Scott v. Harris,* 550 U.S. 372, 380, 127 S.Ct. 1769, 167 L.Ed.2d 686 (2007). "Where the record taken as a whole could not lead a rational trier of fact to find for the nonmoving party, there is no genuine issue for trial." *Matsushita Elec. Industrial Co. v. Zenith Radio Corp.,* 475 U.S. 574, 587, 106 S.Ct. 1348, 89 L.Ed.2d 538 (1986) (internal quotation marks omitted). In this Court, the City's only defense is that it acted to comply with Title VII's disparate-impact provision. To succeed on their motion, then, petitioners must demonstrate that there can be no genuine dispute that there was no strong basis in evidence for the City to conclude it would face disparate-impact liability if it certified the examination results. See *Celotex Corp. v. Catrett,* 477 U.S. 317, 324, 106 S.Ct. 2548, 91 L.Ed.2d 265 (1986) (where the nonmoving party "will bear the burden of proof at trial on a dispositive issue," the nonmoving party bears the burden

of production under Rule 56 to "designate specific facts showing that there is a genuine issue for trial" (internal quotation marks omitted)).

The racial adverse impact here was significant, and petitioners do not dispute that the City was faced with a prima facie case of disparate-impact liability. On the *2678 captain exam, the pass rate for white candidates was 64 percent but was 37.5 percent for both black and Hispanic candidates. On the lieutenant exam, the pass rate for white candidates was 58.1 percent; for black candidates, 31.6 percent; and for Hispanic candidates, 20 percent. The pass rates of minorities, which were approximately one-half the pass rates for white candidates, fall well below the 80-percent standard set by the EEOC to implement the disparate-impact provision of Title VII. See 29 CFR § 1607.4(D) (2008) (selection rate that is less than 80 percent "of the rate for the group with the highest rate will generally be regarded by the Federal enforcement agencies as evidence of adverse impact"); *Watson,* 487 U.S., at 995-996, n. 3, 108 S.Ct. 2777 (plurality opinion) (EEOC's 80-percent standard is "a rule of thumb for the courts"). Based on how the passing candidates ranked and an application of the "rule of three," certifying the examinations would have meant that the City could not have considered black candidates for any of the then-vacant lieutenant or captain positions.

[19] ☑ Based on the degree of adverse impact reflected in the results, respondents were compelled to take a hard look at the examinations to determine whether certifying the results would have had an impermissible disparate impact. The problem for respondents is that a prima facie case of disparate-impact liability— essentially, a threshold showing of a significant statistical disparity, *Connecticut v. Teal,* 457 U.S. 440, 446, 102 S.Ct. 2525, 73 L.Ed.2d 130 (1982), and nothing more—is far from a strong basis in evidence that the City would have been liable under Title VII had it certified the results. That is because the City could be liable for disparate-impact discrimination only if the examinations were not job related and consistent with business necessity, or if there existed an equally valid, less discriminatory alternative that served the City's needs but that the City refused to adopt. § 2000e-2(k)(1)(A), (C). We conclude there is no strong basis in evidence to establish that the test was deficient in either of these respects. We address each of the two points in turn, based on the record developed by the parties through discovery—a record that concentrates in substantial part on the statements various witnesses made to the CSB.

1

[20] ☑ There is no genuine dispute that the examinations were job related and consistent with business necessity. The City's assertions to the contrary are "blatantly contradicted by the record" *Scott, supra,* at 380, 127 S.Ct. 1769. The CSB heard statements from Chad Legel (the IOS vice president) as well as city officials outlining the detailed steps IOS took to develop and administer the examinations. IOS devised the written examinations, which were the focus of the CSB's inquiry, after painstaking analyses of the captain and lieutenant positions—analyses in which IOS made sure that minorities were overrepresented. And IOS drew the questions from source material approved by the Department. Of the outside witnesses who appeared before the CSB, only one, Vincent Lewis,

had reviewed the examinations in any detail, and he was the only one with any firefighting experience. Lewis stated that the "questions were relevant for both exams." CA2 App. A1053. The only other witness who had seen any part of the examinations, Christopher Hornick (a competitor of IOS's), criticized the fact that no one within the Department had reviewed the tests—a condition imposed by the City to protect the integrity of the exams in light of past alleged security breaches. But Hornick stated that the exams "appea[r] to be ... reasonably good" and recommended *2679 that the CSB certify the results. *Id.*, at A1041.

Arguing that the examinations were not job related, respondents note some candidates' complaints that certain examination questions were contradictory or did not specifically apply to firefighting practices in New Haven. But Legel told the CSB that IOS had addressed those concerns—that it entertained "a handful" of challenges to the validity of particular examination questions, that it "reviewed those challenges and provided feedback [to the City] as to what we thought the best course of action was," and that he could remember at least one question IOS had thrown out ("offer[ing] credit to everybody for that particular question"). *Id.*, at A955-A957. For his part, Hornick said he "suspect[ed] that some of the criticisms ... [leveled] by candidates" were not valid. *Id.*, at A1035.

The City, moreover, turned a blind eye to evidence that supported the exams' validity. Although the City's contract with IOS contemplated that IOS would prepare a technical report consistent with EEOC guidelines for examination-validity studies, the City made no request for its report. After the January 2004 meeting between Legel and some of the city-official respondents, in which Legel defended the examinations, the City sought no further information from IOS, save its appearance at a CSB meeting to explain how it developed and administered the examinations. IOS stood ready to provide respondents with detailed information to establish the validity of the exams, but respondents did not accept that offer.

2

[21] ☑ Respondents also lacked a strong basis in evidence of an equally valid, less discriminatory testing alternative that the City, by certifying the examination results, would necessarily have refused to adopt. Respondents raise three arguments to the contrary, but each argument fails. First, respondents refer to testimony before the CSB that a different composite-score calculation weighting the written and oral examination scores 30/70—would have allowed the City to consider two black candidates for then-open lieutenant positions and one black candidate for then-open captain positions. (The City used a 60/40 weighting as required by its contract with the New Haven firefighters' union.) But respondents have produced no evidence to show that the 60/40 weighting was indeed arbitrary. In fact, because that formula was the result of a union-negotiated collective-bargaining agreement, we presume the parties negotiated that weighting for a rational reason. Nor does the record contain any evidence that the 30/70 weighting would be an equally valid way to determine whether candidates possess the proper mix of job knowledge and situational skills to earn promotions. Changing the weighting formula, moreover, could well have violated Title VII's

prohibition of altering test scores on the basis of race. See § 2000e-2(*l*). On this record, there is no basis to conclude that a 30/70 weighting was an equally valid alternative the City could have adopted.

Second, respondents argue that the City could have adopted a different interpretation of the "rule of three" that would have produced less discriminatory results. The rule, in the New Haven city charter, requires the City to promote only from "those applicants with the three highest scores" on a promotional examination. New Haven, Conn., Code of Ordinances, Tit. I, Art. XXX, § 160 (1992). A state court has interpreted the charter to prohibit so-called "banding"- the City's previous practice of rounding scores to the nearest whole number and considering all *2680 candidates with the same whole-number score as being of one rank. Banding allowed the City to consider three ranks of candidates (with the possibility of multiple candidates filling each rank) for purposes of the rule of three. See *Kelly v. New Haven,* No. CV000444614, 2004 WL 114377, *3 (Conn. Super.Ct., Jan. 9, 2004). Respondents claim that employing banding here would have made four black and one Hispanic candidates eligible for then-open lieutenant and captain positions.

A state court's prohibition of banding, as a matter of municipal law under the charter, may not eliminate banding as a valid alternative under Title VII. See 42 U.S.C. § 2000e-7. We need not resolve that point, however. Here, banding was not a valid alternative for this reason: Had the City reviewed the exam results and then adopted banding to make the minority test scores appear higher, it would have violated Title VII's prohibition of adjusting test results on the basis of race. § 2000e-2(*l*); see also *Chicago Firefighters Local 2 v. Chicago,* 249 F.3d 649, 656 (C.A.7 2001) (Posner, J.) ("We have no doubt that if banding were adopted in order to make lower black scores seem higher, it would indeed be ... forbidden"). As a matter of law, banding was not an alternative available to the City when it was considering whether to certify the examination results.

Third, and finally, respondents refer to statements by Hornick in his telephone interview with the CSB regarding alternatives to the written examinations. Hornick stated his "belie[f]" that an "assessment center process," which would have evaluated candidates' behavior in typical job tasks, "would have demonstrated less adverse impact." CA2 App. A1039. But Hornick's brief mention of alternative testing methods, standing alone, does not raise a genuine issue of material fact that assessment centers were available to the City at the time of the examinations and that they would have produced less adverse impact. Other statements to the CSB indicated that the Department could not have used assessment centers for the 2003 examinations. *Supra,* at 2670. And although respondents later argued to the CSB that Hornick had pushed the City to reject the test results, *supra,* at 2671–2672, the truth is that the essence of Hornick's remarks supported its certifying the test results. See *Scott,* 550 U.S., at 380, 127 S.Ct. 1769. Hornick stated that adverse impact in standardized testing "has been in existence since the beginning of testing," CA2 App. A1037, and that the disparity in New Haven's test results was "somewhat higher but generally in the range that we've seen professionally." *Id.,* at A1030-A1031. He told the CSB he was "not suggesting" that IOS "somehow created a test that had adverse impacts that it should not have had." *Id.,* at A1038. And he suggested that the CSB should "certify the list as it exists." *Id.,* at A1041.

Especially when it is noted that the strong-basis-in-evidence standard applies, respondents cannot create a genuine issue of fact based on a few stray (and contradictory) statements in the record. And there is no doubt respondents fall short of the mark by relying entirely on isolated statements by Hornick. Hornick had not "stud[ied] the test at length or in detail." *Id.,* at A1030. And as he told the CSB, he is a "direct competitor" of IOS's. *Id.,* at A1029. The remainder of his remarks showed that Hornick's primary concern—somewhat to the frustration of CSB members— was marketing his services for the future, not commenting on the results of the tests the City had already administered. See, e.g., *id.,* at A1026, A1027, A1032, A1036, A1040, A1041. Hornick's hinting had its intended effect: The City has since hired him as a consultant. As for the other outside witnesses who spoke *2681 to the CSB, Vincent Lewis (the retired fire captain) thought the CSB should certify the test results. And Janet Helms (the Boston College professor) declined to review the examinations and told the CSB that, as a society, "we need to develop a new way of assessing people." *Id.,* at A1073. That task was beyond the reach of the CSB, which was concerned with the adequacy of the test results before it.

3

[22] ☑ On the record before us, there is no genuine dispute that the City lacked a strong basis in evidence to believe it would face disparate-impact liability if it certified the examination results. In other words, there is no evidence—let alone the required strong basis in evidence—that the tests were flawed because they were not job related or because other equally valid and less discriminatory tests were available to the City. Fear of litigation alone cannot justify an employer's reliance on race to the detriment of individuals who passed the examinations and qualified for promotions. The City's discarding the test results was impermissible under Title VII, and summary judgment is appropriate for petitioners on their disparate-treatment claim.

* * *

The record in this litigation documents a process that, at the outset, had the potential to produce a testing procedure that was true to the promise of Title VII: No individual should face workplace discrimination based on race. Respondents thought about promotion qualifications and relevant experience in neutral ways. They were careful to ensure broad racial participation in the design of the test itself and its administration. As we have discussed at length, the process was open and fair.

The problem, of course, is that after the tests were completed, the raw racial results became the predominant rationale for the City's refusal to certify the results. The injury arises in part from the high, and justified, expectations of the candidates who had participated in the testing process on the terms the City had established for the promotional process. Many of the candidates had studied for months, at considerable personal and financial expense, and thus the injury caused by the City's reliance on raw racial statistics at the end of the process was all the more severe. Confronted with arguments both for and against certifying the test results—and threats of a lawsuit either way—the City was required to make a difficult inquiry. But its hearings produced no strong evidence of a disparate-impact violation, and the City was not entitled to disregard the tests based solely on the racial disparity in the results.

Our holding today clarifies how Title VII applies to resolve competing expectations under the disparate-treatment and disparate-impact provisions. If, after it certifies the test results, the City faces a disparate-impact suit, then in light of our holding today it should be clear that the City would avoid disparate-impact liability based on the strong basis in evidence that, had it not certified the results, it would have been subject to disparate-treatment liability.

Petitioners are entitled to summary judgment on their Title VII claim, and we therefore need not decide the underlying constitutional question. The judgment of the Court of Appeals is reversed, and the cases are remanded for further proceedings consistent with this opinion.

It is so ordered.

JUSTICE **SCALIA,** CONCURRING.

I join the Court's opinion in full, but write separately to observe that its resolution *2682 of this dispute merely postpones the evil day on which the Court will have to confront the question: Whether, or to what extent, are the disparate-impact provisions of Title VII of the Civil Rights Act of 1964 consistent with the Constitution's guarantee of equal protection? The question is not an easy one. See generally Primus, Equal Protection and Disparate Impact: Round Three, 117 Harv. L.Rev. 493 (2003).

The difficulty is this: Whether or not Title VII's disparate-treatment provisions forbid "remedial" race-based actions when a disparate-impact violation would *not* otherwise result—the question resolved by the Court today—it is clear that Title VII not only permits but affirmatively *requires* such actions when a disparate-impact violation *would* otherwise result. See *ante,* at 2674. But if the Federal Government is prohibited from discriminating on the basis of race, *Bolling v. Sharpe,* 347 U.S. 497, 500, 74 S.Ct. 693, 98 L.Ed. 884 (1954), then surely it is also prohibited from enacting laws mandating that third parties—e.g., employers, whether private, State, or municipal—discriminate on the basis of race. See *Buchanan v. Warley,* 245 U.S. 60, 78-82, 38 S.Ct. 16, 62 L.Ed. 149 (1917). As the facts of these cases illustrate, Title VII's disparate-impact provisions place a racial thumb on the scales, often requiring employers to evaluate the racial outcomes of their policies, and to make decisions based on (because of) those racial outcomes. That type of racial decision-making is, as the Court explains, discriminatory. See *ante,* at 2673; *Personnel Administrator of Mass. v. Feeney,* 442 U.S. 256, 279, 99 S.Ct. 2282, 60 L.Ed.2d 870 (1979).

To be sure, the disparate-impact laws do not mandate imposition of quotas, but it is not clear why that should provide a safe harbor. Would a private employer not be guilty of unlawful discrimination if he refrained from establishing a racial hiring quota but intentionally designed his hiring practices to achieve the same end? Surely he would. Intentional discrimination is still occurring, just one step up the chain. Government compulsion of such design would therefore seemingly

violate equal protection principles. Nor would it matter that Title VII requires
consideration of race on a wholesale, rather than retail, level. "[T]he Government
must treat citizens as individuals, not as simply components of a racial, religious,
sexual or national class." *Miller v. Johnson,* 515 U.S. 900, 911, 115 S.Ct. 2475,
132 L.Ed.2d 762 (1995) (internal quotation marks omitted). And, of course, the
purportedly benign motive for the disparate-impact provisions cannot save the
statute. See *Adarand Constructors, Inc. v. Pena,* 515 U.S. 200, 227, 115 S.Ct.
2097, 132 L.Ed.2d 158 (1995).

It might be possible to defend the law by framing it as simply an evidentiary
tool used to identify genuine, intentional discrimination—to "smoke out," as it
were, disparate treatment. See Primus, *supra,* at 498-499, 520-521. Disparate
impact is sometimes (though not always; see *Watson v. Fort Worth Bank &
Trust,* 487 U.S. 977, 992, 108 S.Ct. 2777, 101 L.Ed.2d 827 (1988) (plurality opinion)) a signal of something illicit, so a regulator might allow statistical disparities
to play some role in the evidentiary process. Cf. *McDonnell Douglas Corp. v.
Green,* 411 U.S. 792, 802-803, 93 S.Ct. 1817, 36 L.Ed.2d 668 (1973). But arguably, the disparate-impact provisions sweep too broadly to be fairly characterized in such a fashion—since they fail to provide an affirmative defense for
good-faith (i.e., nonracially motivated) conduct, or perhaps even for good faith
plus hiring standards that are entirely reasonable. See *post,* at 2697–2698, and
n. 1 (GINSBURG, J., dissenting) (describing the demanding *2683 nature of the
"business necessity" defense). This is a question that this Court will have to consider in due course. It is one thing to free plaintiffs from proving an employer's
illicit intent, but quite another to preclude the employer from proving that its
motives were pure and its actions reasonable.

The Court's resolution of these cases makes it unnecessary to resolve these
matters today. But the war between disparate impact and equal protection will
be waged sooner or later, and it behooves us to begin thinking about how—and
on what terms—to make peace between them.

JUSTICE **ALITO,** WITH WHOM JUSTICE **SCALIA** AND
JUSTICE **THOMAS** JOIN, CONCURRING.

I join the Court's opinion in full. I write separately only because the dissent,
while claiming that "[t]he Court's recitation of the facts leaves out important
parts of the story," *post,* at 2690 (opinion of GINSBURG, J.), provides an incom-
plete description of the events that led to New Haven's decision to reject the
results of its exam. The dissent's omissions are important because, when all of
the evidence in the record is taken into account, it is clear that, even if the legal
analysis in Parts II and III-A of the dissent were accepted, affirmance of the
decision below is untenable.

I

When an employer in a disparate-treatment case under Title VII of the Civil
Rights Act of 1964 claims that an employment decision, such as the refusal to
promote, was based on a legitimate reason, two questions—one objective and one
subjective—must be decided. The first, objective question is whether the reason

given by the employer is one that is legitimate under Title VII. See *St. Mary's Honor Center v. Hicks,* 509 U.S. 502, 506-507, 113 S.Ct. 2742, 125 L.Ed.2d 407 (1993). If the reason provided by the employer is not legitimate on its face, the employer is liable. *Id.,* at 509, 113 S.Ct. 2742. The second, subjective question concerns the employer's intent. If an employer offers a facially legitimate reason for its decision but it turns out that this explanation was just a pretext for discrimination, the employer is again liable. See *id.,* at 510-512, 113 S.Ct. 2742.

The question on which the opinion of the Court and the dissenting opinion disagree concerns the objective component of the determination that must be made when an employer justifies an employment decision, like the one made in this litigation, on the ground that a contrary decision would have created a risk of disparate-impact liability. The Court holds—and I entirely agree—that concern about disparate-impact liability is a legitimate reason for a decision of the type involved here only if there was a "substantial basis in evidence to find the tests inadequate." *Ante,* at 2677. The Court ably demonstrates that in this litigation no reasonable jury could find that the city of New Haven (City) possessed such evidence and therefore summary judgment for petitioners is required. Because the Court correctly holds that respondents cannot satisfy this objective component, the Court has no need to discuss the question of the respondents' actual intent. As the Court puts it, "[e]ven if respondents were motivated as a subjective matter by a desire to avoid committing disparate-impact discrimination, the record makes clear there is no support for the conclusion that respondents had an objective, substantial basis in evidence to find the tests inadequate." *Ibid.*

The dissent advocates a different objective component of the governing standard. According to the dissent, the objective *2684 component should be whether the evidence provided "good cause" for the decision, *post,* at 2699, and the dissent argues—incorrectly, in my view—that no reasonable juror could fail to find that such evidence was present here. But even if the dissent were correct on this point, I assume that the dissent would not countenance summary judgment for respondents if respondents' professed concern about disparate-impact litigation was simply a pretext. Therefore, the decision below, which sustained the entry of summary judgment for respondents, cannot be affirmed unless no reasonable jury could find that the City's asserted reason for scrapping its test-concern about disparate-impact liability was a pretext and that the City's real reason was illegitimate, namely, the desire to placate a politically important racial constituency.

II

A

As initially described by the dissent, see *post,* at 2690–2695, the process by which the City reached the decision not to accept the test results was open, honest, serious, and deliberative. But even the District Court admitted that "a jury could rationally infer that city officials worked behind the scenes to sabotage the promotional examinations because they knew that, were the exams certified, the Mayor would incur the wrath of [Rev. Boise] Kimber and other influential leaders of New Haven's African-American community." 554 F.Supp.2d 142, 162 (Conn.2006), summarily aff'd, 530 F.3d 87 (C.A.2 2008) *(per curiam)*.

This admission finds ample support in the record. Reverend Boise Kimber, to whom the District Court referred, is a politically powerful New Haven pastor and a self-professed " 'kingmaker.' " App. to Pet. for Cert. in No. 07-1428, p. 906a; see also *id.,* at 909a. On one occasion, "[i]n front of TV cameras, he threatened a race riot during the murder trial of the black man arrested for killing white Yalie Christian Prince. He continues to call whites racist if they question his actions." *Id.,* at 931a.

Reverend Kimber's personal ties with seven-term New Haven Mayor John DeStefano (Mayor) stretch back more than a decade. In 1996, for example, Mayor DeStefano testified for Rev. Kimber as a character witness when Rev. Kimber—then the manager of a funeral home—was prosecuted and convicted for stealing prepaid funeral expenses from an elderly woman and then lying about the matter under oath. See *id.,* at 126a, 907a. "Reverend Kimber has played a leadership role in all of Mayor DeStefano's political campaigns, [and] is considered a valuable political supporter and vote-getter." *Id.,* at 126a. According to the Mayor's former campaign manager (who is currently his executive assistant), Rev. Kimber is an invaluable political asset because "[h]e's very good at organizing people and putting together field operations, as a result of his ties to labor, his prominence in the religious community, and his long-standing commitment to roots." *Id.,* at 908a (internal quotation marks and alteration omitted).

In 2002, the Mayor picked Rev. Kimber to serve as the Chairman of the New Haven Board of Fire Commissioners (BFC), "despite the fact that he had no experience in the profession, fire administration, [or] municipal management." *Id.,* at 127a; see also *id.,* at 928a-929a. In that capacity, Rev. Kimber told firefighters that certain new recruits would not be hired because "they just have too many vowels in their name[s]." Thanawala, New Haven Fire Panel Chairman Steps Down Over Racial Slur, Hartford Courant, June 13, 2002, p. B2. After protests about *2685 this comment, Rev. Kimber stepped down as chairman of the BFC, *ibid.;* see also App. to Pet. for Cert. in No. 07-1428, at 929a, but he remained on the BFC and retained "a direct line to the mayor," *id.,* at 816a.

Almost immediately after the test results were revealed in "early January" 2004, Rev. Kimber called the City's Chief Administrative Officer, Karen Dubois-Walton, who "acts 'on behalf of the Mayor.'" *Id.,* at 221a, 812a. Dubois-Walton and Rev. Kimber met privately in her office because he wanted "to express his opinion" about the test results and "to have some influence" over the City's response. *Id.,* at 815a-816a. As discussed in further detail below, Rev. Kimber adamantly opposed certification of the test results—a fact that he or someone in the Mayor's office eventually conveyed to the Mayor. *Id.,* at 229a.

B

On January 12, 2004, Tina Burgett (the director of the City's Department of Human Resources) sent an e-mail to Dubois-Walton to coordinate the City's response to the test results. Burgett wanted to clarify that the City's executive officials would meet "sans the Chief, and that once we had a better fix on the next steps we would meet with the Mayor (possibly) and then the two Chiefs." *Id.,* at 446a. The "two Chiefs" are Fire Chief William Grant (who is white) and Assistant Fire Chief Ronald Dumas (who is African-American). Both chiefs believed that the test results should be certified. *Id.,* at 228a, 817a. Petitioners allege, and the record suggests, that the Mayor and his staff colluded "sans the Chief[s]" because "the defendants did not want Grant's or Dumas' views to be publicly known; accordingly both men were prevented by the Mayor and his staff from making any statements regarding the matter." *Id.,* at 228a.[FN1]

FN1. Although the dissent disputes it, see *post,* at 2707, n. 17, the record certainly permits the inference that petitioners' allegation is true. See App. to Pet. for Cert. in No. 07-1428, pp. 846a-851a (deposition of Dubois-Walton).

The next day, on January 13, 2004, Chad Legel, who had designed the tests, flew from Chicago to New Haven to meet with Dubois-Walton, Burgett, and Thomas Ude, the City's corporate counsel. *Id.,* at 179a. "Legel outlined the merits of the examination and why city officials should be confident in the validity of the results." *Ibid.* But according to Legel, Dubois-Walton was "argumentative" and apparently had already made up her mind that the tests were " 'discriminatory.' " *Id.,* at 179a-180a. Again according to Legel, "[a] theme" of the meeting was "the political and racial overtones of what was going on in the City." *Id.,* at 181a. "Legel came away from the January 13, 2004 meeting with the impression that defendants were already leaning toward discarding the examination results." *Id.,* at 180a.

On January 22, 2004, the Civil Service Board (CSB or Board) convened its first public meeting. Almost immediately, Rev. Kimber began to exert political pressure on the CSB. He began a loud, minutes-long outburst that required the CSB Chairman to shout him down and hold him out of order three times. See *id.,* at 187a, 467a-468a; see also App. in No. 06-4996-cv (CA2), pp. A703-A705. Reverend Kimber protested the public meeting, arguing that he and the other fire commissioners should first be allowed to meet with the CSB in private. App. to Pet. for Cert. in No. 07-1428, at 188a.

Four days after the CSB's first meeting, Mayor DeStefano's executive aide sent an e-mail to Dubois-Walton, Burgett, and *2686 Ude. *Id.,* at 190a. The message clearly indicated that the Mayor had made up his mind to oppose certification of the test results (but nevertheless wanted to conceal that fact from the public):

> "I wanted to make sure we are all on the same page for this meeting tomorrow .... *[L]et's remember, that these folks are not against certification yet. So we can't go in and tell them that is our position;* we have to deliberate and arrive there as the fairest and most cogent outcome." *Ibid.*

On February 5, 2004, the CSB convened its second public meeting. Reverend Kimber again testified and threatened the CSB with political recriminations if they voted to certify the test results:

> "I look at this [Board] tonight. I look at three whites and one Hispanic and no blacks .... I would hope that you would not put yourself in this type of position, *a political ramification that may come back upon you* as you

sit on this [Board] and decide the future of a department and the future
of those who are being promoted.

"(APPLAUSE)." *Id.*, at 492a (emphasis added).

One of the CSB members "t[ook] great offense" because he believed that
Rev. Kimber "consider[ed][him] a bigot because [his] face is white." *Id.*, at 496a.
The offended CSB member eventually voted not to certify the test results. *Id.*,
at 586a-587a.

One of Rev. Kimber's "friends and allies," Lieutenant Gary Tinney, also
exacerbated racial tensions before the CSB. *Id.*, at 129a. After some firefighters applauded in support of certifying the test results, "Lt. Tinney exclaimed,
'Listen to the Klansmen behind us.'" *Id.*, at 225a.

Tinney also has strong ties to the Mayor's office. See, e.g., *id.*, at 129a-130a,
816a-817a. After learning that he had not scored well enough on the captain's
exam to earn a promotion, Tinney called Dubois-Walton and arranged a meeting
in her office. *Id.*, at 830a-831a, 836a. Tinney alleged that the white firefighters
had cheated on their exams—an accusation that Dubois-Walton conveyed to the
Board without first conducting an investigation into its veracity. *Id.*, at 837a-
838a; see also App. 164 (statement of CSB Chairman, noting the allegations of
cheating). The allegation turned out to be baseless. App. to Pet. for Cert. in No.
07-1428, at 836a.

Dubois-Walton never retracted the cheating allegation, but she and other
executive officials testified several times before the CSB. In accordance with
directions from the Mayor's office to make the CSB meetings appear deliberative, see *id.*, at 190a, executive officials remained publicly uncommitted about
certification—while simultaneously "work[ing] as a team" behind closed doors
with the secretary of the CSB to devise a political message that would convince
the CSB to vote against certification, see *id.*, at 447a. At the public CSB meeting
on March 11, 2004, for example, Corporation Counsel Ude bristled at one board
member's suggestion that City officials were recommending against certifying
the test results. See *id.*, at 215a ("Attorney Ude took offense, stating, 'Frankly,
because I would never make a recommendation; I would not have made a recommendation like that'"). But within days of making that public statement, Ude
privately told other members of the Mayor's team "the ONLY way we get to a
decision not to certify is" to focus on something other than "a big discussion re:
adverse impact" law. *Id.*, at 458a-459a.

*2687 As part of its effort to deflect attention from the specifics of the test,
the City relied heavily on the testimony of Dr. Christopher Hornick, who is one
of Chad Legel's competitors in the test-development business. Hornick never
"stud[ied] the test [that Legel developed] at length or in detail," *id.*, at 549a;
see also *id.*, at 203a, 553a, but Hornick did review and rely upon literature
sent to him by Burgett to criticize Legel's test. For example, Hornick "noted
in the literature that [Burgett] sent that the test was not customized to the New
Haven Fire Department." *Id.*, at 551a. The Chairman of the CSB immediately
corrected Hornick. *Id.*, at 552a ("Actually, it was, Dr. Hornick"). Hornick also
relied on newspaper accounts—again, sent to him by Burgett—pertaining to

the controversy surrounding the certification decision. See *id.*, at 204a, 557a. Although Hornick again admitted that he had no knowledge about the actual test that Legel had developed and that the City had administered, see *id.*, at 560a-561a, the City repeatedly relied upon Hornick as a testing "guru" and, in the CSB Chairman's words, "the City ke[pt] quoting him as a person that we should rely upon more than anybody else [to conclude that there] is a better way—a better mousetrap."[FN2] App. in No. 06-4996-cv (CA2), at A1128. Dubois-Walton later admitted that the City rewarded Hornick for his testimony by hiring him to develop and administer an alternative test. App. to Pet. for Cert. in No. 07-1428, at 854a; see also *id.*, at 562a-563a (Hornick's plea for future business from the City on the basis of his criticisms of Legel's tests).

FN2. The City's heavy reliance on Hornick's testimony makes the two chiefs' silence all the more striking. See *supra*, at 2685. While Hornick knew little or nothing about the tests he criticized, the two chiefs were involved "during the lengthy process that led to the devising of the administration of these exams," App. to Pet. for Cert. in No. 07-1428, at 847a, including "collaborating with City officials on the extensive job analyses that were done," "selection of the oral panelists," and selection of "the proper content and subject matter of the exams," *id.*, at 847a-848a.

At some point prior to the CSB's public meeting on March 18, 2004, the Mayor decided to use his executive authority to disregard the test results—*even if* the CSB ultimately voted to certify them. *Id.*, at 819a-820a. Accordingly, on the evening of March 17th, Dubois-Walton sent an e-mail to the Mayor, the Mayor's executive assistant, Burgett, and attorney Ude, attaching two alternative press releases. *Id.*, at 457a. The first would be issued if the CSB voted not to certify the test results; the second would be issued (and would explain the Mayor's invocation of his executive authority) if the CSB voted to certify the test results. *Id.*, at 217a-218a, 590a-591a, 819a-820a. Half an hour after Dubois-Walton circulated the alternative drafts, Burgett replied: "[W]ell, that seems to say it all. Let's hope draft # 2 hits the shredder tomorrow nite." *Id.*, at 457a.

Soon after the CSB voted against certification, Mayor DeStefano appeared at a dinner event and "took credit for the scu[tt]ling of the examination results." *Id.*, at 230a.

C

Taking into account all the evidence in the summary judgment record, a reasonable jury could find the following. Almost as soon as the City disclosed the racial makeup of the list of firefighters who scored the highest on the exam, the City administration was lobbied by an influential community leader to scrap the test results, and the City administration decided on that course of action before making *2688 any real assessment of the possibility of a disparate-impact violation. To achieve that end, the City administration concealed its internal decision but worked—as things turned out, successfully—to persuade the CSB that acceptance of the test results would be illegal and would expose the City to disparate-impact liability. But in the event that the CSB was not persuaded, the Mayor, wielding ultimate decision-making authority, was prepared to overrule the CSB immediately. Taking this view of the evidence, a reasonable jury could easily find that the City's real reason for scrapping the test results was not a concern about violating the disparate-impact provision of Title VII but a simple desire to please a politically important racial constituency. It is noteworthy that the Solicitor General—whose position on the principal legal issue in this case is largely aligned with the dissent—concludes that "[n]either the district court nor the court of appeals ... adequately considered whether, viewing the evidence in the light most favorable to petitioners, a genuine issue of material fact remained whether respondents' claimed purpose to comply with Title VII was a pretext for intentional racial discrimination ...." Brief for United States as *Amicus Curiae* 6; see also *id.,* at 32-33.

III

I will not comment at length on the dissent's criticism of my analysis, but two points require a response.

The first concerns the dissent's statement that I "equat[e] political considerations with unlawful discrimination." *Post,* at 2708–2709. The dissent misrepresents my position: I draw no such equation. Of course "there are many ways in which a politician can attempt to win over a constituency-including a racial constituency-without engaging in unlawful discrimination." *Post,* at 2708–2709. But—as I assume the dissent would agree—there are some things that a public official cannot do, and one of those is engaging in intentional racial discrimination when making employment decisions.

The second point concerns the dissent's main argument—that efforts by the Mayor and his staff to scuttle the test results are irrelevant because the ultimate decision was made by the CSB. According to the dissent, "[t]he relevant decision was made by the CSB," *post,* at 2708–2709, and there is "scant cause to suspect" that anything done by the opponents of certification, including the Mayor and his staff, "prevented the CSB from evenhandedly assessing the reliability of the exams and rendering an independent, good-faith decision on certification," *post,* at 2708.

Adoption of the dissent's argument would implicitly decide an important question of Title VII law that this Court has never resolved—the circumstances in which an employer may be held liable based on the discriminatory intent of subordinate employees who influence but do not make the ultimate employment decision. There is a large body of court of appeals case law on this issue, and these cases disagree about the proper standard. See *EEOC v. BCI Coca-Cola Bottling Co. of Los Angeles,* 450 F.3d 476, 484-488 (C.A.10 2006) (citing cases and describing the approaches taken in different Circuits). One standard is whether the subordinate "exerted influenc [e] over the titular decision maker." *Russell v. McKinney Hosp. Venture,* 235 F.3d 219, 227 (C.A.5 2000); see also *Poland v. Chertoff,* 494 F.3d 1174, 1182 (C.A.9 2007) (A subordinate's bias is imputed to the employer where the subordinate "influenced or was involved in the decision or decision-making process"). Another is whether the discriminatory input "caused the adverse employment action." *2689 See *BCI Coca-Cola Bottling Co. of Los Angeles, supra,* at 487.

In the present cases, a reasonable jury could certainly find that these standards were met. The dissent makes much of the fact that members of the CSB swore under oath that their votes were based on the good-faith belief that certification of the results would have violated federal law. See *post,* at 2707–2708. But the good faith of the CSB members would not preclude a finding that the presentations engineered by the Mayor and his staff influenced or caused the CSB decision.

The least employee-friendly standard asks only whether "the actual decision maker" acted with discriminatory intent, see *Hill v. Lockheed Martin Logistics Management, Inc.,* 354 F.3d 277, 291 (C.A.4 2004) (en banc), and it is telling that, even under this standard, summary judgment for respondents would not be proper. This is so because a reasonable jury could certainly find that in New Haven, the Mayor—not the CSB—wielded the final decision-making power. After all, the Mayor claimed that authority and was poised to use it in the event that the CSB decided to accept the test results. See *supra,* at 2687. If the Mayor had the authority to overrule a CSB decision *accepting* the test results, the Mayor also presumably had the authority to overrule the CSB's decision *rejecting* the test results. In light of the Mayor's conduct, it would be quite wrong to throw out petitioners' case on the ground that the CSB was the ultimate decision maker.

* * *

Petitioners are firefighters who seek only a fair chance to move up the ranks in their chosen profession. In order to qualify for promotion, they made personal sacrifices. Petitioner Frank Ricci, who is dyslexic, found it necessary to "hir[e] someone, at considerable expense, to read onto audiotape the content of the books and study materials." App. to Pet. for Cert. in No. 07-1428, at 169a. He "studied an average of eight to thirteen hours a day ..., even listening to audio tapes while driving his car." *Ibid.* Petitioner Benjamin Vargas, who is Hispanic, had to "give up a part-time job," and his wife had to "take leave from her own job in order to take care of their three young children while Vargas studied." *Id.,* at 176a. "Vargas devoted countless hours to study ..., missed two of his children's birthdays and over two weeks of vacation time," and "incurred significant financial expense" during the three-month study period. *Id.,* at 176a-177a.

Petitioners were denied promotions for which they qualified because of the race and ethnicity of the firefighters who achieved the highest scores on the City's exam. The District Court threw out their case on summary judgment, even though that court all but conceded that a jury could find that the City's asserted justification was pretextual. The Court of Appeals then summarily affirmed that decision.

The dissent grants that petitioners' situation is "unfortunate" and that they "understandably attract this Court's sympathy." *Post,* at 2690, 2710. But "sympathy" is not what petitioners have a right to demand. What they have a right to demand is even-handed enforcement of the law—of Title VII's prohibition against discrimination based on race. And that is what, until today's decision, has been denied them.

Justice **GINSBURG**, with whom Justice **STEVENS**, Justice
SOUTER, and Justice **BREYER** join, dissenting.

In assessing claims of race discrimination, "[c]ontext matters." *2690 *Grutter v.
Bollinger,* 539 U.S. 306, 327, 123 S.Ct. 2325, 156 L.Ed.2d 304 (2003). In 1972,
Congress extended Title VII of the Civil Rights Act of 1964 to cover public
employment. At that time, municipal fire departments across the country, including New Haven's, pervasively discriminated against minorities. The extension
of Title VII to cover jobs in firefighting effected no overnight change. It took
decades of persistent effort, advanced by Title VII litigation, to open firefighting
posts to members of racial minorities.

The white firefighters who scored high on New Haven's promotional exams
understandably attract this Court's sympathy. But they had no vested right to
promotion. Nor have other persons received promotions in preference to them.
New Haven maintains that it refused to certify the test results because it believed,
for good cause, that it would be vulnerable to a Title VII disparate-impact suit if
it relied on those results. The Court today holds that New Haven has not demonstrated "a strong basis in evidence" for its plea. *Ante,* at 2664. In so holding, the
Court pretends that "[t]he City rejected the test results solely because the higher
scoring candidates were white." *Ante,* at 2674. That pretension, essential to the
Court's disposition, ignores substantial evidence of multiple flaws in the tests
New Haven used. The Court similarly fails to acknowledge the better tests used
in other cities, which have yielded less racially skewed outcomes.[FN1]

FN1. Never mind the flawed tests New Haven used and the better selection
methods used elsewhere, Justice ALITO's concurring opinion urges. Overriding
all else, racial politics, fired up by a strident African-American pastor, were at
work in New Haven. See *ante,* at 2665–2668. Even a detached and disinterested
observer, however, would have every reason to ask: Why did such racially skewed
results occur in New Haven, when better tests likely would have produced less
disproportionate results?

By order of this Court, New Haven, a city in which African-Americans
and Hispanics account for nearly 60 percent of the population, must today be
served—as it was in the days of undisguised discrimination—by a fire department in which members of racial and ethnic minorities are rarely seen in command positions. In arriving at its order, the Court barely acknowledges the
pathmarking decision in *Griggs v. Duke Power Co.,* 401 U.S. 424, 91 S.Ct. 849,
28 L.Ed.2d 158 (1971), which explained the centrality of the disparate-impact
concept to effective enforcement of Title VII. The Court's order and opinion, I
anticipate, will not have staying power.

I

A

The Court's recitation of the facts leaves out important parts of the story.
Firefighting is a profession in which the legacy of racial discrimination casts
an especially long shadow. In extending Title VII to state and local government
employers in 1972, Congress took note of a U.S. Commission on Civil Rights

(USCCR) report finding racial discrimination in municipal employment even "more pervasive than in the private sector." H.R.Rep. No. 92-238, p. 17 (1971). According to the report, overt racism was partly to blame, but so too was a failure on the part of municipal employers to apply merit-based employment principles. In making hiring and promotion decisions, public employers often "rel[ied] on criteria unrelated to job performance," including nepotism or political patronage. 118 Cong. Rec. 1817 (1972). Such flawed selection methods served to entrench preexisting racial hierarchies. The USCCR report singled out police and fire departments for having "[b]arriers to equal employment ... greater ... than in *2691 any other area of State or local government," with African-Americans "hold[ing] almost no positions in the officer ranks." *Ibid.* See also National Commission on Fire Prevention and Control, America Burning 5 (1973) ("Racial minorities are under-represented in the fire departments in nearly every community in which they live.").

The City of New Haven (City) was no exception. In the early 1970s, African-Americans and Hispanics composed 30 percent of New Haven's population, but only 3.6 percent of the City's 502 firefighters. The racial disparity in the officer ranks was even more pronounced: "[O]f the 107 officers in the Department only one was black, and he held the lowest rank above private." *Firebird Soc. of New Haven, Inc. v. New Haven Bd. of Fire Comm'rs,* 66 F.R.D. 457, 460 (Conn.1975).

Following a lawsuit and settlement agreement, see *ibid.,* the City initiated efforts to increase minority representation in the New Haven Fire Department (Department). Those litigation-induced efforts produced some positive change. New Haven's population includes a greater proportion of minorities today than it did in the 1970s: Nearly 40 percent of the City's residents are African-American, and more than 20 percent are Hispanic. Among entry-level firefighters, minorities are still underrepresented, but not starkly so. As of 2003, African-Americans and Hispanics constituted 30 percent and 16 percent of the City's firefighters, respectively. In supervisory positions, however, significant disparities remain. Overall, the senior officer ranks (captain and higher) are nine percent African-American and nine percent Hispanic. Only 1 of the Department's 21 fire captains is African-American. See App. in No. 06-4996-cv (CA2), p. A1588 (hereinafter CA2 App.). It is against this backdrop of entrenched inequality that the promotion process at issue in this litigation should be assessed.

B

By order of its charter, New Haven must use competitive examinations to fill vacancies in fire officer and other civil-service positions. Such examinations, the City's civil service rules specify, "shall be practical in nature, shall relate to matters which fairly measure the relative fitness and capacity of the applicants to discharge the duties of the position which they seek, and shall take into account character, training, experience, physical and mental fitness." *Id.,* at A331. The City may choose among a variety of testing methods, including written and oral exams and "[p]erformance tests to demonstrate skill and ability in performing actual work." *Id.,* at A332.

New Haven, the record indicates, did not closely consider what sort of "practical" examination would "fairly measure the relative fitness and capacity of the applicants to discharge the duties" of a fire officer. Instead, the City simply adhered to the testing regime outlined in its two-decades-old contract with the local firefighters' union: a written exam, which would account for 60 percent of an applicant's total score, and an oral exam, which would account for the remaining 40 percent. *Id.,* at A1045. In soliciting bids from exam development companies, New Haven made clear that it would entertain only "proposals that include a written component that will be weighted at 60%, and an oral component that will be weighted at 40%." *Id.,* at A342. Chad Legel, a representative of the winning bidder, Industrial/Organizational Solutions, Inc. (IOS), testified during his deposition that the City never asked whether alternative methods might better measure the qualities of a successful fire officer, including leadership *2692 skills and command presence. See *id.,* at A522 ("I was under contract and had responsibility only to create the oral interview and the written exam.").

Pursuant to New Haven's specifications, IOS developed and administered the oral and written exams. The results showed significant racial disparities. On the lieutenant exam, the pass rate for African-American candidates was about one-half the rate for Caucasian candidates; the pass rate for Hispanic candidates was even lower. On the captain exam, both African-American and Hispanic candidates passed at about half the rate of their Caucasian counterparts. See App. 225-226. More striking still, although nearly half of the 77 lieutenant candidates were African-American or Hispanic, none would have been eligible for promotion to the eight positions then vacant. The highest scoring African-American candidate ranked 13th; the top Hispanic candidate was 26th. As for the seven then-vacant captain positions, two Hispanic candidates would have been eligible, but no African-Americans. The highest scoring African-American candidate ranked 15th. See *id.,* at 218-219.

These stark disparities, the Court acknowledges, sufficed to state a prima facie case under Title VII's disparate-impact provision. See *ante,* at 2678 ("The pass rates of minorities ... f[e]ll well below the 80-percent standard set by the [Equal Employment Opportunity Commission (EEOC)] to implement the disparate-impact provision of Title VII."). New Haven thus had cause for concern about the prospect of Title VII litigation and liability. City officials referred the matter to the New Haven Civil Service Board (CSB), the entity responsible for certifying the results of employment exams.

Between January and March 2004, the CSB held five public meetings to consider the proper course. At the first meeting, New Haven's Corporation Counsel, Thomas Ude, described the legal standard governing Title VII disparate-impact claims. Statistical imbalances alone, Ude correctly recognized, do not give rise to liability. Instead, presented with a disparity, an employer "has the opportunity and the burden of proving that the test is job-related and consistent with business necessity." CA2 App. A724. A Title VII plaintiff may attempt to rebut an employer's showing of job-relatedness and necessity by identifying alternative selection methods that would have been at least as valid but with "less of an adverse or disparate or discriminatory effect." *Ibid.* See also *id.,* at A738. Accordingly, the CSB Commissioners understood, their principal task was to decide whether they were confident about the reliability of the exams: Had the exams fairly measured the qualities of a successful fire officer despite their disparate results? Might an alternative examination process have identified the most qualified candidates without creating such significant racial imbalances?

Seeking a range of input on these questions, the CSB heard from test takers, the test designer, subject-matter experts, City officials, union leaders, and community members. Several candidates for promotion, who did not yet know their exam results, spoke at the CSB's first two meetings. Some candidates favored certification. The exams, they emphasized, had closely tracked the assigned study materials. Having invested substantial time and money to prepare themselves for the test, they felt it would be unfair to scrap the results. See, e.g., *id.,* at A772-A773, A785-A789.

Other firefighters had a different view. A number of the exam questions, they pointed out, were not germane to New Haven's practices and procedures. See, e.g., *id.,* at A774-A784. At least two candidates *2693 opposed to certification noted unequal access to study materials. Some individuals, they asserted, had the necessary books even before the syllabus was issued. Others had to invest substantial sums to purchase the materials and "wait a month and a half for some of the books because they were on back-order." *Id.,* at A858. These disparities, it was suggested, fell at least in part along racial lines. While many Caucasian applicants could obtain materials and assistance from relatives in the fire service, the overwhelming majority of minority applicants were "first-generation firefighters" without such support networks. See *id.,* at A857-A861, A886-A887.

A representative of the Northeast Region of the International Association of Black Professional Firefighters, Donald Day, also spoke at the second meeting. Statistical disparities, he told the CSB, had been present in the Department's previous promotional exams. On earlier tests, however, a few minority candidates had fared well enough to earn promotions. *Id.,* at A828. See also App. 218-219. Day contrasted New Haven's experience with that of nearby Bridgeport, where minority firefighters held one-third of lieutenant and captain positions. Bridgeport, Day observed, had once used a testing process similar to New Haven's, with a written exam accounting for 70 percent of an applicant's score, an oral exam for 25 percent, and seniority for the remaining five percent. CA2 App. A830. Bridgeport recognized, however, that the oral component, more so than the written component, addressed the sort of "real-life scenarios" fire officers encounter on the job. *Id.,* at A832. Accordingly, that city "changed the relative weights" to give primacy to the oral exam. *Ibid.* Since that time, Day reported, Bridgeport had seen minorities "fairly represented" in its exam results. *Ibid.*

The CSB's third meeting featured IOS representative Legel, the leader of the team that had designed and administered the exams for New Haven. Several City officials also participated in the discussion. Legel described the exam development process in detail. The City, he recounted, had set the "parameters" for the exams, specifically, the requirement of written and oral components with a 60/40 weighting. *Id.,* at A923, A974. For security reasons, Department officials had not been permitted to check the content of the questions prior to their administration. Instead, IOS retained a senior fire officer from Georgia to review the exams "for content and fidelity to the source material." *Id.,* at

A936. Legel defended the exams as "facially neutral," and stated that he "would stand by the[ir] validity." *Id.*, at A962. City officials did not dispute the neutrality of IOS's work. But, they cautioned, even if individual exam questions had no intrinsic bias, the selection process as a whole may nevertheless have been deficient. The officials urged the CSB to consult with experts about the "larger picture." *Id.*, at A1012.

At its fourth meeting, CSB solicited the views of three individuals with testing-related expertise. Dr. Christopher Hornick, an industrial/organizational psychology consultant with 25 years' experience with police and firefighter testing, described the exam results as having "relatively high adverse impact." *Id.*, at A1028. Most of the tests he had developed, Hornick stated, exhibited "significantly and dramatically less adverse impact." *Id.*, at A1029. Hornick downplayed the notion of "facial neutrality." It was more important, he advised the CSB, to consider "the broader issue of how your procedures and your rules and the types of tests that you are using are contributing to the adverse impact." *Id.*, at A1038.

*2694 Specifically, Hornick questioned New Haven's union-prompted 60/40 written/oral examination structure, noting the availability of "different types of testing procedures that are much more valid in terms of identifying the best potential supervisors in [the] fire department." *Id.*, at A1032. He suggested, for example, "an assessment center process, which is essentially an opportunity for candidates ... to demonstrate how they would address a particular problem as opposed to just verbally saying it or identifying the correct option on a written test." *Id.*, at A1039-A1040. Such selection processes, Hornick said, better "identif[y] the best possible people" and "demonstrate dramatically less adverse impacts." *Ibid.* Hornick added:

> I've spoken to at least 10,000, maybe 15,000 firefighters in group settings in my consulting practice and I have never one time ever had anyone in the fire service say to me, "Well, the person who gets the highest score on a written job knowledge/multiple-guess test makes the best company officer." We know that it's not as valid as other procedures that exist. *Id.*, at A1033.

See also *id.*, at A1042-A1043 ("I think a person's leadership skills, their command presence, their interpersonal skills, their management skills, their tactical skills could have been identified and evaluated in a much more appropriate way.").

Hornick described the written test itself as "reasonably good," *id.*, at A1041, but he criticized the decision not to allow Department officials to check the content. According to Hornick, this "inevitably" led to "test[ing] for processes and procedures that don't necessarily match up into the department." *Id.*, at A1034-A1035. He preferred "experts from within the department who have signed confidentiality agreements ... to make sure that the terminology and equipment that's being identified from standardized reading sources apply to the department." *Id.*, at A1035.

Asked whether he thought the City should certify the results, Hornick hedged: "There is adverse impact in the test. That will be identified in any proceeding that you have. You will have industrial psychology experts, if it goes to court, on both sides. And it will not be a pretty or comfortable position for anyone to be in." *Id.,* at A1040-A1041. Perhaps, he suggested, New Haven might certify the results but immediately begin exploring "alternative ways to deal with these issues" in the future. *Id.,* at A1041.

The two other witnesses made relatively brief appearances. Vincent Lewis, a specialist with the Department of Homeland Security and former fire officer in Michigan, believed the exams had generally tested relevant material, although he noted a relatively heavy emphasis on questions pertaining to being an "apparatus driver." He suggested that this may have disadvantaged test takers "who had not had the training or had not had an opportunity to drive the apparatus." *Id.,* at A1051. He also urged the CSB to consider whether candidates had, in fact, enjoyed equal access to the study materials. *Ibid.* Cf. *supra,* at 2693.

Janet Helms, a professor of counseling psychology at Boston College, observed that two-thirds of the incumbent fire officers who submitted job analyses to IOS during the exam design phase were Caucasian. Members of different racial groups, Helms told the CSB, sometimes do their jobs in different ways, "often because the experiences that are open to white male firefighters are not open to members of these other under-represented groups." CA2 App. A1063-A1064. The heavy reliance on job analyses from white firefighters, *2695 she suggested, may thus have introduced an element of bias. *Id.,* at A1063.

The CSB's fifth and final meeting began with statements from City officials recommending against certification. Ude, New Haven's counsel, repeated the applicable disparate-impact standard:

> [A] finding of adverse impact is the beginning, not the end, of a review of testing procedures. Where a procedure demonstrates adverse impact, you look to how closely it is related to the job that you're looking to fill and you also look at whether there are other ways to test for those qualities, those traits, those positions that are equally valid with less adverse impact. *Id.,* at A1100-A1101.

New Haven, Ude and other officials asserted, would be vulnerable to Title VII liability under this standard. Even if the exams were "facially neutral," significant doubts had been raised about whether they properly assessed the key attributes of a successful fire officer. *Id.,* at A1103. See also *id.,* at A1125 ("Upon close reading of the exams, the questions themselves would appear to test a candidate's ability to memorize textbooks but not necessarily to identify solutions to real problems on the fire ground."). Moreover, City officials reminded the CSB, Hornick and others had identified better, less discriminatory selection methods—such as assessment centers or exams with a more heavily weighted oral component. *Id.,* at A1108-A1109, A1129-A1130.

After giving members of the public a final chance to weigh in, the CSB voted on certification, dividing 2 to 2. By rule, the result was noncertification. Voting no, Commissioner Webber stated, "I originally was going to vote to certify. ... But I've heard enough testimony here to give me great doubts about the test itself and ... some of the procedures. And I believe we can do better." *Id.,* at A1157. Commissioner Tirado likewise concluded that the "flawed" testing process counseled against certification. *Id.,* at A1158. Chairman Segaloff and Commissioner

Caplan voted to certify. According to Segaloff, the testimony had not "compelled [him] to say this exam was not job-related," and he was unconvinced that alternative selection processes would be "less discriminatory." *Id.*, at A1159-A1160. Both Segalhoff and Caplan, however, urged the City to undertake civil service reform. *Id.*, at A1150-A1154.

C

Following the CSB's vote, petitioners—17 white firefighters and one Hispanic firefighter, all of whom had high marks on the exams—filed suit in the United States District Court for the District of Connecticut. They named as defendants—respondents here—the City, several City officials, a local political activist, and the two CSB members who voted against certifying the results. By opposing certification, petitioners alleged, respondents had discriminated against them in violation of Title VII's disparate-treatment provision and the Fourteenth Amendment's Equal Protection Clause. The decision not to certify, respondents answered, was a lawful effort to comply with Title VII's disparate-impact provision and thus could not have run afoul of Title VII's prohibition of disparate treatment. Characterizing respondents' stated rationale as a mere pretext, petitioners insisted that New Haven would have had a solid defense to any disparate-impact suit.

In a decision summarily affirmed by the Court of Appeals, the District Court granted summary judgment for respondents. 554 F.Supp.2d 142 (Conn.2006), aff'd, 530 F.3d 87 (C.A.2 2008) *(per curiam)*. Under Second Circuit precedent, the District Court explained, "the intent to remedy the *2696 disparate impact" of a promotional exam "is not equivalent to an intent to discriminate against non-minority applicants." 554 F.Supp.2d, at 157 (quoting *Hayden v. County of Nassau,* 180 F.3d 42, 51 (C.A.2 1999)). Rejecting petitioners' pretext argument, the court observed that the exam results were sufficiently skewed "to make out a prima facie case of discrimination" under Title VII's disparate-impact provision. 554 F.Supp.2d, at 158. Had New Haven gone forward with certification and been sued by aggrieved minority test takers, the City would have been forced to defend tests that were presumptively invalid. And, as the CSB testimony of Hornick and others indicated, overcoming that presumption would have been no easy task. *Id.*, at 153-156. Given Title VII's preference for voluntary compliance, the court held, New Haven could lawfully discard the disputed exams even if the City had not definitively "pinpoint[ed]" the source of the disparity and "ha[d] not yet formulated a better selection method." *Id.*, at 156.

Respondents were no doubt conscious of race during their decision-making process, the court acknowledged, but this did not mean they had engaged in racially disparate treatment. The conclusion they had reached and the action thereupon taken were race-neutral in this sense: "[A]ll the test results were discarded, no one was promoted, and firefighters of every race will have to participate in another selection process to be considered for promotion." *Id.*, at 158. New Haven's action, which gave no individual a preference, "was 'simply not analogous to a quota system or a minority set-aside where candidates, on the basis of their race, are not treated uniformly.' " *Id.*, at 157 (quoting *Hayden,* 180 F.3d, at 50). For these and other reasons, the court also rejected petitioners' equal protection claim.

II

A

Title VII became effective in July 1965. Employers responded to the law by eliminating rules and practices that explicitly barred racial minorities from "white" jobs. But removing overtly race-based job classifications did not usher in genuinely equal opportunity. More subtle—and sometimes unconscious—forms of discrimination replaced once undisguised restrictions.

In *Griggs v. Duke Power Co.,* 401 U.S. 424, 91 S.Ct. 849, 28 L.Ed.2d 158 (1971), this Court responded to that reality and supplied important guidance on Title VII's mission and scope. Congress, the landmark decision recognized, aimed beyond "disparate treatment"; it targeted "disparate impact" as well. Title VII's original text, it was plain to the Court, "proscribe [d] not only overt discrimination but also practices that are fair in form, but discriminatory in operation." *Id.,* at 431, 91 S.Ct. 849.[FN2] Only by ignoring *Griggs* *2697 could one maintain that intentionally disparate treatment alone was Title VII's "original, foundational prohibition," and disparate impact a mere afterthought. Cf. *ante,* at 2675.

FN2. The Court's disparate-impact analysis rested on two provisions of Title VII: § 703(a)(2), which made it unlawful for an employer "to limit, segregate, or classify his employees in any way which would deprive or tend to deprive any individual of employment opportunities or otherwise adversely affect his status as an employee, because of such individual's race, color, religion, sex, or national origin"; and § 703(h), which permitted employers "to act upon the results of any professionally developed ability test provided that such test, its administration or action upon the results is not designed, intended or used to discriminate because of race, color, religion, sex or national origin." *Griggs v. Duke Power Co.,* 401 U.S. 424, 426, n. 1, 91 S.Ct. 849, 28 L.Ed.2d 158 (1971) (quoting 78 Stat. 255, 42 U.S.C. § 2000e-2(a)(2), (h) (1964 ed.)). See also 401 U.S., at 433-436, 91 S.Ct. 849 (explaining that § 703(h) authorizes only tests that are "demonstrably a reasonable measure of job performance").

Griggs addressed Duke Power Company's policy that applicants for positions, save in the company's labor department, be high school graduates and score satisfactorily on two professionally prepared aptitude tests. "[T]here was no showing of a discriminatory purpose in the adoption of the diploma and test requirements." 401 U.S., at 428, 91 S.Ct. 849. The policy, however, "operated to render ineligible a markedly disproportionate number of [African-Americans]." *Id.,* at 429, 91 S.Ct. 849. At the time of the litigation, in North Carolina, where the Duke Power plant was located, 34 percent of white males, but only 12 percent of African-American males, had high school diplomas. *Id.,* at 430, n. 6, 91 S.Ct. 849. African-Americans also failed the aptitude tests at a significantly higher rate than whites. *Ibid.* Neither requirement had been "shown to bear a demonstrable relationship to successful performance of the jobs for which it was used." *Id.,* at 431, 91 S.Ct. 849.

The Court unanimously held that the company's diploma and test requirements violated Title VII. "[T]o achieve equality of employment opportunities," the Court comprehended, Congress "directed the thrust of the Act to the *consequences* of employment practices, not simply the motivation." *Id.,* at 429, 432, 91 S.Ct. 849. That meant "unnecessary barriers to employment" must fall, even if "neutral on their face" and "neutral in terms of intent." *Id.*, at 430, 431, 91 S.Ct. 849. "The touchstone" for determining whether a test or qualification meets Title VII's measure, the Court said, is not "good intent or the absence of discriminatory intent"; it is "business necessity." *Id.*, at 431, 432, 91 S.Ct. 849. Matching procedure to substance, the *Griggs* Court observed, Congress "placed on the employer the burden of showing that any given requirement ... ha[s] a manifest relationship to the employment in question." *Id.*, at 432, 91 S.Ct. 849.

In *Albemarle Paper Co. v. Moody,* 422 U.S. 405, 95 S.Ct. 2362, 45 L.Ed.2d 280 (1975), the Court, again without dissent, elaborated on *Griggs.* When an employment test "select[s] applicants for hire or promotion in a racial pattern significantly different from the pool of applicants," the Court reiterated, the employer must demonstrate a "manifest relationship" between test and job. 422 U.S., at 425, 95 S.Ct. 2362. Such a showing, the Court cautioned, does not necessarily mean the employer prevails: "[I]t remains open to the complaining party to show that other tests or selection devices, without a similarly undesirable racial effect, would also serve the employer's legitimate interest in 'efficient and trustworthy workmanship.' " *Ibid.*

Federal trial and appellate courts applied *Griggs* and *Albemarle* to disallow a host of hiring and promotion practices that "operate[d] as 'built in headwinds' for minority groups." *Griggs,* 401 U.S., at 432, 91 S.Ct. 849. Practices discriminatory in effect, courts repeatedly emphasized, could be maintained only upon an employer's showing of "an overriding and compelling business purpose." *Chrisner v. Complete Auto Transit, Inc.,* 645 F.2d 1251, 1261, n. 9 (C.A.6 1981).[FN3] That a practice served *2698 "legitimate management functions" did not, it was generally understood, suffice to establish business necessity. *Williams v. Colorado Springs, Colo., School Dist.,* 641 F.2d 835, 840-841 (C.A.10 1981) (internal quotation marks omitted). Among selection methods cast aside for lack of a "manifest relationship" to job performance were a number of written hiring and promotional examinations for firefighters.[FN4]

FN3. See also *Dothard v. Rawlinson,* 433 U.S. 321, 332, n. 14, 97 S.Ct. 2720, 53 L.Ed.2d 786 (1977) ("a discriminatory employment practice must be shown to be necessary to safe and efficient job performance to survive a Title VII challenge"); *Williams v. Colorado Springs, Colo., School Dist.,* 641 F.2d 835, 840-841 (C.A.10 1981) ("The term 'necessity' connotes that the exclusionary practice must be shown to be of great importance to job performance."); *Kirby v. Colony Furniture Co.,* 613 F.2d 696, 705, n. 6 (C.A.8 1980) ("the proper standard for determining whether 'business necessity' justifies a practice which has a racially discriminatory result is not whether it is justified by routine business considerations but whether there is a *compelling* need for the employer to maintain that practice and whether the employer can prove there is *no* alternative to the challenged practice"); *Pettway v. American Cast Iron Pipe Co.,* 494 F.2d 211, 244, n. 87 (C.A.5 1974) ("this doctrine of business necessity ... connotes an irresistible demand" (internal quotation marks omitted)); *United States v. Bethlehem Steel Corp.,* 446 F.2d 652, 662 (C.A.2 1971) (an exclusionary practice "must not only directly foster safety and efficiency of a plant, but also be essential to those goals"); *Robinson v. Lorillard Corp.,* 444 F.2d 791, 798 (C.A.4 1971) ("The test is whether there exists an overriding legitimate business purpose such that the practice is necessary to the safe and efficient operation of the business.").

FN4. See, e.g., *Nash v. Jacksonville,* 837 F.2d 1534 (C.A.11 1988), vacated, 490 U.S. 1103, 109 S.Ct. 3151, 104 L.Ed.2d 1015 (1989), opinion reinstated, 905 F.2d 355 (C.A.11 1990); *Vulcan Pioneers, Inc. v. New Jersey Dept. of Civil Serv.,* 832 F.2d 811 (CA3 1987); *Guardians Assn. of N.Y. City Police Dept. v. Civil Serv. Comm'n,* 630 F.2d 79 (C.A.2 1980); *Ensley Branch of NAACP v. Seibels,* 616 F.2d 812 (C.A.5 1980); *Firefighters Inst. for Racial Equality v. St. Louis,* 616 F.2d 350 (C.A.8 1980); *Boston Chapter, NAACP v. Beecher,* 504 F.2d 1017 (C.A.1 1974).

Moving in a different direction, in *Wards Cove Packing Co. v. Atonio,* 490 U.S. 642, 109 S.Ct. 2115, 104 L.Ed.2d 733 (1989), a bare majority of this Court significantly modified the *Griggs-Albemarle* delineation of Title VII's disparate-impact proscription. As to business necessity for a practice that disproportionately excludes members of minority groups, *Wards Cove* held, the employer bears only the burden of production, not the burden of persuasion. 490 U.S., at 659-660, 109 S.Ct. 2115. And in place of the instruction that the challenged practice "must have a manifest relationship to the employment in question," *Griggs,* 401 U.S., at 432, 91 S.Ct. 849, *Wards Cove* said that the practice would be permissible as long as it "serve[d], in a significant way, the legitimate employment goals of the employer." 490 U.S., at 659, 109 S.Ct. 2115.

In response to *Wards Cove* and "a number of [other] recent decisions by the United States Supreme Court that sharply cut back on the scope and effectiveness of [civil rights] laws," Congress enacted the Civil Rights Act of 1991. H.R.Rep. No. 102-40, pt. 2, p. 2 (1991). Among the 1991 alterations, Congress formally codified the disparate-impact component of Title VII. In so amending the statute, Congress made plain its intention to restore "the concepts of 'business necessity' and 'job related' enunciated by the Supreme Court in Griggs v. Duke Power Co. ... and in other Supreme Court decisions prior to Wards Cove Packing Co. v. Atonio." § 3(2), 105 Stat. 1071. Once a complaining party demonstrates that an employment practice causes a disparate impact, amended Title VII states, the burden is on the employer "to demonstrate that the challenged practice is job related for the position in question and consistent with business necessity." 42 U.S.C. § 2000e-2(k)(1)(A)(i). If the employer carries that substantial burden, the complainant may respond by identifying "an alternative employment *2699 practice" which the employer "refuses to adopt." § 2000e-2(k)(1)(A)(ii), (C).

B

Neither Congress' enactments nor this Court's Title VII precedents (including the now-discredited decision in *Wards Cove*) offer even a hint of "conflict" between an employer's obligations under the statute's disparate-treatment and disparate-impact provisions. Cf. *ante,* at 2673–2674. Standing on an equal footing, these twin pillars of Title VII advance the same objectives: ending workplace discrimination and promoting genuinely equal opportunity. See *McDonnell Douglas Corp. v. Green,* 411 U.S. 792, 800, 93 S.Ct. 1817, 36 L.Ed.2d 668 (1973).

Yet the Court today sets at odds the statute's core directives. When an employer changes an employment practice in an effort to comply with Title VII's disparate-impact provision, the Court reasons, it acts "because of race"-something Title VII's disparate-treatment provision, see § 2000e-2(a)(1), generally forbids. *Ante,* at 2673–2674. This characterization of an employer's compliance-directed action shows little attention to Congress' design or to the *Griggs* line of cases Congress recognized as pathmarking.

"[O]ur task in interpreting separate provisions of a single Act is to give the Act the most harmonious, comprehensive meaning possible in light of the legislative policy and purpose." *Weinberger v. Hynson, Westcott & Dunning, Inc.,* 412 U.S. 609, 631-632, 93 S.Ct. 2469, 37 L.Ed.2d 207 (1973) (internal quotation marks omitted). A particular phrase need not "extend to the outer limits of its definitional possibilities" if an incongruity would result. *Dolan v. Postal Service,* 546 U.S. 481, 486, 126 S.Ct. 1252, 163 L.Ed.2d 1079 (2006). Here, Title VII's disparate-treatment and disparate-impact proscriptions must be read as complementary.

In codifying the *Griggs* and *Albemarle* instructions, Congress declared unambiguously that selection criteria operating to the disadvantage of minority group members can be retained only if justified by business necessity.[FN5] In keeping with Congress' design, employers who reject such criteria due to reasonable doubts about their reliability can hardly be held to have engaged in discrimination "because of" race. A reasonable endeavor to comply with the law and to ensure that qualified candidates of all races have a fair opportunity to compete is simply not what Congress meant to interdict. I would therefore hold that an employer who jettisons a selection device when its disproportionate racial impact becomes apparent does not violate Title VII's disparate-treatment bar automatically or at all, subject to this key condition: The employer must have good cause to believe the device would not withstand examination for business necessity. Cf. *Faragher v. Boca Raton,* 524 U.S. 775, 806, 118 S.Ct. 2275, 141 L.Ed.2d 662 (1998) (observing that it accords with "clear statutory policy" for employers "to prevent violations" and "make reasonable efforts to discharge their duty" under Title VII).

FN5. What was the "business necessity" for the tests New Haven used? How could one justify, e.g., the 60/40 written/oral ratio, see *supra,* at 2665–2666, 2667–2668, under that standard? Neither the Court nor the concurring opinions attempt to defend the ratio.

EEOC's interpretative guidelines are corroborative. "[B]y the enactment of title VII," the guidelines state, "Congress did not intend to expose those who comply with the Act to charges that they are violating the very statute they are seeking to implement." 29 CFR § 1608.1(a) (2008). Recognizing EEOC's "enforcement *2700 responsibility" under Title VII, we have previously accorded the Commission's position respectful consideration. See, e.g., *Albemarle,* 422 U.S., at 431, 95 S.Ct. 2362; *Griggs,* 401 U.S., at 434, 91 S.Ct. 849. Yet the Court today does not so much as mention EEOC's counsel.

Our precedents defining the contours of Title VII's disparate-treatment prohibition further confirm the absence of any intra-statutory discord. In *Johnson v. Transportation Agency, Santa Clara Cty.,* 480 U.S. 616, 107 S.Ct. 1442, 94 L.Ed.2d 615 (1987), we upheld a municipal employer's voluntary affirmative-action plan against a disparate-treatment challenge. Pursuant to the plan, the employer selected a woman for a road-dispatcher position, a job category traditionally regarded as "male." A male applicant who had a slightly higher interview score brought suit under Title VII. This Court rejected his claim and approved the plan, which allowed consideration of gender as "one of numerous factors." *Id.*, at 638, 107 S.Ct. 1442. Such consideration, we said, is "fully consistent with Title VII" because plans of that order can aid "in eliminating the vestiges of discrimination in the workplace." *Id.*, at 642, 107 S.Ct. 1442.

This litigation does not involve affirmative action. But if the voluntary affirmative action at issue in *Johnson* does not discriminate within the meaning of Title VII, neither does an employer's reasonable effort to comply with Title VII's disparate-impact provision by refraining from action of doubtful consistency with business necessity.

C

To "reconcile" the supposed "conflict" between disparate treatment and disparate impact, the Court offers an enigmatic standard. *Ante,* at 2673–2674. Employers may attempt to comply with Title VII's disparate-impact provision, the Court declares, only where there is a "strong basis in evidence" documenting the necessity of their action. *Ante,* at 2662. The Court's standard, drawn from inapposite equal protection precedents, is not elaborated. One is left to wonder what cases would meet the standard and why the Court is so sure this case does not.

1

In construing Title VII, I note preliminarily, equal protection doctrine is of limited utility. The Equal Protection Clause, this Court has held, prohibits only intentional discrimination; it does not have a disparate-impact component. See *Personnel Administrator of Mass. v. Feeney,* 442 U.S. 256, 272, 99 S.Ct. 2282, 60 L.Ed.2d 870 (1979); *Washington v. Davis,* 426 U.S. 229, 239, 96 S.Ct. 2040, 48 L.Ed.2d 597 (1976). Title VII, in contrast, aims to eliminate all forms of employment discrimination, unintentional as well as deliberate. Until today, cf. *ante,* at 2664; *ante,* p. 2664 (SCALIA, J., concurring), this Court has never questioned the constitutionality of the disparate-impact component of Title VII, and for good reason. By instructing employers to avoid needlessly exclusionary selection processes, Title VII's disparate-impact provision calls for a "race-neutral means to increase minority ... participation"—something this Court's equal protection precedents also encourage. See *Adarand Constructors, Inc. v. Pena,* 515 U.S. 200, 238, 115 S.Ct. 2097, 132 L.Ed.2d 158 (1995) (quoting *Richmond v. J.A. Croson Co.,* 488 U.S. 469, 507, 109 S.Ct. 706, 102 L.Ed.2d 854 (1989)). "The very radicalism of holding disparate impact doctrine unconstitutional as a matter of equal protection," moreover, "suggests that only a very uncompromising court would issue such a decision." Primus, *2701 Equal Protection and Disparate Impact: Round Three, 117 Harv. L.Rev. 493, 585 (2003).

The cases from which the Court draws its strong-basis-in-evidence standard are particularly inapt; they concern the constitutionality of absolute racial preferences. See *Wygant v. Jackson Bd. of Ed.,* 476 U.S. 267, 277, 106 S.Ct. 1842, 90 L.Ed.2d 260 (1986) (plurality opinion) (invalidating a school district's plan to lay off nonminority teachers while retaining minority teachers with less seniority); *Croson,* 488 U.S., at 499-500, 109 S.Ct. 706 (rejecting a set-aside program for minority contractors that operated as "an unyielding racial quota"). An employer's effort to avoid Title VII liability by repudiating a suspect selection method scarcely resembles those cases. Race was not merely a relevant consideration in *Wygant* and *Croson;* it was the decisive factor. Observance of Title VII's disparate-impact provision, in contrast, calls for no racial preference, absolute or otherwise. The very purpose of the provision is to ensure that individuals are hired and promoted based on qualifications manifestly necessary to successful performance of the job in question, qualifications that do not screen out members of any race.[FN6]

FN6. Even in Title VII cases involving race-conscious (or gender-conscious) affirmative-action plans, the Court has never proposed a strong-basis-in-evidence standard. In *Johnson v. Transportation Agency, Santa Clara Cty.,* 480 U.S. 616, 107 S.Ct. 1442, 94 L.Ed.2d 615 (1987), the Court simply examined the municipal employer's action for reasonableness: "Given the obvious imbalance in the Skilled Craft category, and given the Agency's commitment to eliminating such imbalances, it was plainly not unreasonable for the Agency ... to consider as one factor the sex of [applicants] in making its decision." *Id.,* at 637, 107 S.Ct. 1442. See also *Firefighters v. Cleveland,* 478 U.S. 501, 516, 106 S.Ct. 3063, 92 L.Ed.2d 405 (1986) ("Title VII permits employers and unions voluntarily to make use of reasonable race-conscious affirmative action.").

2

The Court's decision in this litigation underplays a dominant Title VII theme. This Court has repeatedly emphasized that the statute "should not be read to thwart" efforts at voluntary compliance. *Johnson,* 480 U.S., at 630, 107 S.Ct. 1442. Such compliance, we have explained, is "the preferred means of achieving [Title VII's] objectives." *Firefighters v. Cleveland,* 478 U.S. 501, 515, 106 S.Ct. 3063, 92 L.Ed.2d 405 (1986). See also *Kolstad v. American Dental Assn.,* 527 U.S. 526, 545, 119 S.Ct. 2118, 144 L.Ed.2d 494 (1999) ("Dissuading employers from [taking voluntary action] to prevent discrimination in the workplace is directly contrary to the purposes underlying Title VII."); 29 CFR § 1608.1(c). The strong-basis-in-evidence standard, however, as barely described in general, and cavalierly applied in this case, makes voluntary compliance a hazardous venture.

As a result of today's decision, an employer who discards a dubious selection process can anticipate costly disparate-treatment litigation in which its chances for success—even for surviving a summary-judgment motion—are highly problematic. Concern about exposure to disparate-impact liability, however well grounded, is insufficient to insulate an employer from attack. Instead, the employer must make a "strong" showing that (1) its selection method was "not job related and consistent with business necessity," or (2) that it refused to adopt "an equally valid, less-discriminatory alternative." *Ante,* at 2778. It is hard to see how these requirements differ from demanding that an employer establish "a provable, actual violation" *against itself.* Cf. *ante,* at 2676. There is indeed a sharp conflict here, but it is not the false one the Court describes between Title VII's core provisions. It is, *2702 instead, the discordance of the Court's opinion with the voluntary compliance ideal. Cf. *Wygant,* 476 U.S., at 290, 106 S.Ct. 1842 (O'Connor, J., concurring in part and concurring in judgment)

("The imposition of a requirement that public employers make findings that they have engaged in illegal discrimination before they [act] would severely undermine public employers' incentive to meet voluntarily their civil rights obligations.").[FN7]

FN7. Notably, prior decisions applying a strong-basis-in-evidence standard have not imposed a burden as heavy as the one the Court imposes today. In *Croson,* the Court found no strong basis in evidence because the City had offered "nothing approaching a prima facie case." *Richmond v. J.A. Croson Co.,* 488 U.S. 469, 500, 109 S.Ct. 706, 102 L.Ed.2d 854 (1989). The Court did not suggest that anything beyond a prima facie case would have been required. In the context of race-based electoral districting, the Court has indicated that a "strong basis" exists when the "threshold conditions" for liability are present. *Bush v. Vera,* 517 U.S. 952, 978, 116 S.Ct. 1941, 135 L.Ed.2d 248 (1996) (plurality opinion).

3

The Court's additional justifications for announcing a strong-basis-in-evidence standard are unimpressive. First, discarding the results of tests, the Court suggests, calls for a heightened standard because it "upset[s] an employee's legitimate expectation." *Ante,* at 2677. This rationale puts the cart before the horse. The legitimacy of an employee's expectation depends on the legitimacy of the selection method. If an employer reasonably concludes that an exam fails to identify the most qualified individuals and needlessly shuts out a segment of the applicant pool, Title VII surely does not compel the employer to hire or promote based on the test, however unreliable it may be. Indeed, the statute's prime objective is to prevent exclusionary practices from "operat[ing] to 'freeze' the status quo." *Griggs,* 401 U.S., at 430, 91 S.Ct. 849.

Second, the Court suggests, anything less than a strong-basis-in-evidence standard risks creating "a *de facto* quota system, in which ... an employer could discard test results ... with the intent of obtaining the employer's preferred racial balance." *Ante,* at 2675. Under a reasonableness standard, however, an employer could not cast aside a selection method based on a statistical disparity alone.[FN8] The employer must have good cause to believe that the method screens out qualified applicants and would be difficult to justify as grounded in business necessity. Should an employer repeatedly reject test results, it would be fair, I agree, to infer that the employer is simply seeking a racially balanced outcome and is not genuinely endeavoring to comply with Title VII.

FN8. Infecting the Court's entire analysis is its insistence that the City rejected the test results "in sole reliance upon race-based statistics." *Ante,* at 2676. See also *ante,* at 2673–2674, 2677–2678. But as the part of the story the Court leaves out, see *supra,* at 2690–2695, so plainly shows—the long history of rank discrimination against African-Americans in the firefighting profession, the multiple flaws in New Haven's test for promotions—"sole reliance" on statistics certainly is not descriptive of the CSB's decision.

D

The Court stacks the deck further by denying respondents any chance to satisfy the newly announced strong-basis-in-evidence standard. When this Court formulates a new legal rule, the ordinary course is to remand and allow the lower courts to apply the rule in the first instance. See, e.g., *Johnson v. California,* 543 U.S. 499, 515, 125 S.Ct. 1141, 160 L.Ed.2d 949 (2005); *2703 *Pullman-Standard v. Swint,* 456 U.S. 273, 291, 102 S.Ct. 1781, 72 L.Ed.2d 66 (1982). I see no good reason why the Court fails to follow that course in this case. Indeed, the sole basis for the Court's peremptory ruling is the demonstrably false pretension that respondents showed "nothing more" than "a significant statistical disparity." *Ante,* at 2677–2678; see *supra,* at 2702, n. 8.[FN9]

FN9. The Court's refusal to remand for further proceedings also deprives respondents of an opportunity to invoke 42 U.S.C. § 2000e-12(b) as a shield against liability. Section 2000e-12(b) provides:

"In any action or proceeding based on any alleged unlawful employment practice, no person shall be subject to any liability or punishment for or on account of (1) the commission by such person of an unlawful employment practice if he pleads and proves that the act or omission complained of was in good faith, in conformity with, and in reliance on any written interpretation or opinion of the [EEOC] .... Such a defense, if established, shall be a bar to the action or proceeding, notwithstanding that (A) after such act or omission, such interpretation or opinion is modified or rescinded or is determined by judicial authority to be invalid or of no legal effect ... ."

Specifically, given the chance, respondents might have called attention to the EEOC guidelines set out in 29 CFR §§ 1608.3 and 1608.4 (2008). The guidelines recognize that employers may "take affirmative action based on an analysis which reveals facts constituting actual or potential adverse impact." § 1608.3(a). If "affirmative action" is in order, so is the lesser step of discarding a dubious selection device.

III

A

Applying what I view as the proper standard to the record thus far made, I would hold that New Haven had ample cause to believe its selection process was flawed and not justified by business necessity. Judged by that standard, petitioners have not shown that New Haven's failure to certify the exam results violated Title VII's disparate-treatment provision.[FN10]

FN10. The lower courts focused on respondents' "intent" rather than on whether respondents in fact had good cause to act. See 554 F.Supp.2d 142, 157 (Conn.2006). Ordinarily, a remand for fresh consideration would be in order. But the Court has seen fit to preclude further proceedings. I therefore explain why, if final adjudication by this Court is indeed appropriate, New Haven should be the prevailing party.

The City, all agree, "was faced with a prima facie case of disparate-impact liability," *ante,* at 2677: The pass rate for minority candidates was half the rate for nonminority candidates, and virtually no minority candidates would have been eligible for promotion had the exam results been certified. Alerted to this stark disparity, the CSB heard expert and lay testimony, presented at public hearings, in an endeavor to ascertain whether the exams were fair and consistent with business necessity. Its investigation revealed grave cause for concern about the exam process itself and the City's failure to consider alternative selection devices.

Chief among the City's problems was the very nature of the tests for promotion. In choosing to use written and oral exams with a 60/40 weighting, the City simply adhered to the union's preference and apparently gave no consideration to whether the weighting was likely to identify the most qualified fire-officer candidates.[FN11] There is strong reason to think it was not.

FN11. This alone would have posed a substantial problem for New Haven in a disparate-impact suit, particularly in light of the disparate results the City's scheme had produced in the past. See *supra,* at 2692–2693. Under the Uniform Guidelines on Employee Selection Procedures (Uniform Guidelines), employers must conduct "an investigation of suitable alternative selection procedures." 29 CFR § 1607.3(B). See also *Officers for Justice v. Civil Serv. Comm'n,* 979 F.2d 721, 728 (C.A.9 1992) ("before utilizing a procedure that has an adverse impact on minorities, the City has an *obligation* pursuant to the *Uniform Guidelines* to explore alternative procedures and to implement them if they have less adverse impact and are substantially equally valid"). It is no answer to "presume" that the two-decades-old 60/40 formula was adopted for a "rational reason" because it "was the result of a union-negotiated collective bargaining agreement." Cf. *ante,* at 2667. That the parties may have been "rational" says nothing about whether their agreed-upon selection process was consistent with business necessity. It is not at all unusual for agreements negotiated between employers and unions to run afoul of Title VII. See, e.g., *Peters v. Missouri-Pacific R. Co.,* 483 F.2d 490, 497 (C.A.5 1973) (an employment practice "is not shielded [from the requirements of Title VII] by the facts that it is the product of collective bargaining and meets the standards of fair representation").

*2704 Relying heavily on written tests to select fire officers is a questionable practice, to say the least. Successful fire officers, the City's description of the position makes clear, must have the "[a]bility to lead personnel effectively, maintain discipline, promote harmony, exercise sound judgment, and cooperate with other officials." CA2 App. A432. These qualities are not well measured by written tests. Testifying before the CSB, Christopher Hornick, an exam-design expert with more than two decades of relevant experience, was emphatic on this point: Leadership skills, command presence, and the like "could have been identified and evaluated in a much more appropriate way." *Id.,* at A1042–A1043.

Hornick's commonsense observation is mirrored in case law and in Title VII' s administrative guidelines. Courts have long criticized written firefighter promotion exams for being "more probative of the test-taker's ability to recall what a particular text stated on a given topic than of his firefighting or supervisory knowledge and abilities." *Vulcan Pioneers, Inc. v. New Jersey Dept. of Civil Serv.,* 625 F.Supp. 527, 539 (NJ 1985). A fire officer's job, courts have observed, "involves complex behaviors, good interpersonal skills, the ability to make decisions under tremendous pressure, and a host of other abilities—none of which is easily measured by a written, multiple choice test." *Firefighters Inst. for Racial Equality v. St. Louis,* 616 F.2d 350, 359 (C.A.8 1980).[FN12] Interpreting the Uniform Guidelines, EEOC and other federal agencies responsible for enforcing equal opportunity employment laws have similarly recognized that, as measures of "interpersonal relations" or "ability to function under danger (e.g., firefighters)," "[p]encil-and-paper tests ... generally are not close enough approximations of work behaviors to show content validity." 44 Fed.Reg. 12007 (1979). See also 29 CFR § 1607.15(C)(4).[FN13]

FN12. See also *Nash,* 837 F.2d, at 1538 ("the examination did not test the one aspect of job performance that differentiated the job of firefighter engineer from fire lieutenant (combat): supervisory skills"); *Firefighters Inst. for Racial Equality v. St. Louis,* 549 F.2d 506, 512 (C.A.8 1977) ("there is no good pen and paper test for evaluating supervisory skills"); *Boston Chapter, NAACP,* 504 F.2d, at 1023 ("[T]here is a difference between memorizing ... fire fighting terminology and being a good fire fighter. If the Boston Red Sox recruited players on the basis of their knowledge of baseball history and vocabulary, the team might acquire [players] who could not bat, pitch or catch.").

FN13. Cf. *Gillespie v. Wisconsin,* 771 F.2d 1035, 1043 (C.A.7 1985) (courts must evaluate "the degree to which the nature of the examination procedure approximates the job conditions"). In addition to "content validity," the Uniform Guidelines discuss "construct validity" and "criterion validity" as means by which an employer might establish the reliability of a selection method. See 29 CFR § 1607.14(B)–(D). Content validity, however, is the only type of validity addressed by the parties and "the only feasible type of validation in these circumstances." Brief for Industrial-Organizational Psychologists as *Amicus Curiae* 7, n. 2 (hereinafter I-O Psychologists Brief).

*2705 Given these unfavorable appraisals, it is unsurprising that most municipal employers do not evaluate their fire-officer candidates as New Haven does. Although comprehensive statistics are scarce, a 1996 study found that nearly two-thirds of surveyed municipalities used assessment centers ("simulations of the real world of work") as part of their promotion processes. P. Lowry, A Survey of the Assessment Center Process in the Public Sector, 25 Public Personnel Management 307, 315 (1996). That figure represented a marked increase over the previous decade, see *ibid.,* so the percentage today may well be even higher. Among municipalities still relying in part on written exams, the median weight assigned to them was 30 percent—half the weight given to New Haven's written exam. *Id.,* at 309.

Testimony before the CSB indicated that these alternative methods were both more reliable and notably less discriminatory in operation. According to Donald Day of the International Association of Black Professional Firefighters, nearby Bridgeport saw less skewed results after switching to a selection process that placed primary weight on an oral exam. CA2 App. A830-A832; see *supra,* at 2692–2693. And Hornick described assessment centers as "demonstrat[ing] dramatically less adverse impacts" than written exams. CA2 App. A1040.[FN14] Considering the prevalence of these proven alternatives, New Haven was poorly positioned to argue that promotions based on its outmoded and exclusionary selection process qualified as a business necessity. Cf. *Robinson v. Lorillard Corp.,* 444 F.2d 791, 798, n. 7 (C.A.4 1971) ("It should go without saying that a practice is hardly 'necessary' if an alternative practice better effectuates its intended purpose or is equally effective but less discriminatory.").[FN15]

FN14. See also G. Thornton & D. Rupp, Assessment Centers in Human Resource Management 15 (2006) ("Assessment centers predict future success, do not cause adverse impact, and are seen as fair by participants."); W. Cascio & H. Aguinis, Applied Psychology in Human Resource Management 372 (6th ed. 2005) ("research has demonstrated that adverse impact is less of a problem in an [assessment center] as compared to an aptitude test"). Cf. *Firefighters Inst. for Racial Equality,* 549 F.2d, at 513 (recommending assessment centers as an alternative to written exams).

FN15. Finding the evidence concerning these alternatives insufficiently developed to "create a genuine issue of fact," *ante,* at 2680–2681, the Court effectively confirms that an employer cannot prevail under its strong-basis-in-evidence standard unless the employer decisively proves a disparate-impact violation against itself. The Court's specific arguments are unavailing. First, the Court suggests, changing the oral/written weighting may have violated Title VII's prohibition on altering test scores. *Ante,* at 2680. No one is arguing, however, that the results of the exams given should have been altered. Rather, the argument is that the City could have availed itself of a better option when it initially decided what selection process to use. Second, with respect to assessment centers, the Court identifies "statements to the CSB indicat [ing] that the Department could not have used [them] for the 2003 examinations." *Ante,* at 2680–2681. The Court comes up with only a single statement on this subject—an offhand remark made by petitioner Ricci, who hardly qualifies as an expert in testing methods. See *ante,* at 2686. Given the large number of municipalities that regularly use assessment centers, it is impossible to fathom why the City, with proper planning, could not have done so as well.

Ignoring the conceptual and other defects in New Haven's selection process, the Court describes the exams as "painstaking[ly]" developed to test "relevant" material and on that basis finds no substantial risk of disparate-impact liability. See *ante,* at 2778. Perhaps such reasoning would have sufficed under *Wards Cove,* which permitted exclusionary practices as long as they advanced an employer's "legitimate"*2706 goals. 490 U.S., at 659, 109 S.Ct. 2115. But Congress repudiated *Wards Cove* and reinstated the "business necessity" rule attended by a "manifest relationship" requirement. See *Griggs,* 401 U.S., at 431-432, 91 S.Ct. 849. See also *supra,* at 2672. Like the chess player who tries to win by sweeping the opponent's pieces off the table, the Court simply shuts from its sight the formidable obstacles New Haven would have faced in defending against a disparate-impact suit. See *Lanning v. Southeastern Pa. Transp. Auth.,* 181 F.3d 478, 489 (C.A.3 1999) ("Judicial application of a standard focusing solely on whether the qualities measured by an ... exam bear some relationship to the job in question would impermissibly write out the business necessity prong of the Act's chosen standard.").

That IOS representative Chad Legel and his team may have been diligent in designing the exams says little about the exams' suitability for selecting fire officers. IOS worked within the City's constraints. Legel never discussed with the City the propriety of the 60/40 weighting and "was not asked to consider the possibility of an assessment center." CA2 App. A522. See also *id.,* at A467. The IOS exams, Legel admitted, had not even attempted to assess "command presence": "[Y]ou would probably be better off with an assessment center if you cared to measure that." *Id.,* at A521. Cf. *Boston Chapter, NAACP v. Beecher,* 504 F.2d 1017, 1021-1022 (C.A.1 1974) ("A test fashioned from materials pertaining to the job ... superficially may seem job-related. But what is at issue is whether it demonstrably selects people who will perform better the required on-the-job behaviors.").

In addition to the highly questionable character of the exams and the neglect of available alternatives, the City had other reasons to worry about its vulnerability to disparate-impact liability. Under the City's ground rules, IOS was not allowed to show the exams to anyone in the New Haven Fire Department prior to their administration. This "precluded [IOS] from being able to engage in [its] normal subject matter expert review process"—something Legel described as "very critical." CA2 App. A477, A506. As a result, some of the exam questions were confusing or irrelevant, and the exams may have over-tested some subject-matter areas while missing others. See, e.g., *id.*, at A1034-A1035, A1051. Testimony before the CSB also raised questions concerning unequal access to study materials, see *id.*, at A857-A861, and the potential bias introduced by relying principally on job analyses from nonminority fire officers to develop the exams, see *id.*, at A1063-A1064.[FN16] See also *supra*, at 2667, 2694.

FN16. The I-O Psychologists Brief identifies still other, more technical flaws in the exams that may well have precluded the City from prevailing in a disparate-impact suit. Notably, the exams were never shown to be suitably precise to allow strict rank ordering of candidates. A difference of one or two points on a multiple-choice exam should not be decisive of an applicant's promotion chances if that difference bears little relationship to the applicant's qualifications for the job. Relatedly, it appears that the line between a passing and failing score did not accurately differentiate between qualified and unqualified candidates. A number of fire-officer promotional exams have been invalidated on these bases. See, e.g., *Guardians Assn.*, 630 F.2d, at 105 ("When a cutoff score unrelated to job performance produces disparate racial results, Title VII is violated."); *Vulcan Pioneers, Inc. v. New Jersey Dept. of Civil Serv.*, 625 F.Supp. 527, 538 (NJ 1985) ("[T]he tests here at issue are not appropriate for ranking candidates.").

The Court criticizes New Haven for failing to obtain a "technical report" from IOS, which, the Court maintains, would have provided "detailed information to establish *2707 the validity of the exams." *Ante*, at 2679. The record does not substantiate this assertion. As Legel testified during his deposition, the technical report merely summarized "the steps that [IOS] took methodologically speaking," and would not have established the exams' reliability. CA2 App. A461. See also *id.*, at A462 (the report "doesn't say anything that other documents that already existed wouldn't say").

In sum, the record solidly establishes that the City had good cause to fear disparate-impact liability. Moreover, the Court supplies no tenable explanation why the evidence of the tests' multiple deficiencies does not create at least a triable issue under a strong-basis-in-evidence standard.

B

Concurring in the Court's opinion, Justice ALITO asserts that summary judgment for respondents would be improper even if the City had good cause for its noncertification decision. A reasonable jury, he maintains, could have found that respondents were not actually motivated by concern about disparate-impact litigation, but instead sought only "to placate a politically important [African-American] constituency." *Ante*, at 2665. As earlier noted, I would not oppose a remand for further proceedings fair to both sides. See *supra*, at 2703, n. 10. It is the Court that has chosen to short-circuit this litigation based on its pretension that the City has shown, and can show, nothing more than a statistical disparity. See *supra,* at 2702, n. 8, 2702–2703. Justice ALITO compounds the Court's error.

Offering a truncated synopsis of the many hours of deliberations undertaken by the CSB, Justice ALITO finds evidence suggesting that respondents' stated desire to comply with Title VII was insincere, a mere "pretext" for discrimination against white firefighters. *Ante,* at 2683–2684. In support of his assertion, Justice ALITO recounts at length the alleged machinations of Rev. Boise Kimber (a local political activist), Mayor John DeStefano, and certain members of the mayor's staff. See *ante,* at 2684–2687.

Most of the allegations Justice ALITO repeats are drawn from petitioners' statement of facts they deem undisputed, a statement displaying an adversarial zeal not uncommonly found in such presentations.[FN17] What cannot credibly be denied, however, is that the decision against certification of the exams was made neither by Kimber nor by the mayor and his staff. The relevant decision was made by the *2708 CSB, an unelected, politically insulated body. It is striking that Justice ALITO's concurrence says hardly a word about the CSB itself, perhaps because there is scant evidence that its motivation was anything other than to comply with Title VII's disparate-impact provision. Notably, petitioners did not even seek to take depositions of the two commissioners who voted against certification. Both submitted uncontested affidavits declaring unequivocally that their votes were "based solely on [their] good faith belief that certification" would have discriminated against minority candidates in violation of federal law. CA2 App. A1605, A1611.

FN17. Some of petitioners' so-called facts find little support in the record, and many others can scarcely be deemed material. Petitioners allege, for example, that City officials prevented New Haven's fire chief and assistant chief from sharing their views about the exams with the CSB. App. to Pet. for Cert. in No. 07-1428, p. 228a. None of the materials petitioners cite, however, "suggests" that this proposition is accurate. Cf. *ante,* at 2685. In her deposition testimony, City official Karen Dubois-Walton specifically denied that she or her colleagues directed the chief and assistant chief not to appear. App. to Pet. for Cert. in No. 07-1428, p. 850a. Moreover, contrary to the insinuations of petitioners and Justice ALITO, the statements made by City officials before the CSB did not emphasize allegations of cheating by test takers. Cf. *ante,* at 2686–2687. In her deposition, Dubois-Walton acknowledged sharing the cheating allegations not with the CSB, but with a different City commission. App. to Pet. for Cert. in No. 07-1428, p. 837a. Justice ALITO also reports that the City's attorney advised the mayor's team that the way to convince the CSB not to certify was "to focus on something other than 'a big discussion re: adverse impact' law." *Ante,* at 2686–2687 (quoting App. to Pet. for Cert. in No. 07-1428, p. 458a). This is a misleading abbreviation of the attorney's advice. Focusing on the exams' defects and on disparate-impact law is precisely what he recommended. See *id.,* at 458a-459a.

Justice ALITO discounts these sworn statements, suggesting that the CSB's deliberations were tainted by the preferences of Kimber and City officials, whether or not the CSB itself was aware of the taint. Kimber and City officials, Justice ALITO speculates, decided early on to oppose certification and then "engineered" a skewed presentation to the CSB to achieve their preferred outcome. *Ante,* at 2683.

As an initial matter, Justice ALITO exaggerates the influence of these actors. The CSB, the record reveals, designed and conducted an inclusive decision-making process, in which it heard from numerous individuals on both sides of the certification question. See, e.g., CA2 App. A1090. Kimber and others no doubt used strong words to urge the CSB not to certify the exam results, but the CSB received "pressure" from supporters of certification as well as opponents. Cf. *ante,* at 2686. Petitioners, for example, engaged counsel to speak on their behalf before the CSB. Their counsel did not mince words: "[I]f you discard these results," she warned, "you will get sued. You will force the taxpayers of the city of New Haven into protracted litigation." CA2 App. A816. See also *id.,* at A788.

The local firefighters union—an organization required by law to represent all the City's firefighters—was similarly outspoken in favor of certification. Discarding the test results, the union's president told the CSB, would be "totally ridiculous." *Id.,* at A806. He insisted, inaccurately, that the City was not at risk of disparate-impact liability because the exams were administered pursuant to "a collective bargaining agreement." *Id.,* at A1137. Cf. *supra,* at 2703–2704, n. 11. Never mentioned by Justice ALITO in his attempt to show testing expert Christopher Hornick's alliance with the City, *ante,* at 2684, the CSB solicited Hornick's testimony at the union's suggestion, not the City's. CA2 App. A1128. Hornick's cogent testimony raised substantial doubts about the exams' reliability. See *supra,* at 2686–2687.[FN18]

FN18. City officials, Justice ALITO reports, sent Hornick newspaper accounts and other material about the exams prior to his testimony. *Ante,* at 2686. Some of these materials, Justice ALITO intimates, may have given Hornick an inaccurate portrait of the exams. But Hornick's testimony before the CSB, viewed in full, indicates that Hornick had an accurate understanding of the exam process. Much of Hornick's analysis focused on the 60/40 weighting of the written and oral exams, something that neither the Court nor the concurrences even attempt to defend. It is, moreover, entirely misleading to say that the City later hired union-proposed Hornick as a "rewar[d]" for his testimony. Cf. *Ante,* at 2687.

There is scant cause to suspect that maneuvering or overheated rhetoric, from either side, prevented the CSB from evenhandedly assessing the reliability of the exams and rendering an independent, good-faith decision on certification. Justice ALITO acknowledges that the CSB had little patience for Kimber's antics. *2709 Ante,* at 2685–2686.[FN19] As to petitioners, Chairman Segaloff—who voted to certify the exam results—dismissed the threats made by their counsel as unhelpful and needlessly "inflammatory." CA2 App. A821. Regarding the views expressed by City officials, the CSB made clear that they were entitled to no special weight. *Id.,* at A1080.[FN20]

FN19. To be clear, the Board of Fire Commissioners on which Kimber served is an entity separate from the CSB. Kimber was *not* a member of the CSB. Kimber, Justice ALITO states, requested a private meeting with the CSB. *Ante,* at 2685. There is not a shred of evidence that a private meeting with Kimber or anyone else took place.

FN20. Justice ALITO points to evidence that the mayor had decided not to make promotions based on the exams even if the CSB voted to certify the results, going so far as to prepare a press release to that effect. *Ante,* at 2687. If anything, this evidence reinforces the conclusion that the CSB—which made the noncertification decision—remained independent and above the political fray. The mayor and his staff needed a contingency plan precisely because they did not control the CSB.

In any event, Justice ALITO's analysis contains a more fundamental flaw: It equates political considerations with unlawful discrimination. As Justice ALITO sees it, if the mayor and his staff were motivated by their desire "to placate a ... racial constituency," *ante,* at 2684, then they engaged in unlawful discrimination against petitioners. But Justice ALITO fails to ask a vital question: "[P]lacate" how? That political officials would have politics in mind is hardly extraordinary, and there are many ways in which a politician can attempt to win over a constituency—including a racial constituency—without engaging in unlawful discrimination. As courts have recognized, "[p]oliticians routinely respond to bad press ..., but it is not a violation of Title VII to take advantage of a situation to gain political favor." *Henry v. Jones,* 507 F.3d 558, 567 (C.A.7 2007).

The real issue, then, is not whether the mayor and his staff were politically motivated; it is whether their attempt to score political points was legitimate (i.e., nondiscriminatory). Were they seeking to exclude white firefighters from promotion (unlikely, as a fair test would undoubtedly result in the addition of white firefighters to the officer ranks), or did they realize, at least belatedly, that their tests could be toppled in a disparate-impact suit? In the latter case, there is no disparate-treatment violation. Justice ALITO, I recognize, would disagree. In his view, an employer's action to avoid Title VII disparate-impact liability qualifies as a presumptively improper race-based employment decision. See *ante,* at 2683. I reject that construction of Title VII. See *supra,* at 2699–2700. As I see it, when employers endeavor to avoid exposure to disparate-impact liability, they do not thereby encounter liability for disparate treatment.

Applying this understanding of Title VII, supported by *Griggs* and the long line of decisions following *Griggs,* see *supra,* at 2697–2698, and nn. 3-4, the District Court found no genuine dispute of material fact. That court noted, particularly, the guidance furnished by Second Circuit precedent. See *supra,* at 2688–2689. Petitioners' allegations that City officials took account of politics, the District Court determined, simply "d[id] not suffice" to create an inference of unlawful discrimination. 554 F.Supp.2d, at 160, n. 12. The noncertification decision, even if undertaken "in a political context," reflected a legitimate "intent not to implement a promotional process based on testing results that had an adverse impact." *Id.,* at 158, 160. Indeed, the District Court perceived *2710 "a total absence of any evidence of discriminatory animus towards [petitioners]." *Id.,* at 158. See also *id.,* at 162 ("Nothing in the record in this case suggests that the City defendants or CSB acted 'because of' discriminatory animus toward [petitioners] or other non-minority applicants for promotion."). Perhaps the District Court could have been more expansive in its discussion of these issues, but its conclusions appear entirely consistent with the record before it.[FN21]

FN21. The District Court, Justice ALITO writes, "all but conceded that a jury could find that the City's asserted justification was pretextual" by "admitt[ing] that 'a jury could rationally infer that city officials worked behind the scenes to sabotage the promotional examinations because they knew that, were the exams certified, the Mayor would incur the wrath of [Rev. Boise] Kimber and other influential leaders of New Haven's African-American community.' " *Ante,* at 2696, 2689 (quoting 554

F.Supp.2d, at 162). The District Court drew the quoted passage from petitioners' lower court brief, and used it in reference to a First Amendment claim not before this Court. In any event, it is not apparent why these alleged political maneuvers suggest an intent to discriminate against petitioners. That City officials may have wanted to please political supporters is entirely consistent with their stated desire to avoid a disparate-impact violation. Cf. *Ashcroft v. Iqbal*, 556 U.S. ----, ----, 129 S.Ct. 1937, 1951-1952, 173 L.Ed.2d 868 (2009) (allegations that senior Government officials condoned the arrest and detention of thousands of Arab Muslim men following the September 11 attacks failed to establish even a "plausible inference" of unlawful discrimination sufficient to survive a motion to dismiss).

It is indeed regrettable that the City's noncertification decision would have required all candidates to go through another selection process. But it would have been more regrettable to rely on flawed exams to shut out candidates who may well have the command presence and other qualities needed to excel as fire officers. Yet that is the choice the Court makes today. It is a choice that breaks the promise of *Griggs* that groups long denied equal opportunity would not be held back by tests "fair in form, but discriminatory in operation." 401 U.S., at 431, 91 S.Ct. 849.

* * *

This case presents an unfortunate situation, one New Haven might well have avoided had it utilized a better selection process in the first place. But what this case does not present is race-based discrimination in violation of Title VII. I dissent from the Court's judgment, which rests on the false premise that respondents showed "a significant statistical disparity," but "nothing more." See *ante,* at 2677–2678.

ENDNOTES

1. 29 U.S.C.A. Section 654(a) (2) and (b (1971).
2. 42 U.S.C.A. Sections 200e-2000e17 (1964).
3. 42 U.S.C. Sections 12101-12102, 12111-12117, 12201-12213 (1990).
4. 42 U.S.C. Sections 1981 and 1981a (1866).
5. 42 U.S.C. Section 1983 (1871).
6. Pub.L.No. 102-166, 105 Stat. 1071(1991).
7. 29 U.S.C. Sections 621-634 (1967).
8. 29 U.S.C. Section 206(d) (1963).
9. Pub.L.No. 110-233, 122 Stat. 881(2008) (Codified at 42 U.S.C. Section 2000ff).
10. Pub.L.No. 110-535, 122 Stat. 3553 (2008).
11. 29 U.S.C. Sections 705, 791, 793-794a.
12. 29 U.S.C. Sections 2601-2619, 2651-26454 (1993).
13. 8 U.S.C. Section 1324b (1986).
14. Pub. L.No. 102-166, 105 Stat. 1081 (1991) (Codified at 42 U.S.C. Section 2000e).
15. Taken from the EEOC web site located at www.EEOC.gov.

2 Title VII of the Civil Rights Act

Don't take the TRY out of a man or a horse.

D. Wayne Lukas, racehorse trainer

There is no such thing as a good excuse.

John P. Grier

LEARNING OBJECTIVES

1. Acquire an understanding of the requirements of Title Vll.
2. Analyze the unlawful employment practices prohibited by Title VII.
3. Acquire an understanding of the protected classes within Title VII.

HYPOTHETICAL SITUATION

A female employee in the sales department comes to you and informs you that her boss, the sales manager, has arbitrarily stopped all of her travel since she announced to the sales group that she was pregnant. The female employee tells you that she loves her job and the best part of her job is the travel, which makes up 75% of her job function. She tells you that no one "officially" informed her that she could not travel; however, she has been unofficially assigned to "desk duty," and no travel is being authorized by her boss. She states that there are no complications with her pregnancy, and her physician told her that she can travel up to the last trimester. She asks you, as the safety professional, what she should do. What are you going to tell this employee?

OVERVIEW

As the primary federal antidiscrimination law encompassing a broad spectrum of protections, Title VII of the Civil Rights Act of 1964[1] (hereinafter referred to as "Title VII") is general in its application and extends protections to all races, including Caucasian, as well as protections in the areas of religion, pregnancy, sex, color, and national origin. At the time of its creation by Congress in 1964, Title VII was the most sweeping civil rights legislation every enacted and contains 11 different titles addressing discrimination in public accommodations, voting, education and, most important to safety professionals, in the employment setting. Safety professionals should be aware that the Civil Rights Act has been modified and amended on several occasions since 1964, the most recent being the Civil Rights Act of 1991.[2] In the Civil Rights Act of 1991,[3] Congress reacted to a series of court decisions by the U.S. Supreme Court changing the landscape of discrimination law and precedent. With the Civil Rights Act of 1991, Congress amended several of the statutes enforced by the EEOC and added jury trials; compensatory and punitive damages in title VII (and ADA) lawsuits involving intentional discrimination and added statutory caps on damages awarded for future losses, pain, and suffering; and punitive damages depending on the size of the employer. Additionally, Congress codified the disparate impact theory of discrimination and expanded the coverage of Title VII to cover employees of American-controlled companies or organizations with operations outside of the United States.

Safety professionals should be aware that most states also possess individual state antidiscrimination laws primarily focused in the employment setting, which can be more encompassing than the federal law. Section 708 of Title VII permits each state to have parallel state legislation and regulation in the employment setting as long as the law does not conflict with Title VII. Safety professionals should be aware that employers with 15 or fewer employees and employees employed for each workday in 20 or more calendar weeks in a calendar year and engaged in interstate commerce can be considered outside of Title VII's jurisdiction.[4]

For most companies or organizations that possess a full-time safety professional, this small employer requirement is not applicable.

Safety professionals should be aware that Title VII identified unlawful employment practices as

- *Fail or refuse to hire or to discharge any individual, or otherwise to discriminate against any individual with respect to his or her compensation, terms, conditions, or privileges of employment, because of such individual's race, color, religion, sex or national origin; or*
- *Limit, segregate, or classify his or her employment or applicants for employment in any way which would deprive or tend to deprive any individual of employment opportunities or otherwise adversely affect his or her status as an employee, because of such individual's race, color, religion, sex or national origin.*[5]

Safety professionals should be aware that in addition to employers, labor unions, employment agencies and joint labor-management training committees are also required to comply with Title VII. For employers, labor organizations and others covered under Title VII, retaliation is prohibited in any manner. Safety professionals should be aware that it is considered an unlawful employment practice "to discriminate against the respective employees, applicants, members, or other related individuals because of their opposition to an unlawful employment practice or because of their filing a charge, testifying, assisting, or participating in any manner in an investigation, proceeding, or hearing under the Act."[6]

Safety professionals should be cognizant of the scope and application of the protections extended to individuals under Title VII and be able to identify situations and issues where the potential for discrimination may be present. Being aware of the protections provided to individuals under Title VII is the first step in preventing discrimination in your workplace.

Within the categories of race and color, safety professionals should be aware that Title VII includes protections to all races, including Caucasians.[7] In the category of national origin, safety professionals should be aware that language requirements are often found to be improper unless job relatedness can be shown[8] and prohibiting employees of foreign descent from using their native language at work has also been found to be an unlawful employment practice unless the business necessity it can be proven by the employer.[9] Title VII prohibits discrimination against male or female individuals based on their sex.[10] Safety professionals should be aware that the prohibition against discrimination based on sex extends beyond the hiring process to include such other areas as life insurance programs, profit sharing plans, bonus programs, and all other plans or programs.[11]

Safety professionals must be cognizant of sexual harassment in their workplace, especially with activities involving managerial or exempt personnel and hourly or nonexempt personnel. The EEOC has established special obligations on companies or organizations to ensure that their supervisory personnel as well as other employees do not engage in the unlawful employment practice of sexually based harassment. Safety professionals should be aware that sexual harassment can include "any unwelcome sexual advances, requests for sexual favor, or other verbal or physical conduct of a sexual nature[12]" Additionally, safety professionals should be aware that sexual advances or requests for sexual favors in the workplace substantially and detrimentally affects an employee's work performance or can create an "intimidating, hostile or offensive work environment."[13]

Safety professionals should also be aware that "petty slights, annoyances, and isolated incidents (unless extremely serious) will not rise to the level of illegality. To be unlawful, the conduct must create a work environment that would be intimidating, hostile, or offensive to reasonable people. Offensive conduct may include, but is not limited to, offensive jokes, slurs, epithets or name calling, physical assaults or threats, intimidation, ridicule or mockery, insults or put-downs, offensive objects or pictures, and interference with work performance. Harassment can occur in a variety of circumstances, including, but not limited to, the following:

- *The harasser can be the victim's supervisor, a supervisor in another area, an agent of the employer, a coworker, or a nonemployee.*
- *The victim does not have to be the person harassed, but can be anyone affected by the offensive conduct.*
- *Unlawful harassment may occur without economic injury to, or discharge of, the victim.*[14]

Safety professionals should be aware that Title VII defines "religion" as including all aspects of religious observances, practices, or beliefs.[15] The EEOC expanded this definition in its regulations to include moral and ethical beliefs not confined to theistic concepts or to traditional precepts that are sincerely held by individuals with the strength of traditional religious views and beliefs.[16] Safety professionals should be aware that Title VII goes beyond simple neutrality in the workplace providing for companies or organizations to provide reasonable accommodation of an employee's religious observance or practice.[17] However, safety professionals should be aware that Title VII also requires the employee seeking to observe his or her religious beliefs to provide proper notification to the company or organization as to their religious needs and an obligation to resolve any conflicts between job requirements and religious observances or practices.[18] Of particular importance for safety professionals, requiring an applicant or employee to wear clothing or other apparel other than the clothing required by the employee's religion may constitute an unfair employment practice.[19]

Safety professionals should be aware that the Pregnancy Discrimination Act was added as an amendment to Title VII. Safety professionals should be aware that unlawful employment practices in this category can include exclusion from medical or insurance programs, denial of a leave of absence, or discrimination based on the time or duration of a leave of absence. Of particular note is that protections under the Pregnancy Discrimination Act extend not only to female employees but also to the spouses of male employees.[20]

As can be seen, Title VII provides a wide spectrum of protections for individuals working within an employment setting. Safety professionals should be cognizant of these protections and be able to identify situations and issues that may have the potential for discrimination and take the appropriate actions to avoid such discrimination. As identified above, Title VII offers protection against discrimination based on race, sex, color, religion, and pregnancy within the employment setting. Most companies and organizations have addressed the requirements of Title VII and strictly adhere to the letter of this law. Safety professionals should not only be aware of the requirements of Title VII but also their internal company policies and procedures when addressing issues and circumstances of potential discrimination in the workplace.

As with many of the activities within the safety function, safety professionals should be aware that Title VII does not function in an isolated manner. Often, issues or circumstances of potential discrimination may be intertwined with workers' compensation claims, training functions, or other routine safety activity. It is imperative that a safety professional be able to identify the potential issue or circumstance in which protections against discrimination may be present and take the appropriate actions as proscribed by company policy or procedures to prevent or address the potential discriminatory action.

Prudent safety professionals may wish to examine each and every aspect of their safety programs to ensure that any potential for discrimination has been eliminated. From the hiring process through the termination process and all safety practices, programs, and procedures in between, a careful analysis and evaluation of issues and situations where discrimination potential could be present should be undertaken, and those identified issues or situations should be addressed to avoid any potential of discrimination within the safety function. Although an applicant or employee cannot win by simply claiming discrimination, proactive measures to avoid any type of discrimination are often the best approach.

CHAPTER QUESTIONS

1. Title VII provides protection against discrimination based on:
   a. Race
   b. Sex
   c. Religion
   d. All of the above

2. The agency responsible for enforcing Title VII is
   a. OSHA
   b. EEOC
   c. DOT
   d. None of the above

3. Title VII has never been amended: True or false.
4. An employer must have ___ employees to be covered under Title VII.
   a. 15
   b. 20
   c. 50
   d. 75

5. Regarding the hypothetical situation described at the chapter opening, what are you going to tell the employee?

Answers: 1—d; 2—b; 3—false; 4—a; 5— varied responses possible

ENDNOTES

1. 42 U.S.C.A. Section 2000E through 2000E-17.
2. Public Law 102-166 (1991).
3. Id.
4. 42 U.S.C.A. Section 2000e(b).
5. 42 U.S.C.A. Section 2000-2(a).
6. 42 U.S.C.A. Section 2000e-2(a).
7. Note: This is often referred to as "reverse discrimination" actions.
8. EEOC decision No. 73-0377 (1972).
9. *Saucedo v. Brothers Well Service, Inc.,* 464 F.Supp. 919 (DC Tex., 19790.
10. 42 U.S.C.A. Section 2000e-2.
11. 29 CFR Section 1604.9.
12. EEOC, Sex Discrimination Guidelines, 29 CFR Section 1604.11.
13. 29 CFR Section 1604.2 and 1604.11.
14. EEOC website located at www.eeoc.gov.
15. 42 U.S.C.A. Section 2000e(j).
16. 29 C.F.R. Section 1605.1.
17. 42 U.S.C.A. Section 2000e(j); 29 CFR Section 1605.2.
18. *Chrysler v. Mann,* 561 F.2d 1282 (CA-8, 1977).
19. EEOC decision No. 71-2620 (1971).
20. *Newport News Shipping & Dry Dock Co. v. EEOC,* 32 FEP 1 (S.Ct. 1983).

3 Proactive Protections to Prevent Discrimination under Title VII

The pessimist complains about the wind; the optimist expects it to change; the realist adjusts the sails.

William Authur Ward

The purpose is the eternal condition of success.

Theodore T. Munger

LEARNING OBJECTIVES

1. Acquire an understanding of the methods of identifying characteristic discrimination.
2. Identify the issues and activities where potential discrimination may be present.
3. Analyze and implement proactive measures to prevent discrimination in the workplace.

HYPOTHETICAL SITUATION

You have a manager who is always hugging everyone, male and female, when they come in contact with him. One of your female safety committee members comes to you and states that she is uncomfortable with the manager repeatedly hugging her and rubbing against her. What are you going to do?

OVERVIEW

"Shooting from the hip" can get a safety professional in hot water, especially when dealing with a potential issue involving discrimination. Safety professionals should always exercise extreme caution whenever an employee claims any type of action or inaction that has even a hint of possible discrimination. The safest response of any safety professional in any situation involving an employee and possible discrimination is to *stop* and *listen* to the employee. Always treat the employee with respect and dignity. There is a reason why the employee is coming to you with his or her issues.

Although the management of a claim of discrimination is usually outside of the safety professional's job function, there are certain activities within the safety domain which have the potential for discrimination. Some of these activities that safety professionals should take special notice of include, but are not limited to, the following activities:

1. **Hiring**—Safety professionals should be aware that discrimination can occur even before the applicant even submits an application to your company. Each and every stage in the hiring process should be carefully scrutinized and a plan designed to ensure that all activities within the hiring process are compliant with the laws. One of the areas in which safety professionals can inadvertently discriminate against a potential candidate is in the job advertisement or job posting. Safety professionals can potentially discriminate against potential candidates by using words that can be construed negatively by certain protected classes or by including criteria, such as height, that are not applicable to the performance of the job. Using the height example, if the safety professional is achieving a forklift job and specifies a height requirement of 6 feet in the job advertisement, this height requirement, although not applicable in the performance of the forklift job, may inadvertently screen out women, Asian men, Latino men, or other protected classes. Safety professionals should always focus any advertisement or posting on objective, job-related criteria.

2. **Applications**—Safety professionals should screen their application documents to ensure that the application does *not* ask any questions about the prospective employee's gender, age, disability, race, color, religion, national origin, pregnancy, or other protected classes. Safety professionals should also analyze any application documents to ascertain whether any questions may elicit a response that identifies the applicant within a protected class. For example, asking whether the applicant is over the age of 18 is appropriate given the child labor and safety laws and regulations. Asking applicants for their age or may identify the individuals as being within a protected class based on their age.

3. **Screening**—Safety professionals are often actively involved in screening of potential applicants. Safety professionals should focus on the applicant's ability to perform the job functions rather than any protected class information that may have been volunteered by the applicant on their application, resumé, or other sources.

4. **Testing**—Safety professionals should be cautious when designing and implementing job-related physical testing. Although job- or skill-related testing is permissible, testing activities that are not job related or related to a job function can be susceptible to challenge.

5. **Alcohol and Controlled Substance Testing**—Safety professionals are often directly responsible for the management of any type of post-offer or other type of alcohol and controlled substance testing. Safety professionals should be cognizant of the specific requirements of the testing program and ensure that each and every facet of the program strictly adheres to the specifics of the alcohol and controlled substance testing program.

6. **Interviewing**—It is relatively easy for a safety professional to ask a question during an interview conversation that may identify the applicant's protected status. A relatively benign question or response such as "Oh, I also went to Central High School … When did you graduate?" may identify the applicant's age, religion, or other protected status. Safety professionals are cautioned to keep all interview questions job related and may wish to develop, analyze, and write the questions down prior to the interview so as not to deviate from the prepared questions.

7. **English-Only Rules**—Safety professionals should be aware that although the OSHA standards are only in English and many employers do not want to go to the expense of interpreting safety signs, programs, policies, and procedures, unless English is absolutely necessary for the performance of the job and a business necessity, most English-Only policies are discriminatory on their face. For most safety professionals, interpretation of safety signs, programs, policies, and procedures into the primary language of the employees is a necessity in order to ensure complete understanding of the safety requirements. Safety professionals should also be aware that this may require translation into several different languages depending on your workforce.

8. **Safety Rules**—Safety professionals are often called upon to develop and draft written safety policies, procedures, and programs to ensure the safety of their employees. Safety professionals should pay special attention to the development and drafting of these written documents so as not to inadvertently discriminate against any of the protected classes. For example, if the company employs a process or chemical that can cause birth defects in unborn children, if the safety professional prohibits women who may be pregnant from working near the process or chemical to protect the women and unborn child, could this prohibition be considered discriminatory?

9. **Dress Codes**—Safety professionals should exercise caution when developing or implementing a dress code within their organization. Although the color, insignia, or other markings on employer-provided uniform, hard hat, or other uniform locations may possess a safety-related meaning within the specific operation (such as green hard hats or forklift operator patches), safety professionals should be cognizant of the cultural differences as well as the potential religious significance of individual employee's dress. Although most courts have permitted dress codes, especially where the dress is employer provided and neutral in nature, prudent safety professionals may want to investigate the potential cultural issues involved in any dress requirements, colors of clothing, and other details.

10. **Grooming Policies**—Safety professionals may want to investigate any issue involving grooming-related policies prior to implementation. Often, the safety professional is called upon to decide the safety of employees with longer hair working around moving machinery, the use of facial jewelry on the job, the use of facial hair, and other related issues. Prudent safety professionals should not only examine the issue from a safety prospective but also from the potential discrimination perspective before deciding the issue. For example, a "no beard" policy may be discriminatory against

African American males with pseudofolliculitis barbae or hair length may be related to the employee's religious beliefs.

11. **Training Activities**—Training is an important aspect of most safety programs. Safety professionals should exercise care in all aspects of the training component to ensure that the training is not discriminatory against any employee in any way. For example, the safety professional does not schedule a pregnant female employee for the required safety training because the safety professional assumes she will be off on maternity leave. Because the employee was not able to acquire the required safety training, she is denied a promotion.

12. **Job Assignments**—Although most safety professionals do not provide specific job assignments for the vast majority of employees involved in production, safety professionals may provide job assignments for their safety staffs, safety interns, or employees of restricted duty following a work-related accident. Safety professionals should exercise diligence to ensure all assignments are appropriate and based on business necessity as well as individual skills, abilities, and/or medical restrictions of the individual.

13. **Performance Evaluations**—Safety professionals should exercise caution when completing performance evaluations or any other type of job performance assessments. Although job performance evaluations are common in most organizations and a good method of communicating how the employee is doing in his or her job as well as the areas for future improvement, safety professionals should be cognizant that what is said during this evaluation as well as what is written in this evaluation document is not discriminatory in any way.

14. **Disciplinary Actions**—Safety professionals are often involved in the disciplinary process in many organizations especially when the alleged infraction involves a safety issue. Most companies possess a form of progressive disciplinary policy or program that provides guidance to the safety professional. It is important that the safety professional ensures that the matter is appropriately investigated and any disciplinary action initiated is fair and in accordance with company policy.

15. **Promotion and Demotion**—Safety professionals should be cognizant that when one employee is promoted, there is another employee who expected to get the job and is disappointed. All promotions should be based on job-related performance, experience, and expertise only. Safety professionals should be aware that any demotion or reduction in rank or pay should also be based solely on the individual's lack of performance, lack of experience, or lack of expertise in the specific job and should always be done in a fair and consistent manner in accordance with company policy.

16. **Termination**—Termination from employment is often perceived as the workplace version of the death penalty. Safety professionals should be aware that terminated employees often search for discriminatory motives by the employer behind the termination. Safety professionals involved in any involuntary termination should be cognizant of the

potential of discriminatory charges or claims being brought by the terminated employee in an attempt to avoid the involuntary termination.

17. **Layoff and Reduction in Force**—Safety professionals should be aware that employees facing a reduction-in-force (RIF) or layoff situation may be more likely to bring a claim of discrimination against the employer if they perceive their selection for the RIF or layoff is done in a discriminatory manner. Safety professionals involved in an RIF or layoff should ensure that their decision-making process is in strict accordance with company policy. (It should be noted that safety professionals should become familiar with the Worker Adjustment and Retraining Notification Act (known as 'WARN") as well as the Older Worker Benefit Protection Act (known as "OWBPA") as well as individual state laws when addressing a layoff or RIF situation.)

18. **Resignation**—Voluntary termination or a resignation by an employee is often on good terms and involves a better opportunity for the employee. Many employees provide their current employer with a minimum of two weeks notice to be able to acquire their replacement. Safety professionals should pay special attention to a situation where an employee, often a managerial-level employee, is provided the option to resign or be involuntarily terminated by the company. If there is any discriminatory treatment of the employee who is being forced to resign, safety professionals should be aware that the courts have recognized the concept of a "constructive termination," where an employee is forced to resign as a result of discriminatory actions by the employer.

19. **Providing a Reference**—Safety professionals should be aware that discriminatory or defamatory information provided to a prospective employer when called or requested upon to provide information regarding a previous employer can result in legal action against the safety professionals or company.

20. **Dating the Boss**—Although many companies possess policies forbidding managerial employees from dating subordinates and/or possess policies where employees who wish to date sign an internal agreement not to bring a discrimination claim against the company as a result of this relationship, interpersonal relationships between employees do happen. Safety professionals should be cognizant of the potential of discrimination resulting from the relationship as well as the employees working for and with the parties involved in the relationship.

Safety professionals should be aware that the fact that an employee has filed a claim of discrimination on the job does not automatically mean that discrimination is present in the workplace. As many safety professionals are aware, once the employee files a claim of discrimination, then they must prove the facts of the discriminatory actions in a court of law. In most cases, the employee will need to prove that he or she was a member of a protected class and the employee or applicant was qualified for the particular job in question. The employee will also need to prove that he or she suffered an adverse action by the employer and prove that the employer had an improper motive for the action. Once the employee has established the above, this does not mean that the employer is automatically at fault. The employer would then be provided an opportunity to prove that the job action taken was for a legitimate business reason. Safety professionals should be aware that proper documentation and adherence to company policies is essential in proving the business reason for the actions.

In many circumstances, safety professionals are given guidance by their human resource department as to the policies and procedures to follow from the hiring process through termination. However, if a safety professional should encounter situations where guidance is not provided, it is essential that he or she refrain from "shooting from the hip" and ask the appropriate managerial team members or legal counsel for assistance with the specific issue. Although safety professionals are usually not directly involved in issues involving alleged discrimination in the workplace, they can easily become the target of an allegation of discrimination, given their visibility and functions within the organization.

CHAPTER QUESTIONS

1. The potential of discrimination exists:
   a. In hiring
   b. In promotion
   c. At termination
   d. Before, during, and after employment
2. English-only rules are applicable for safety rules. True or false?
3. A claim of discrimination with the EEOC automatically means discrimination is present. True or False?
4. A voluntary termination is the same as quitting a job. True or false?
5. In the hypothetical situation at the start of the chapter, what are you going to do?

Answers: 1—d; 2—false; 3—false; 4—true; 5—varied responses possible.

4 Defending a Claim of Discrimination under Title VII

When you come to a fork in the road—take it.

Yogi Berra

In many lines of work, it isn't how much you do that counts,
but how much you do well and how often you decide right.

William Feather

LEARNING OBJECTIVES

1. Acquire an understanding of the process of filing and defending a claim of discrimination.
2. Identify and analyze the potential defenses to a claim of discrimination.
3. Identify and analyze the proactive measures which can be taken to prepare for a claim of discrimination.

HYPOTHETICAL SITUATION

A female safety professional at plant level has an affair with one of her male safety staff member for a number of months. After the relationship had ended, the safety professional evaluated the safety staff member's work performance, which was previously identified as excellent, and graded him as poor. As a result of this evaluation, the safety staff member was demoted and transferred to another less desirable facility. The safety staff member files a claim of discrimination. As the corporate safety director, what are you going to do?

OVERVIEW

In order to defend against a claim of discrimination under Title VII, safety professionals must possess a thorough and complete understanding of the claims process as identified by the Equal Employment Opportunity Commission (EEOC). Although many states permit employees to proceed directly to state court for claims of discrimination utilizing individual state antidiscrimination laws, many claims are initiated at the federal level with the Equal Employment Opportunity Commission.

Safety professionals should be aware that it is relatively easy for an employee or applicant to file a charge of discrimination with the EEOC. As can be seen in Appendix A, companies and organizations are required to post notices of employee rights in every private employer operations, state and local government buildings, educational institutions, employment agencies, and labor organization buildings. On these posters, employees are provided several methods through which to file a charge of discrimination including a toll free telephone number, toll free TTY number (for hearing impairments), and in person at EEOC field offices. The EEOC does also offer an online assessment tool on their website and provides an intake questionnaire which can be printed, completed, and returned to one of the EEOC offices. The poster is offered in English as well as Spanish and Chinese.

On average, the EEOC receives between 50,000 and 70,000 claims of discrimination in any given year.* Safety professionals should be aware that, from the data provided by the EEOC, the largest percentage of claims was found to possess no reasonable cause.† A smaller percentage of these claims are closed by administrative closure, withdrawn, or resolved by settlement.‡ Safety professionals should be aware, as noted above, that it is relatively easy to file a charge of discrimination, however, the data identifies that it is more difficult to prove the charge of discrimination for employees than for applicants to prove the required elements of the charge.

Safety professionals should also be aware of the time limitations in which an employee or applicant can file a charge of discrimination. On the federal level with the EEOC, the time limitation is 180 days from the last date of discrimination. However, safety professionals should be aware that under many state antidiscrimination laws, employees or applicants can be provided 300 days from the date of discrimination to file the charge of discrimination. Charges of discrimination that have exceeded the time limitations are usually discharged.

All EEOC charges must be submitted in writing to an EEOC location. The written charge identifies the company, organization, or individual to which the allegations are being directed, a general statement of the alleged unlawful conduct, and other general information. The written charge must be signed and verified (by sworn oath to be truthful) by the EEOC.§ It is important for safety professionals to recognize that verbal allegations, absent a written and sworn charge filed with the EEOC, are usually not valid.

Safety professionals should be aware that the company or organization is usually notified in writing from the EEOC within a few weeks following the filing of the charge. The written correspondence often identifies the allegations of discrimination and requests identified documents or data related to the charge from the company or organization.¶ The EEOC can schedule a fact-finding, face-to-face conference with

---

* Retrieved from EEOC website at www.eeoc.gov. The claim receipts as identified by year: FY 1997—58,615; FY 1998—58.124; FY 1999—57582; FY 2000—59,588; FY 2001 -59,631; FY 2002—61,459; FY 2003—59.075; FY 2004—58,328; FY 2005—55, 976 ; FY 2006—56,155; FY 2007—61,159; FY 2008—69,064; FY 2009—68,710.
† Id.
‡ Id.
§ 29 CFR Section 1601.9
¶ EEOC Compliance Manual, Section 14.5.

the parties or their representatives and the EEOC records the fact-finding conference for use for determination or settlement purposes. Safety professionals should be
aware that an attorney may attend the fact-finding conference; however, the attorney
may not speak on behalf of their client or direct- or cross-examine any witnesses.
The attorney may only serve as an advisor to the company or employee.*

Safety professionals should be aware that the EEOC can make an on-site investigation at the company or organization's business site.† Safety professionals should be
aware that most on-site investigations are scheduled in advanced and utilized by the
EEOC investigator to become acquainted with the worksite as well as any specific
job which is at issue.‡ Safety professionals involved in an on-site visit should, as
with an OSHA inspection, exercise caution thoughout the visit and especially when
providing statements or explanations about the jobsite or job functions to the EEOC
investigator.

When the EEOC investigator or agent has concluded his or her investigation, the
EEOC investigator will submit their findings to the commission. The commission,
based upon the finding of the investigator or agent, will determine whether it believes
that the charge by the employee or applicant is true.§ The commission has two possible determinations, namely, a "reasonable cause" determination (the commission
believes the alleged discrimination did occur) or a "no reasonable cause" determination (the commission believes that the alleged discrimination did not occur).¶

Safety professionals should be aware that if the commission issues a determination finding no cause, the EEOC will send the employee or applicant a "right to sue"
letter advising the employee or applicant that their charge of discrimination will be
dismissed by the EEOC.** The "right to sue" letter or notice informs the applicant
or employee that if he or she wishes to initiate a lawsuit in district court under Title
VII, the employee or applicant will have 90 days after receipt of the letter to file suit.††
Safety professionals should be aware that if the employee or applicant pursues legal
action in district court, a complaint will be received within the 90 day time period.

If the EEOC determines that there is reasonable cause to believe discrimination has taken place, safety professionals should expect a determination letter from
the EEOC with a request to join in mediation or conciliation discussions.‡‡ If the
employee or applicant and the company or organization agree to conciliation and
the discussions yield an agreement, the agreement is reduced to written form, signed
by the employee or applicant and the company or organization and also signed by
the EEOC conciliator. The written agreement is then sent to the EEOC for approval
with copies to the employee or applicant, the company or organization, and a copy to
the EEOC office in Washington, D.C.§§ Safety professionals should be aware that the

---

* Id. at Section 14.7.
† Id. at Section 25.2.
‡ Id.
§ Id. at Section 40.
¶ Id.
** 42 USCA Section 2000e-1-5; 29 CFR Section 1601.19(f).
†† 29 CFR Section 1601.29(e).
‡‡ EEOC Compliance Manual Section 64.
§§ Id. at Section 64 (b) and (f).

EEOC will review every conciliation agreement and has up to 18 months to complete this review.*

Safety professionals should be aware that the EEOC will insist on very specific conditions and language, such as that a waiver of the right to sue be included in any settlement agreement.† If the employee or applicant and the company or organization cannot reach an agreement or either side refuses conciliation, safety professionals should be aware that the EEOC will send a letter to the employee or applicant that the conciliation efforts are terminated and provide the employee or applicant with a right-to-sue notice.‡

Safety professionals should be aware that the EEOC has a duty to notify the employee or applicant bring the charge of his or her right to sue. If the EEOC is unable to complete their investigation with 180 days or if the employee or applicant request a right-to-sue letter, the EEOC usually will issue the right-to-sue notice.§ Given the volume of charges being filed with the EEOC, safety professionals should not be surprised if the EEOC is not able to investigate the charge within 180 days and a right-to-sue letter or notice is provided to the employee or applicant.

If the EEOC, in their investigation, determines that there is reasonable cause that discrimination has taken place, Title VII provides that the EEOC, or the individual applicant or employee with a right-to-sue notice, can begin a legal action in federal district court if specific conditions are met including jurisdiction, time limitations, complaint drafting requirement and other specific requirements. Safety professionals should also be aware that class action lawsuits are permitted under Title VII, however, there are very specific requirements under Rule 23 of the Federal Rules of Civil Procedure, as well as other requirements which must be met to certify the class action.

Safety professionals should be aware that prior to the Civil Rights Act of 1991¶, Title VII trials were de novo, or a trial before a judge only. However, after the passage of the Civil Rights Act of 1991**, jury trials, as well as compensatory and punitive damages, were allowed when intentional employment discrimination can be shown. Additionally, safety professionals should be aware that the Civil Rights Act of 1991 also permits the applicant or employee who is found to be the victim of discrimination to recover their attorney's fees, expert witness fees, court costs, other out-of-pocket expenses (such as job search and medical costs) and compensation for emotional hard (such as mental anguish, inconvenience, and loss of enjoyment of life.††

The reason for Congress making changes in the Civil Rights Act of 1991‡‡ was in response "to seven Supreme Court decisions" or which two, namely *Wards Cove*

---

* Id. at Section 80.3.
† EEOC Compliance Manual, Forward.
‡ Id at Section 66.4.
§ 29 CFR Section 1601(a)(1) & (2).
¶ 42 U.S.C Section 1981a and 42 U.S.C. Section 2000-e-2(k)-(n).
**Id.
†† EEOC website located at www.eeoc.gov.
‡‡ Supra note 19.

Packing Co. v. Atonio and *Price Waterhouse v. Hopkins* getting the most attention.* "In Wards Cove, the Court had reformulated the standards and burdens of proof for a disparate impact claim, making it more difficult for plaintiffs to prevail. In Price Waterhouse, the Court held that even when a plaintiff proves that an adverse decision was made for discriminatory reasons, the employer can escape liability by proving it would have made the same decision even if it had not been motivated by discriminatory animus."† The Civil Rights Act of 1991 responded to these decisions by the U.S. Supreme Court "by codifying the disparate impact theory of discrimination as originally articulated in the *Griggs v. Duke Power* decision, and clarifying that whenever a plaintiff in a "mixed motive" case proves that discrimination motivated an employment decision, she has established a violation of Title VII."‡

Safety professionals should also be aware that the Civil Rights Act of 1991 also placed limits on compensatory and punitive damages based upon the size of the employer. These limits on compensatory and punitive damages are as follows:

For employers with 15-100 employees, the limit is $50,000.00.
For employers with 101-200 employees, the limit is $100,000.00.
For employers with 201-500 employees, the limit is $200,000.00.
For employers with more than 500 employees, the limit is $300,000.00.§

As can be seen, the human as well as monetary costs of a discriminatory action can be very expensive. Safety professionals with responsibilities within the area should consider a proactive policy which prohibits any type of discrimination in the workplace. Generally, an effective policy includes, but is not limited to, a policy statement that the company or organization is committed to preventing harassment and discrimination in their workplace and must

Inform employees of prohibited conduct
Establish the company standard as to how employees are to be treated
Identify disciplinary measures for noncompliance
Establish a confidential internal method for employees to report incidents of
    alleged discrimination or harassment.

Safety professionals should also consider the development of an investigative methodology through which to investigate and respond to any allegation of harassment or discrimination where there is no fear of retaliation by the employee.

In the event that a charge is filed, safety professionals should be aware that no stone should be left unturned when preparing to defend your company or organization against a charge of discrimination. Although the best action is to never have the charge in the first place, once the charge is filed, it is imperative that all possible and appropriate defenses to the circumstances be appropriately evaluated and assessed.

---

* Taken from 40th Anniversary Panel, EEOC website located at www.eeoc.gov.
† Id.
‡ Id.
§ EEOC website.

Possible defenses to a charge of discrimination can include, but are not limited to, a business necessity. This defense if often utilized by employers and has been described as an overriding legitimate, nondiscriminatory business purpose for the action of the employer.* Safety professionals should be aware that the courts require that an employer prove the following three elements:

1. There must exist an overriding legitimate business purpose such that the practice is necessary to the safe and efficient operations of the business;
2. The challenged practice must effectively carry the business purpose it is alleged to serve
3. There must exist no acceptable alternative policies or practices which would accomplish the business purpose advanced, or accomplish it equally well with lesser differential impact.[†]

Further defenses include the following:

- "For Cause" Defense—In this defense, the employer can overcome the employee or applicant's prima facie case of discrimination by showing it has good cause or nondiscriminatory reasons for the action taken by the employer.[‡]
- Reasonable Accommodation—As with ADA, the employer can utilize proof of attempting to make reasonable accommodation in cases alleging religious discrimination.[§]
- Reliance on EEOC Opinion—Safety professionals should be aware that good faith reliance upon a written interpretation of opinion by the EEOC can be utilized as a defense against a charge of discrimination. This defense is similar to the good faith reliance argument often utilized within the OSHA area.
- National Security—For safety professionals working for companies or organizations with national security requirements, the company or organization may utilize national security as a defense. This defense may arise where applicants are subject to certain national security or homeland requirements.
- Bona Fide Occupational Qualifications—The "BFOQ" defense is vary narrow in scope and is primarily utilized in cases of sex discrimination. In essence, the BFOQ defense identified occupational qualifications which are reasonably necessary to the normal operations of a particular business or operation.[¶] As identified by the EEOC, "job descriptions, properly prepared, can support the goal of eradicating unlawful employment discrimination. Racial requirements are never lawful in job descriptions and should not be used under any circumstances. Job requirements based on

---

* United Papermakers & Paperworkers v. U.S., 416 F.2d 980 (Ca-5, 1969).
† Robinson v. Lorillard Corp., 444 F.2d 791 (CA-4, 1971).
‡ EEOC v. AT&T, 556 F.2d 167 (1977).
§ 42 U.S.C.A. Section 2000-12(b).
¶ See, 42 U.S.C.A. Section 200003-2(e).

an employee's gender, national origin, religion or age can be used in very limited circumstances. Job requirements based on these protected characteristics are lawful only when an employer can demonstrate that they are bona fide occupational qualifications ("BFOQs") reasonably necessary to the normal operation of business. ...Furthermore, the Commission encourages employers to carefully assess whether their job requirements or duties, although neutral or evenly applied, indirectly impact employees based on their protected characteristics. If an adverse impact exists based on an employee's race, sex, color, national origin or religion, then the particular job requirement or duty should only be included in the job description if it is job-related, consistent with business necessity and the least discriminatory alternative available. ...In addition, employers should explicitly divide job requirements into the major and minor functions of the position.... Employers should also be careful to focus on the major or minor functions of a position as opposed to recommended methods of performance when more than one acceptable method exists....Employers are not required to compose job descriptions. Yet, to the extent they are developed, they should not be discriminatory. Also, employers must preserve records of job descriptions made in the regular course of business for a two-year period.*

- Employment Testing—Of importance to safety professionals is the defense of employment testing. Once the charge of discrimination is filed, the employer has a duty to prove that the testing procedures and requirements have a manifest relationship to the specific job at issue.
- Jurisdictional Issues—As previously discussed, there are time limitations and jurisdictional requirements that must be met in order for the action to move forward. It is important that safety professionals carefully check each and every requirement to ensure that the action meets all jurisdictional requirements.
- Statistical Data—Statistical data can be utilized as a defense to prove or rebut an allegation of discrimination. Safety professionals utilizing this defense should ensure that the data is "relevant, meaningful and not segmented and particularized so as to obtain a desired result."†

In summation, safety professionals are usually not expected to be experts in the area of employment discrimination and harassment, however, it is important that safety professionals have a firm grasp on the concepts and procedures involved in addressing issues and situations within the company or business. Safety professionals are often the "eyes and ears" of the company and frequently see or hear potential issues while performing their various activities within the operations. The best method of avoiding charges of discrimination and harassment in the workplace is a carefully crafted and strictly enforced policy that prohibits any type of possible discrimination or harassment to exist in the first place. However, in the event of an allegation of discrimination or harassment by an applicant or employee, safety professionals

---

* EEOC Interpretation Letter, EEOC Website.
† EEOC vc. Datapoint Corp., 570 F.2d 1264 (1978).

may find themselves part of the team that investigates the allegations or assists the human resources or legal professionals define the issues and defenses. A rudimentary knowledge of the policies, procedures, and laws addressing discrimination in the workplace should be a tool in every safety professional's toolbox.

CHAPTER QUESTIONS

1. Title VII is designed to prevent discrimination based on:
    a. Race
    b. Sex
    c. Religion
    d. All of the above
2. The primary agency tasked with enforcing title VII is:
    a. EEOC
    b. OSHA
    c. Department of Defense
    d. None of the above
3. The maximum limit on compensatory and punitive damages for an employer with 499 employees is:
    a. $50,000.00
    b. $100,000.00
    c. $300,000.00
    d. None of the above.
4. On average, the EEOC received how many claims in a year?
    a. 10,000-20,000
    b. 50,000-70,000
    c. 100,000–150,000
    d. None of the above
5. In the hypothetical situation provided at the beginning of this chapter, what are you going to do?

Answers: 1—d; 2—a; 3—d; 4—b; 5—varied responses possible.

5 Americans with Disabilities Act

We may not imagine how our lives could be more frustrating and complex—but Congress can.

Cullen Hightower

Congress does from a third to a half of what I think is the minimum that it ought to do, and I am profoundly grateful that I get as much.

Theodore Roosevelt

LEARNING OBJECTIVES

1. Acquire an understanding of the requirements and protections afforded under the Americans with Disabilities Act.
2. Identify the interaction of the Americans with Disabilities Act and the safety function.
3. Acquire an understanding of the overall purpose and scope of the Americans with Disabilities Act.
4. Acquire an understanding of the ADA Amendment Act and regulations.

HYPOTHETICAL SITUATION

An employee returning to work with a full medical release after a work-related injury tells you, " I don't think I can do my old job anymore…. I would like to be moved to the forklift job." What are you going to do?

OVERVIEW

Outside of the Occupational Safety and Health Act, the Americans with Disabilities Act* (ADA) often impacts and intersects with the safety function more than any other law. It is important that safety professionals acquire a working knowledge of the ADA as well as key areas in which the ADA and safety function may intersect in order to be able to recognize when the ADA may be applicable to the situation. Safety professionals should be able to recognize when a potentially qualified individual is requesting an accommodation and appropriately address the situation appropriately in order to comply with the ADA. Within the safety function, the ADA can impact and intersect in the areas of workers' compensation, restricted duty programs, facil-

---

* 42 U.S.C. Section 12101–12102, 12111–12117, 12201–12213.

ity modifications, safety training, equipment design, PPE and other safety functions creating duties and responsibilities for the safety professional as well as potential liabilities for the company or organization. Additionally, safety professionals should continuously educate themselves in the changing requirements of the ADA as well as the ADA Amendment Act and EEOC regulations and decisions in order to ensure that the safety function as well as the company or organization is in compliance with these antidiscrimination protections provided to qualified individuals with disabilities.

Safety professionals should be aware that the ADA is one of the more extensive laws which can provide protections to a vastly large number of applicants, employees and others. In a nutshell, the ADA prohibits discriminating against qualified individuals with physical or mental disabilities in all employment settings. In theory, this can sound relatively simple however the ADA can be a minefield for safety professionals who are uninformed in the requirements of this law.

So where do we start? Mr. David Fram, Director of ADA and EEO Services from the National Employment Law Institute in his 2010 presentation, utilizes the very simple question "How can I help you?" This initial response can open the doors to communications between the safety professional and employee as well as starting down the right path to ensure compliance with the ADA.* Unlike other areas of the safety functions where action is the order of the day, safety professionals confronting ADA issues must stop and listen to the individual and assess the facts and the ADA requirements and decisions in order to make an informed decision. Shooting from the hip or failing to listen to the employee or applicant can create potential liabilities for the safety professional and his or her company or organization. Remember ... "How can I help you?!"

Safety professionals should be aware that although the ADA became law in 1990, the ADA is still transitioning through court decisions, agency regulations, and interpretations and was substantially amended by the ADA Amendments Act of 2008[†] (ADAAA) and the Civil Rights Act of 1991.[‡]

Safety professionals should be aware that effective the ADAAA makes significant changes to the term "disability" as defined in the ADA by rejecting several Supreme Court decisions and portions of previous EEOC's ADA regulations. The ADAAA retains the basic definition of "disability" as defined in the ADA as being an impairment that substantially limits one or more major life activities, a record of such an impairment, or being regarded as possessing an impairment. More specifically, the ADAAA provided the following:

1. *The Act requires the EEOC to revise the section of their regulations which define the term "substantially limits";*
2. *The Act expands the definition of "major life activities" by including two non-exhaustive lists of which list 1 includes many activities that the EEOC has previously recognized (such as walking) as well as activities that the*

---

* Presentation by David Fram at the National Employment Law Conference (2010).
† Public Law 110–325.
‡ 42 U.S.C. Section 1981(a) & 42 U.S.C. Section 2000e(2)(k)–(n).

EEOC previously did not specifically recognize (such as reading, bending and communicating) and the second list includes major bodily functions (such as functions of the immune system, normal cell growth, digestive system, bowel, bladder, neurological, brain, respiratory, circulatory, endocrine and reproductive functions.

3. *The Act states that mitigating measures such as "ordinary eyeglasses or contact lenses" shall not be considered in assessing whether an individual possesses a disability;*
4. *The Act clarified that an impairment that is episodic or in remission is a disability if it would substantially limit a major life activity when active; the Act provides that an individual subjected to an action prohibited by the ADA (such as failure to hire) because of an actual or perceived impairment will meet the "regarded as" definition of disability, unless the impairment is transitory or minor;*
5. *The Act provides that individuals covered only under the "regarded as" prong of the ADA test are not entitled to reasonable accommodation; and*
6. *The Act emphasizes that the definition of "disability" should be interpreted broadly.**

Safety professionals should acquire a firm grasp of the requirements of the ADA as well as the interpretations and changes resulting from the ADAAA, the Civil Rights Act of 1991, and the EEOC interpretations. The initial place to start for safety professionals is with the federal or state agency responsible for administration and enforcement of the law. For the ADA, safety professionals should be aware that the federal agency responsible for administration and enforcement of the ADA is the Equal Employment Opportunity Commission (hereinafter "EEOC") and information can be found on their website located at www.eeoc.gov. Safety professionals are encouraged to monitor the EEOC website as well as recent case decisions, regulations, and interpretations to ensure the most current information regarding the ADA. Additionally, safety professionals should also be aware that individual states may also possess laws which parallel or are more stringent than the federal level ADA and information can usually be found on the individual state agency's website.

Safety professionals should be aware that the ADA is divided into five titles, and all titles possess the potential of substantially impacting the safety function in covered public or private sector organizations. Title I contains the employment provisions that protect all individuals with disabilities who are in the United States, regardless of their national origin or immigration status. Title II prohibits discriminating against qualified individuals with disabilities or excluding them from the services, programs, or activities provided by public entities. Title II contains the transportation provisions of the Act. Title III, entitled "Public Accommodations," requires that

---

* Notice Concerning The Americans with Disabilities (ADA) Amendment Act of 2008, EEOC website located at www.eeoc.gov.

goods, services, privileges, advantages, and facilities of any public place be offered "in the most integrated setting appropriate to the needs of the individual."*

Title IV also covers transportation offered by private entities and addresses telecommunications. Title IV requires that telephone companies provide telecommunication relay services and that public service television announcements that are produced or funded with federal money include closed caption. Title V includes the miscellaneous provisions. This Title notes that the ADA does not limit or invalidate other federal and state laws providing equal or greater protection for the rights of individuals with disabilities, and addresses related insurance, alternate dispute, and congressional coverage issues.

Safety professionals in the private sector should provide careful attention to the scope and potential impact of Title I of the ADA on safety functions. Title I prohibits covered employers from discriminating against a "qualified individual with a disability" with regard to job applications, hiring, advancement, discharge, compensation, training, and other terms, conditions, and privileges of employment.[†]

Of particular importance for safety professionals is Section 101 (8). This section defines a "qualified individual with a disability" as any person who, with or without reasonable accommodation, can perform the essential functions of the employment position that such individual holds or desires . . . consideration shall be given to the employer's judgment as to what functions of a job are essential, and if an employer has prepared a written description before advertising or interviewing applicants for the job, this description shall be considered evidence of the essential function of the job.[‡] The Equal Employment Opportunity Commission (EEOC) provides additional clarification of this definition by stating, "an individual with a disability who satisfies the requisite skill, experience and educational requirements of the employment position such individual holds or desires, and who, with or without reasonable accommodation, can perform the essential functions of such position."[§]

Congress did not provide a specific list of disabilities covered under the ADA because "of the difficulty of ensuring the comprehensiveness of such a list."[1] Under the ADA, individuals have a disability if they

- *Have a physical or mental impairment that substantially limits one or more of the major life activities of such individual,*
- *Can produce a record of such an impairment, or*
- *Are regarded as having such an impairment.*[¶]

Safety professionals should be aware that the ADA utilizes the broader language of "disability" rather than the term "handicapped" adopted under the Rehabilitation Act. For an individual to be considered "disabled" under the ADA, the physical or mental impairment must limit one or more "major life activities." Under the U.S.

---

* ADA Section 305.
† ADA Section 102(a); 42 U.S.C. Section 12122.
‡ ADA Section 101(8).
§ EEOC Interpretive Rules, 56 Fed. Reg. 35 (July 26, 1991).
¶ ADA Section 101(8).

Justice Department's regulation issued for Section 504 of the Rehabilitation Act, "major life activities" are defined as, "functions such as caring for one's self, performing manual tasks, walking, seeing, hearing, speaking, breathing, learning and working."[2] Congress clearly intended to have the term "disability" broadly construed. However, this definition does not include simple physical characteristics, nor limitations based on environmental, cultural, or economic disadvantages.[3]

The second prong of this definition is "a record of such an impairment disability." The Senate Report and the House Judiciary Committee Report each stated:

> *This provision is included in the definition in part to protect individuals who have recovered from a physical or mental impairment which previously limited them in a major life activity. Discrimination on the basis of such a past impairment would be prohibited under this legislation. Frequently occurring examples of the first group (i.e., those who have a history of an impairment) are people with histories of mental or emotional illness, heart disease or cancer; examples of the second group (i.e., those who have been misclassified as having an impairment) are people who have been misclassified as mentally retarded.*[4]

The third prong of the statutory definition of a disability extends coverage to individuals who are "being regarded as having a disability." The ADA has adopted the same "regarded as" test that is used in Section 504 of the Rehabilitation Act:

> *"Is regarded as having an impairment" means (A) has a physical or mental impairment that does not substantially limit major life activities but is treated . . . as constituting such a limitation; (B) has a physical or mental impairment that substantially limits major life activities only as a result of the attitudes of others toward such impairment; (C) has none of the impairments defined (in the impairment paragraph of the Department of Justice regulations) but is treated . . . as having such an impairment.**

Safety professionals should be aware that a "qualified individual with a disability" under the ADA is any individual who can perform the essential or vital functions of a particular job with or without the employer accommodating the particular disability. Safety professionals should be aware that companies or organizations are provided the opportunity to determine the "essential functions" of the particular job before offering the position through the development of a written job description. This written job description will be considered evidence to which functions of the particular job are essential and which are peripheral. In deciding the "essential functions" of a particular position, the EEOC will consider the company or organization's judgment, whether the written job description was developed prior to advertising or beginning the interview process, the amount of time spent performing the job, the past and current experience of the individual to be hired, relevant collective bargaining agreements, and other factors.[5]

The EEOC defines the term "essential function" of a job as meaning "primary job duties that are intrinsic to the employment position the individual holds or desires" and precludes any marginal or peripheral functions which may be incidental to the

---

* EEOC Interpretive Rules, 56 Fed. Reg. 35 (July 26, 1991).

primary job function. The factors provided by the EEOC in evaluating the "essential functions" of a particular job include the reason that the position exists, the number of employees available, and the degree of specialization required to perform the job. This determination is especially important to safety professionals who may be required to develop the written job descriptions or to determine the "essential functions" of a given position.*

Safety professionals should recognize that they may be placed in a difficult position involving the issue involved is whether or not the qualified individual creates a direct threat to the safety and health of themselves or others in the workplace. This issue may require the safety professional to evaluate and render a decision which will not only impact the individual with a disability but also the company or organization. Safety professionals should be aware that the ADA does identify that any individual who poses a direct threat to the health and safety of others that cannot be eliminated by reasonable accommodation may be disqualified from the particular job.† The term "direct threat" to others is defined by the EEOC as creating "a significant risk of substantial harm to the health and safety of the individual or others that cannot be eliminated by reasonable accommodation."[6] The determining factors that safety professionals should consider in making this determination include the duration of the risk, the nature and severity of the potential harm, and the likelihood that the potential harm will occur.[7] Safety professionals when addressing this issue should also consider the EEOC's Interpretive Guidelines, which state:

> *[If] an individual poses a direct threat as a result of a disability, the employer must determine whether a reasonable accommodation would either eliminate the risk or reduce it to an acceptable level. If no accommodation exists that would either eliminate the risk or reduce the risk, the employer may refuse to hire an applicant or may discharge an employee who poses a direct threat.*[8]

Safety professionals should note that Title I provides that if a company or organization does not make the reasonable accommodations for the known limitations of a qualified individual with disabilities, this action or inaction is to be considered discrimination. However, if the company or organization can prove that providing the accommodation would place an undue hardship on the operation of the business the claim of discrimination be disproved. Section 101(9) defines a "reasonable accommodation" as:

> (a) *making existing facilities used by employees readily accessible to and usable by the qualified individual with a disability and includes:*
> (b) *job restriction, part-time or modified work schedules, reassignment to a vacant position, acquisition or modification of equipment or devices, appropriate adjustments or modification of examinations, training materials, or policies, the provisions of qualified readers or interpreters and other similar accommodations for . . . the QID (qualified individual with a disability)*‡

The EEOC further defines "reasonable accommodation" as:

---

* Id.
† ADA Section 103(b).
‡ ADA Section 101(9).

1. *Any modification or adjustment to a job application process that enables a qualified individual with a disability to be considered for the position such qualified individual with a disability desires, and which will not impose an undue hardship on the . . . business; or*
2. *Any modification or adjustment to the work environment, or to the manner or circumstances which the position held or desired is customarily performed, that enables the qualified individual with a disability to perform the essential functions of that position and which will not impose an undue hardship on the . . . business; or*
3. *Any modification or adjustment that enables the qualified individual with a disability to enjoy the same benefits and privileges of employment that other employees enjoy and does not impose an undue hardship on the ... business.**

Of particular importance for safety professionals is the area of "reasonable accommodation" for qualified individuals with disabilities. Safety professionals should be aware that the company or organization would be required to make "reasonable accommodations" for any/all known physical or mental limitations of the qualified individual with a disability, unless the employer can demonstrate that the accommodations would impose an "undue hardship" on the business, or that the particular disability directly affects the safety and health of that individual or others. Safety professionals should also be aware that included under this section is the prohibition against the use of qualification standards, employment tests, and other selection criteria that can be used to screen out individuals with disabilities, unless the employer can demonstrate that the procedure is directly related to the job function. In addition to the modifications to facilities, work schedules, equipment, and training programs, the company or organization is required to initiate an "informal interactive (communication) process" with the qualified individual to promote voluntary disclosure of his or her specific limitations and restrictions to enable the employer to make appropriate accommodations that will compensate for the limitation.†

Additionally, safety professionals should pay careful attention to Title I, Section 102(c)(1). This section prohibits discrimination through medical screening, employment inquiries, and similar scrutiny. Safety professionals should be aware that underlying this section was Congress's conclusion that information obtained from employment applications and interviews "was often used to exclude individuals with disabilities—particularly those with so-called hidden disabilities such as epilepsy, diabetes, emotional illness, heart disease, and cancer—before their ability to perform the job was even evaluated."‡

Under Title I, section 102(c)(2), safety professionals should be aware that conducting preemployment physical examinations of applicants and asking prospective employees if they are qualified individuals with disabilities is prohibited. Employers are further prohibited from inquiring as to the nature or severity of the disability, even if the disability is visible or obvious. Safety professionals should also be aware that individuals may ask whether any candidates for transfer or promotion who have a known disability can perform the required tasks of the new position if the tasks are

---

* EEOC Interpretive Rules, supra. note 11.

† Id.

‡ S. Comm. on Labor and Human Resources rep. at 38; H. Comm. on Jud. Rep. at 42.

job related and consistent with business necessity. An employer is also permitted to inquire about the applicant's ability to perform the essential job functions prior to employment. The employer should use the written job description as evidence of the essential functions of the position.*

Safety professionals may require medical examinations of employees only if the medical examination is specifically job related and is consistent with business necessity. Medical examinations are permitted only after the applicant with a disability has been offered the job position. The medical examination may be given before the applicant starts the particular job, and the job offer may be contingent upon the results of the medical examination if all employees are subject to the medical examinations and information obtained from the medical examination is maintained in separate, confidential medical files. Companies or organizations are permitted to conduct voluntary medical examinations for current employees as part of an ongoing medical health program, but again, the medical files must be maintained separately and in a confidential manner. The ADA does not prohibit safety professionals or their medical staff from making inquiries or requiring medical or "fit for duty" examinations when there is a need to determine whether or not an employee is still able to perform the essential functions of the job, or where periodic physical examinations are required by medical standards or federal, state, or local law.[9]

In the area of medical testing, safety professionals should pay careful attention to the area of controlled substance testing. Under the ADA, the company or organization is permitted to test job applicants for alcohol and controlled substances prior to an offer of employment under Section 104(d). The testing procedure for alcohol and illegal drug use is not considered a medical examination as defined under the ADA. Companies or organizations may additionally prohibit the use of alcohol and illegal drugs in the workplace and may require that employees not be under the influence while on the job. Companies and organizations are permitted to test current employees for alcohol and controlled substance use in the workplace to the limits permitted by current federal and state law. The ADA requires all employers to conform to the requirements of the Drug-Free Workplace Act of 1988. Thus, safety professionals should be aware that most existing preemployment and postemployment alcohol and controlled substance programs which are not part of the preemployment medical examination or ongoing medical screening program will be permitted in their current form.† Individual employees who choose to use alcohol and illegal drugs are afforded no protection under the ADA. However, employees who have successfully completed a supervised rehabilitation program and are no longer using or addicted are offered the protection of a qualified individual with a disability under the ADA.‡

Safety professionals should be note that Title III also requires that "auxiliary aids and services" be provided for the qualified individual with a disability including, but not limited to, interpreters, readers, amplifiers, and other devices (not limited or specified under the ADA) to provide that individual with an equal opportunity for

---

* Id.
† ADA Section 102(c).
‡ ADA Section 511(b).

employment, promotion, etc.* Congress did, however, provide that auxiliary aids and services do not need to be offered to customers, clients, and other members of the public if the auxiliary aid or service creates an undue hardship on the business. Safety professionals may want to consider alternative methods of accommodating the qualified individual with a disability. This section also addresses the modification of existing facilities to provide access to the individual, and requires that all new facilities be readily accessible and usable by the individual.

Under Title V, safety professionals should note that the ADA does not limit or invalidate other federal or state laws that provide equal or greater protection for the rights of individuals with disabilities. Safety professionals should become knowledgeable with regards to their individual state laws addressing disability or handicap discrimination in the workplace which may be more restrictive than the ADA.

Congress wrote the ADA in an all-encompassing manner and is substantially broad in nature. Safety professionals should note that the ADA provides protections to all individuals associated with or having a relationship to the qualified individual with a disability. This inclusion is unlimited in nature, including family members, individuals living together, and an unspecified number of others. The ADA extends coverage to all "individuals," legal or illegal, documented or undocumented, living within the boundaries of the United States, regardless of their status.† Under Section 102(b)(4), unlawful discrimination includes "excluding or otherwise denying equal jobs or benefits to a qualified individual because of the known disability of the individual with whom the qualified individual is known to have a relationship or association." Therefore, the protections afforded under this section are not limited to only familial relationships. There appears to be no limits regarding the kinds of relationships or associations that are afforded protection. Of particular note is the inclusion of unmarried partners of persons with AIDS or other qualified disabilities.‡

Safety professionals note that, similar to the OSHA requirements, that the ADA requires that employers post notices of the pertinent provisions of the ADA in an accessible format in a conspicuous location within the employer's facilities. To further ensure ADA compliance, safety professional may wish to consider providing additional notification on job applications and other pertinent documents.

Under the ADA, safety professionals should be aware that it is unlawful for an employer to "discriminate on the basis of disability against a qualified individual with a disability" in all areas, including

- *recruitment, advertising, and job application procedures*
- *hiring, upgrading, promoting, awarding tenure, demotion, transfer, layoff, termination, the right to return from layoff, and rehiring*
- *rate of pay or other forms of compensation and changes in compensation*
- *job assignments, job classifications, organization structures, position descriptions, lines of progression, and seniority lists*
- *leaves of absence, sick leave, or other leaves*

---

* ADA Section 3(1).
† H. Rep. 101-485, Part 2, 51.
‡ Id.

- *fringe benefits available by virtue of employment, whether or not administered by the employer*
- *selection and financial support for training, including apprenticeships, professional meetings, conferences and other related activities, and selection for leave of absence to pursue training*
- *activities sponsored by the employer, including social and recreational programs*
- *any other term, condition, or privilege of employment.**

Safety professionals should be aware that the enforcement procedures adopted by the ADA mirror those of Title VII of the Civil Rights Act. A claimant under the ADA must file a claim with the EEOC within 180 days from the alleged discriminatory event, or within 300 days in states with approved enforcement agencies such as the Human Rights Commission. These are commonly called dual agency states or Section 706 agencies. The EEOC has 180 days to investigate the allegation and sue the employer or to issue a right-to-sue notice to the employee. The employee will have 90 days to file a civil action from the date of this notice.[†]

Safety professionals should be aware that the governing federal agency for the ADA is the Equal Employment Opportunity Commission. Enforcement of the ADA is also permitted by the attorney general or by private lawsuit. Remedies, as identified below, can include the ordered modification of a facility, and civil penalties. Section 505 permits reasonable attorney fees and litigation costs for the prevailing party in an ADA action but, under section 513, Congress encourages the use of arbitration to resolve disputes arising under the ADA.[‡]

> *Compensatory and punitive damages may be awarded in cases involving intentional discrimination based on a person's race, color, national origin, sex (including pregnancy), religion, disability, or genetic information.*
>
> *Compensatory damages pay victims for out-of-pocket expenses caused by the discrimination (such as costs associated with a job search or medical expenses) and compensate them for any emotional harm suffered (such as mental anguish, inconvenience, or loss of enjoyment of life).*
>
> *Punitive damages may be awarded to punish an employer who has committed an especially malicious or reckless act of discrimination.*
>
> *There are limits on the amount of compensatory and punitive damages a person can recover. These limits vary depending on the size of the employer:*
> - *For employers with 15-100 employees, the limit is $50,000.*
> - *For employers with 101-200 employees, the limit is $100,000.*
> - *For employers with 201-500 employees, the limit is $200,000.*
> - *For employers with more than 500 employees, the limit is $300,000.[§]*

---

* EEOC Interpretive Guidelines, supra. note 11.
† S. Rep. 101-116,21;H. Rep. 101-485; Part 2, 51; Part 3, 28. Also see EEOC website at www.eeoc.gov.
‡ ADA Section 505 and 513.
§ Taken from the EEOC website located at www.eeoc.gov.

As noted above, the remedies provided under the ADA were modified from their original version with the passage of the Civil Rights Act of 1991. Employment discrimination (whether intentional or by practice) that has a discriminatory effect on qualified individuals may include hiring, reinstatement, promotion, back pay, front pay, reasonable accommodation, or other actions that will make an individual "whole." Payment of attorney fees, expert witness fees, and court fees are still permitted, and jury trials also allowed.

Compensatory and punitive damages were also made available if intentional discrimination is found. Damages may be available to compensate for actual monetary losses, future monetary losses, mental anguish, and inconvenience. Punitive damages are also available if an employer acted with malice or reckless indifference. The total amount of punitive and compensatory damages for future monetary loss and emotional injury for each individual is limited, and is based upon the size of the employer.

As can be seen, the potential entanglements and interactions between the safety function and the ADA can be extensive. As noted above, safety professionals may want to stop and simply ask, "How can I help you?" and LISTEN to the applicant or employee before acting in any manner. Careful assessment and evaluation, along with guidance from human resources or legal counsel, can ensure compliance with the ADA, as well as individual state disability laws and regulations, by the safety professional.

CHAPTER QUESTIONS

1. To qualify for protection under the ADA, the employee must:
   a. Possess a permanent mental or physical disability
   b. Possess a record of the disability
   c. Be treated as being disabled even if he/she is not disabled
   d. All of the above.
2. Damages under the ADA include:
   a. Compensatory damages
   b. Punitive damages
   c. Attorney fees
   d. All of the above.
3. ADA can impact the safety function:
   a. In the hiring function
   b. In the training function
   c. In the testing function
   d. All of the above.
4. The ADA possesses how many titles?
   a. One
   b. Three
   c. Five
   d. Ten
5. In the hypothetical situation provided above, what are you going to do?

Answers: 1—d; 2—d; 3—d; 4—c; 5—varied responses possible.

Supreme Court of the United States

CHEVRON U.S.A. INC., Petitioner, v. Mario ECHAZABAL*

No. 00-1406.

Argued Feb. 27, 2002.
Decided June 10, 2002.

Applicant for employment at oil refinery brought state court action against employer under Americans with Disabilities Act (ADA), and employer removed action to federal court. The United States District Court for the Central District of California, Lourdes G. Baird, J., entered summary judgment in favor of employer, and applicant appealed. The United States Court of Appeals for the Ninth Circuit, Reinhardt, Circuit Judge, 226 F.3d 1063, reversed in part, vacated in part, and remanded. Upon grant of certiorari, the Supreme Court, Justice Souter, held that Equal Employment Opportunity Commission (EEOC) regulation authorizing refusal to hire an individual because his performance on the job would endanger his own health owing to a disability did not exceed the scope of permissible rulemaking under the ADA.

Reversed and remanded.

Equal Employment Opportunity Commission (EEOC) regulation authorizing refusal to hire an individual because his performance on the job would endanger his own health owing to a disability did not exceed the scope of permissible rulemaking under Americans with Disabilities Act (ADA). Americans with Disabilities Act of 1990, § 2 et seq., 42 U.S.C.A § 12101 et seq.; 29 C.F.R. § 1630.15(b)(2).

**2045 *73 SYLLABUS [FN*]

FN* The syllabus constitutes no part of the opinion of the Court but has been prepared by the Reporter of Decisions for the convenience of the reader. See *United States v. Detroit Timber & Lumber Co.,* 200 U.S. 321, 337, 26 S.Ct. 282, 50 L.Ed. 499.

Respondent Echazabal worked for independent contractors at one of petitioner **2046 Chevron U.S.A. Inc.'s oil refineries until Chevron refused to hire him because of a liver condition—which its doctors said would be exacerbated by continued exposure to toxins at the refinery—and the contractor employing him laid him off in response to Chevron's request that it reassign him to a job without exposure to toxins or remove him from the refinery. Echazabal filed suit, claiming, among other things, that Chevron's actions violated the Americans with Disabilities Act of 1990 (ADA). Chevron defended under an Equal Employment Opportunity Commission (EEOC) regulation permitting the

---

* Case from Westlaw and modified for the purposes of this text.

defense that a worker's disability on the job would pose a direct threat to his health. The District Court granted Chevron summary judgment, but the Ninth Circuit reversed, finding that the regulation exceeded the scope of permissible rulemaking under the ADA.

Held: The ADA permits the EEOC's regulation. Pp. 2048-2053.

(a) The ADA's discrimination definition covers a number of things an employer might do to block a disabled person from advancing in the workplace, such as "using qualification standards ... that screen out or tend to screen out [such] an individual," 42 U.S.C. § 12112(b)(6). And along with § 12113(a), the definition creates an affirmative defense for action under a qualification standard "shown to be job-related and consistent with business necessity," which "may include a requirement that an individual shall not pose a direct threat to the health or safety of other individuals in the workplace," § 12113(b). The EEOC's regulation carries the defense one step further, allowing an employer to screen out a potential worker with a disability for risks on the job to his own health or safety. Pp. 2048-2049.

(b) Echazabal relies on the canon *expressio unius exclusio alterius*— expressing one item of an associated group excludes another left unmentioned—for his argument that the ADA, by recognizing only threats to others, precludes the regulation as a matter of law. The first strike against the expression-exclusion rule here is in the statute, which includes the threat-to-others provision as an example of legitimate qualifications that are "job-related and consistent with business necessity." These spacious defensive categories seem to give an agency a good deal *74 of discretion in setting the limits of permissible qualification standards. And the expansive "may include" phrase points directly away from the sort of exclusive specifications that Echazabal claims. Strike two is the failure to identify any series of terms or things that should be understood to go hand in hand, which are abridged in circumstances supporting a sensible inference that the term left out must have been meant to be excluded. Echazabal claims that Congress's adoption only of the threat-to-others exception in the ADA was a deliberate omission of the threat-to-self exception included in the EEOC's regulation implementing the precursor Rehabilitation Act of 1973, which has language identical to that in the ADA. But this is not an unequivocal implication of congressional intent. Because the EEOC was not the only agency interpreting the Rehabilitation Act, its regulation did not establish a clear, standard pairing of threats to self and others. And, it is likely that Congress used such language in the ADA knowing what the EEOC had made of that language under the earlier statute. The third strike is simply that there is no apparent stopping point to the argument that, by specifying a threat-to-others defense, Congress intended a negative implication about those whose safety could be considered. For example, Congress could not have meant that an employer could not defend a refusal to hire when a worker's disability would threaten others outside the workplace. Pp. 2049-2051.

**2047 (c) Since Congress has not spoken exhaustively on threats to a worker's own health, the regulation can claim adherence under the rule in *Chevron U.S.A. Inc. v. Natural Resources Defense Council, Inc.,* 467 U.S. 837, 843, 104 S.Ct. 2778, 81 L.Ed.2d 694, so long as it makes sense of the statutory defense for qualification standards that are "job-related and consistent with business necessity." Chevron's reasons for claiming that the regulation is reasonable include, *inter alia,* that it allows Chevron to avoid the risk of violating the Occupational Safety and Health Act of 1970 (OSHA). Whether an employer would be liable under OSHA for hiring an individual who consents to a job's particular dangers is an open question, but the employer would be courting trouble under OSHA. The EEOC's resolution exemplifies the substantive choices that agencies are expected to make when Congress leaves the intersection of competing objectives both imprecisely marked and subject to administrative leeway. Nor can the EEOC's resolution be called unreasonable as allowing the kind of workplace paternalism the ADA was meant to outlaw. The ADA was trying to get at refusals to give an even break to classes of disabled people, while claiming to act for their own good in reliance on untested and pretextual stereotypes. This sort of sham protection is just what the regulation disallows, by demanding a particularized enquiry into the harms an employee would probably face. Finally, that *75 the threat-to-self defense reasonably falls within the general "job related" and "business necessity" standard does not reduce the "direct threat" language to surplusage. The provision made a conclusion clear that might otherwise have been fought over in litigation or administrative rulemaking. Pp. 2051-2053.

226 F.3d 1063, reversed and remanded.

SOUTER, J., delivered the opinion for a unanimous Court.

*76 JUSTICE SOUTER DELIVERED THE OPINION OF THE COURT.

A regulation of the Equal Employment Opportunity Commission authorizes refusal to hire an individual because his performance on the job would endanger his own health, owing to a disability. The question in this case is whether the Americans with Disabilities Act of 1990, 104 Stat. 328, 42 U.S.C. § 12101 *et seq.* (1994 ed. and Supp. V), permits the regulation.[FN1] We hold that it does.

FN1. We do not consider the further issue passed upon by the Ninth Circuit, which held that the respondent is a " 'qualified individual' " who "can perform the essential functions of the employment position," 42 U.S.C. § 12111(8) (1994 ed.). 226 F.3d 1063, 1072 (C.A.9 2000). That issue will only resurface if the Circuit concludes that the decision of respondent's employer to exclude him was not based on the sort of individualized medical enquiry required by the regulation, an issue on which the District Court granted summary judgment for petitioner and which we leave to the Ninth Circuit for initial appellate consideration if warranted.

I

Beginning in 1972, respondent Mario Echazabal worked for independent contractors at an oil refinery owned by petitioner Chevron U.S.A. Inc. Twice he applied for a job directly with Chevron, which offered to hire him if he could pass the company's physical examination. See **2048 42 U.S.C. § 12112(d)(3) (1994 ed.). Each time, the exam showed liver abnormality or damage, the cause eventually being identified as Hepatitis C, which Chevron's doctors said would be aggravated by continued exposure to toxins at Chevron's refinery. In each instance, the company withdrew the offer, and the second time it asked the contractor employing Echazabal either to reassign him to a job without exposure to harmful chemicals or to remove him from the refinery altogether. The contractor laid him off in early 1996.

Echazabal filed suit, ultimately removed to federal court, claiming, among other things, that Chevron violated the Americans with Disabilities Act (ADA or Act) in refusing to *77 hire him, or even to let him continue working in the plant, because of a disability, his liver condition.[FN2] Chevron defended under a regulation of the Equal Employment Opportunity Commission (EEOC) permitting the defense that a worker's disability on the job would pose a "direct threat" to his health, see 29 CFR § 1630.15(b)(2) (2001). Although two medical witnesses disputed Chevron's judgment that Echazabal's liver function was impaired and subject to further damage under the job conditions in the refinery, the District Court granted summary judgment for Chevron. It held that Echazabal raised no genuine issue of material fact as to whether the company acted reasonably in relying on its own doctors' medical advice, regardless of its accuracy.

FN2. Chevron did not dispute for purposes of its summary-judgment motion that Echazabal is "disabled" under the ADA, and Echazabal did not argue that Chevron could have made a " 'reasonable accommodation.' " App. 184, n. 6.

On appeal, the Ninth Circuit asked for briefs on a threshold question not raised before, whether the EEOC's regulation recognizing a threat-to-self defense, *ibid.*, exceeded the scope of permissible rulemaking under the ADA. 226 F.3d 1063, 1066, n. 3 (C.A.9 2000). The Circuit held that it did and reversed the summary judgment. The court rested its position on the text of the ADA itself in explicitly recognizing an employer's right to adopt an employment qualification barring anyone whose disability would place others in the workplace at risk, while saying nothing about threats to the disabled employee himself. The majority opinion reasoned that "by specifying only threats to 'other individuals in the workplace,' the statute makes it clear that threats to other persons—including the disabled individual himself—are not included within the scope of the [direct threat] defense," *id.*, at 1066-1067, and it indicated that any such regulation would unreasonably conflict with congressional policy against paternalism in the workplace, *id.*, at 1067-1070. The court went on to reject Chevron's further argument that Echazabal *78 was not " 'otherwise qualified' " to perform the job, holding that the ability to perform a job without risk to one's health or safety is not an " 'essential function' " of the job. *Id.,* at 1070.

The decision conflicted with one from the Eleventh Circuit, *Moses v. American Nonwovens, Inc.,* 97 F.3d 446, 447 (1996), and raised tension with the Seventh Circuit case of *Koshinski v. Decatur Foundry, Inc.,* 177 F.3d 599, 603 (1999). We granted certiorari, 534 U.S. 991, 122 S.Ct. 456, 151 L.Ed.2d 375 (2001), and now reverse.

II

Section 102 of the ADA, 104 Stat. 328, 42 U.S.C. § 12101 *et seq.* , prohibits "discriminat[ion] against a qualified individual with a disability because of the disability ... in regard to" a number of actions by an employer, including "hiring." 42 U.S.C. § 12112(a). The statutory definition of "discriminat[ion]" covers a number of things an employer might do to block a disabled person from advancing in the workplace, such as "using qualification standards ... that screen out or tend to **2049 screen out an individual with a disability." § 12112(b)(6). By that same definition, *ibid.,* as well as by separate provision, § 12113(a), the Act creates an affirmative defense for action under a qualification standard "shown to be job-related for the position in question and ... consistent with business necessity." Such a standard may include "a requirement that an individual shall not pose a direct threat to the health or safety of other individuals in the workplace," § 12113(b), if the individual cannot perform the job safely with reasonable accommodation, § 12113(a). By regulation, the EEOC carries the defense one step further, in allowing an employer to screen out a potential worker with a disability not only for risks that he would pose to others in the workplace but for risks on the job to his own health or safety as well: "The term 'qualification standard' may include a requirement that an individual shall not pose *79 a direct threat to the health or safety of the individual or others in the workplace." 29 CFR § 1630.15(b)(2) (2001).

Chevron relies on the regulation here, since it says a job in the refinery would pose a "direct threat" to Echazabal's health. In seeking deference to the agency, it argues that nothing in the statute unambiguously precludes such a defense, while the regulation was adopted under authority explicitly delegated by Congress, 42 U.S.C. § 12116, and after notice-and-comment rulemaking. See *United States v. Mead Corp.,* 533 U.S. 218, 227, 121 S.Ct. 2164, 150 L.Ed.2d 292 (2001); *Chevron U.S.A. Inc. v. Natural Resources Defense Council, Inc.,* 467 U.S. 837, 842-844, 104 S.Ct. 2778, 81 L.Ed.2d 694 (1984). Echazabal, on the contrary, argues that as a matter of law the statute precludes the regulation, which he claims would be an unreasonable interpretation even if the agency had leeway to go beyond the literal text.

A

As for the textual bar to any agency action as a matter of law, Echazabal says that Chevron loses on the threshold question whether the statute leaves a gap for the EEOC to fill. See *id.,* at 843-844, 104 S.Ct. 2778. Echazabal recognizes the generality of the language providing for a defense when a plaintiff is screened out by "qualification standards" that are "job-related and consistent with business necessity" (and reasonable accommodation would not cure the difficulty posed by employment). 42 U.S.C. § 12113(a). Without more, those provisions would allow an employer to turn away someone whose work would pose a serious risk to himself. That possibility is said to be eliminated, however, by the further specification that " 'qualification standards' may include a requirement that an individual shall not pose a direct threat to the health or safety of other individuals in the workplace." § 12113(b); see also § 12111(3) (defining "direct threat" in terms of risk to others). Echazabal contrasts this provision with an EEOC regulation under the Rehabilitation Act of 1973, 87 Stat. 357, as amended, 29 U.S.C. § 701 *et seq.* , antedating *80 the ADA, which recognized an employer's right to consider threats both to other workers and to the threatening employee himself. Because the ADA defense provision recognizes threats only if they extend to another, Echazabal reads the statute to imply as a matter of law that threats to the worker himself cannot count.

The argument follows the reliance of the Ninth Circuit majority on the interpretive canon, *expressio unius est exclusio alterius,* "expressing one item of [an] associated group or series excludes another left unmentioned." *United States v. Vonn,* 535 U.S. 55, 65, 122 S.Ct. 1043, 152 L.Ed.2d 90 (2002). The rule is fine when it applies, but this case joins some others in showing when it does not. See, e.g., *id.,* at 66, 122 S.Ct. 1043; *United Dominion Industries, Inc. v. United States,* 532 U.S. 822, 836, 121 S.Ct. 1934, 150 L.Ed.2d 45 (2001); **2050 *Pauley v. BethEnergy Mines, Inc.,* 501 U.S. 680, 703, 111 S.Ct. 2524, 115 L.Ed.2d 604 (1991).

The first strike against the expression-exclusion rule here is right in the text that Echazabal quotes. Congress included the harm-to-others provision as an example of legitimate qualifications that are "job-related and consistent with business necessity." These are spacious defensive categories, which seem to give an agency (or in the absence of agency action, a court) a good deal of discretion in setting the limits of permissible qualification standards. That discretion is confirmed, if not magnified, by the provision that "qualification standards" falling within the limits of job relation and business necessity "may include" a veto on those who would directly threaten others in the workplace. Far from supporting Echazabal's position, the expansive phrasing of "may include" points directly away from the sort of exclusive specification he claims. *United States v. New York Telephone Co.,* 434 U.S. 159, 169, 98 S.Ct. 364, 54 L.Ed.2d 376 (1977); *Federal Land Bank of St. Paul v. Bismarck Lumber Co.,* 314 U.S. 95, 100, 62 S.Ct. 1, 86 L.Ed. 65 (1941).[FN3]

FN3. In saying that the expansive textual phrases point in the direction of agency leeway we do not mean that the defense provisions place no limit on agency rulemaking. Without deciding whether all safety-related qualification standards must satisfy the ADA's direct-threat standard, see *Albertson's, Inc. v. Kirkingburg,* 527 U.S. 555, 569-570, n. 15, 119 S.Ct. 2162, 144 L.Ed.2d 518 (1999), we assume that some such regulations are implicitly precluded by the Act's specification of a direct-threat defense, such as those allowing "indirect" threats of "insignificant" harm. This is so because the definitional and defense provisions describing the defense in terms of "direct" threats of "significant" harm, 42 U.S.C. §§ 12113(b), 12111(3), are obviously intended to forbid qualifications that screen out by reference to general categories pretextually applied. See *infra,* at 2052-2053, and n. 5. Recognizing the "indirect" and "insignificant" would simply reopen the door to pretext by way of defense.

*81 Just as statutory language suggesting exclusiveness is missing, so is that essential extrastatutory ingredient of an expression-exclusion demonstration, the series of terms from which an omission bespeaks a negative implication. The canon depends on identifying a series of two or more terms or things that should be understood to go hand in hand, which is abridged in circumstances supporting a sensible inference that the term left out must have been meant to be excluded. E. Crawford, Construction of Statutes 337 (1940) (*expressio unius* " 'properly applies only when in the natural association of ideas in the mind of the reader that which is expressed is so set over by way of strong contrast to that which is omitted that the contrast enforces the affirmative inference' " (quoting *State ex rel. Curtis v. De Corps,* 134 Ohio St. 295, 299, 16 N.E.2d 459, 462 (1938))); *United States v. Vonn, supra.*

Strike two in this case is the failure to identify any such established series, including both threats to others and threats to self, from which Congress appears to have made a deliberate choice to omit the latter item as a signal of the affirmative defense's scope. The closest Echazabal comes is the EEOC's rule interpreting the Rehabilitation Act of 1973, 87 Stat. 357, as amended, 29 U.S.C. § 701 *et seq.* , a precursor of the ADA. That statute excepts from the definition of a protected "qualified individual with a handicap" anyone who would pose a "direct threat to the health or safety of other individuals," but, like the later ADA, the Rehabilitation *82 Act says nothing about threats to self that particular employment might pose. 42 U.S.C. § 12113(b). The EEOC nonetheless extended the exception to cover threat-to-self employment, 29 CFR § 1613.702(f) (1990), and Echazabal argues that Congress's adoption only of the threat-to-others exception in the ADA must have been a deliberate omission of the Rehabilitation Act regulation's tandem term of threat-to-self, with intent to exclude it.

But two reasons stand in the way of treating the omission as an unequivocal **2051 implication of congressional intent. The first is that the EEOC was not the only agency interpreting the Rehabilitation Act, with the consequence that its regulation did not establish a clear, standard pairing of threats to self and others. While the EEOC did amplify upon the text of the Rehabilitation Act exclusion by recognizing threats to self along with threats to others, three other agencies adopting regulations under the Rehabilitation Act did not. See 28 CFR § 42.540(*l*)(1) (1990) (Department of Justice), 29 CFR § 32.3 (1990) (Department of Labor), and 45 CFR § 84.3(k)(1) (1990) (Department of Health and Human Services).FN4 It would be a stretch, then, to say that there was a standard usage, with its source in agency practice or elsewhere, that connected threats to others so closely to threats to self that leaving out one was like ignoring a twin.

FN4. In fact, we have said that the regulations issued by the Department of Health and Human Services, which had previously been the regulations of the

Department of Health, Education, and Welfare, are of "particular significance" in interpreting the Rehabilitation Act because "HEW was the agency responsible for coordinating the implementation and enforcement of § 504 of the Rehabilitation Act, 29 U.S.C. § 794," prohibiting discrimination against individuals with disabilities by recipients of federal funds. *Toyota Motor Mfg., Ky., Inc. v. Williams*, 534 U.S. 184, 195, 122 S.Ct. 681, 151 L.Ed.2d 615 (2002). Unfortunately for Echazabal's argument, the congruence of the ADA with the HEW regulations does not produce an unequivocal statement of congressional intent.

Even if we put aside this variety of administrative experience, however, and look no further than the EEOC's Rehabilitation *83 Act regulation pairing self and others, the congressional choice to speak only of threats to others would still be equivocal. Consider what the ADA reference to threats to others might have meant on somewhat different facts. If the Rehabilitation Act had spoken only of "threats to health" and the EEOC regulation had read that to mean threats to self or others, a congressional choice to be more specific in the ADA by listing threats to others but not threats to self would have carried a message. The most probable reading would have been that Congress understood what a failure to specify could lead to and had made a choice to limit the possibilities. The statutory basis for any agency rulemaking under the ADA would have been different from its basis under the Rehabilitation Act and would have indicated a difference in the agency's rulemaking discretion. But these are not the circumstances here. Instead of making the ADA different from the Rehabilitation Act on the point at issue, Congress used identical language, knowing full well what the EEOC had made of that language under the earlier statute. Did Congress mean to imply that the agency had been wrong in reading the earlier language to allow it to recognize threats to self, or did Congress just assume that the agency was free to do under the ADA what it had already done under the earlier Act's identical language? There is no way to tell. Omitting the EEOC's reference to self-harm while using the very language that the EEOC had read as consistent with recognizing self-harm is equivocal at best. No negative inference is possible.

There is even a third strike against applying the expression-exclusion rule here. It is simply that there is no apparent stopping point to the argument that by specifying a threat-to-others defense Congress intended a negative implication about those whose safety could be considered. When Congress specified threats to others in the workplace, for example, could it possibly have meant that an employer could not defend a refusal to hire when a worker's disability *84 would threaten others outside the workplace? If Typhoid Mary had come under the ADA, would a meat packer have been defenseless if Mary had sued after being turned away? See 42 U.S.C. § 12113(d). *Expressio unius* just fails to work here.

B

Since Congress has not spoken exhaustively on threats to a worker's own health, **2052 the agency regulation can claim adherence under the rule in *Chevron*, 467 U.S., at 843, 104 S.Ct. 2778, so long as it makes sense of the statutory defense for qualification standards that are "job-related and consistent with business necessity." 42 U.S.C. § 12113(a). Chevron's reasons for calling the regulation reasonable are unsurprising: moral concerns aside, it wishes to avoid time lost to sickness, excessive turnover from medical retirement or death, litigation under state tort law, and the risk of violating the national Occupational Safety and Health Act of 1970, 84 Stat. 1590, as amended, 29 U.S.C. § 651 *et seq.* Although Echazabal claims that none of these reasons is legitimate, focusing on the concern with OSHA will be enough to show that the regulation is entitled to survive.

Echazabal points out that there is no known instance of OSHA enforcement, or even threatened enforcement, against an employer who relied on the ADA to hire a worker willing to accept a risk to himself from his disability on the job. In Echazabal's mind, this shows that invoking OSHA policy and possible OSHA liability is just a red herring to excuse covert discrimination. But there is another side to this. The text of OSHA itself says its point is "to assure so far as possible every working man and woman in the Nation safe and healthful working conditions," § 651(b), and Congress specifically obligated an employer to "furnish to each of his employees employment and a place of employment which are free from recognized hazards that are causing or are likely to cause death or serious physical harm to his employees," § 654(a)(1). Although there may be an open question *85 whether an employer would actually be liable under OSHA for hiring an individual who knowingly consented to the particular dangers the job would pose to him, see Brief for United States et al. as *Amici Curiae* 19, n. 7, there is no denying that the employer would be asking for trouble: his decision to hire would put Congress's policy in the ADA, a disabled individual's right to operate on equal terms within the workplace, at loggerheads with the competing policy of OSHA, to ensure the safety of "each" and "every" worker. Courts would, of course, resolve the tension if there were no agency action, but the EEOC's resolution exemplifies the substantive choices that agencies are expected to make when Congress leaves the intersection of competing objectives both imprecisely marked but subject to the administrative leeway found in 42 U.S.C. § 12113(a).

Nor can the EEOC's resolution be fairly called unreasonable as allowing the kind of workplace paternalism the ADA was meant to outlaw. It is true that Congress had paternalism in its sights when it passed the ADA, see § 12101(a)(5) (recognizing "overprotective rules and policies" as a form of discrimination). But the EEOC has taken this to mean that Congress was not aiming at an employer's refusal to place disabled workers at a specifically demonstrated risk, but was trying to get at refusals to give an even break to classes of disabled people, while claiming to act for their own good in reliance on untested and pretextual stereotypes.[FN5] Its **2053 regulation *86 disallows just this sort of sham protection, through demands for a particularized enquiry into the harms the employee would probably face. The direct threat defense must be "based on a reasonable medical judgment that relies on the most current medical knowledge and/or the best available objective evidence," and upon an expressly "individualized assessment of the individual's present ability to safely perform the essential functions of the job," reached after considering, among other things, the imminence of the risk and the severity of the harm portended. 29 CFR § 1630.2(r) (2001). The EEOC was certainly acting within the reasonable zone when it saw a difference between rejecting workplace paternalism and ignoring specific and documented risks to the employee himself, even if the employee would take his chances for the sake of getting a job.[FN6]

FN5. Echazabal's contention that the Act's legislative history is to the contrary is unpersuasive. Although some of the comments within the legislative history decry paternalism in general terms, see, e.g., H.R.Rep. No. 101-485, pt. 2, p. 72 (1990), *U.S. Code Cong. & Admin.News* 1990, pp. 303, 354 ("It is critical that paternalistic concerns for the disabled person's own safety not be used to disqualify an otherwise qualified applicant"); ADA Conf. Rep., 136 Cong. Rec. 17377 (1990) (statement of Sen. Kennedy) ("[A]n employer could not use as an excuse for not hiring a person with HIV disease the claim that the employer was simply 'protecting the individual' from opportunistic diseases to which the individual might be exposed"), those comments that elaborate actually express the more pointed concern that such justifications are usually pretextual, rooted in generalities and misperceptions about disabilities. See, e.g., H.R.Rep. No. 101-485, at 74, *U.S. Code Cong. & Admin.News* 1990, pp. 303, 356 ("Generalized fear about risks from the employment environment, such as exacerbation of the disability caused by stress, cannot be used by an employer to disqualify a person with a disability"); S.Rep. No. 101-116, p. 28 (1989) ("It would also be a violation to deny employment to an applicant based on generalized fears about the safety of the applicant .... By definition, such fears are based on averages and group-based predictions. This legislation requires individualized assessments").

Similarly, Echazabal points to several of our decisions expressing concern under Title VII, which like the ADA allows employers to defend otherwise discriminatory practices that are "consistent with business necessity," 42 U.S.C. § 2000e-2(k), with employers adopting rules that exclude women from jobs that are seen as too risky. See, e.g., *Dothard v. Rawlinson,* 433 U.S. 321, 335, 97 S.Ct. 2720, 53 L.Ed.2d 786 (1977); *Automobile Workers v. Johnson Controls, Inc.,* 499 U.S. 187, 202, 111 S.Ct. 1196, 113 L.Ed.2d 158 (1991). Those cases, however, are beside the point, as they, like Title VII generally, were concerned with paternalistic judgments based on the broad category of gender, while the EEOC has required that judgments based on the direct threat provision be made on the basis of individualized risk assessments.

FN6. Respect for this distinction does not entail the requirement, as Echazabal claims, that qualification standards be "neutral," stating what the job requires, as distinct from a worker's disqualifying characteristics. Brief for Respondent 26. It is just as much business necessity for skyscraper contractors to have steelworkers without vertigo as to have well-balanced ones. See 226 F.3d, at 1074 (Trott, J., dissenting). Reasonableness does not turn on formalism. We have no occasion, however, to try to describe how acutely an employee must exhibit a disqualifying condition before an employer may exclude him from the class of the generally qualified. See Brief for Respondent 31. This is a job for the trial courts in the first instance.

*87 Finally, our conclusions that some regulation is permissible and this one is reasonable are not open to Echazabal's objection that they reduce the direct threat provision to "surplusage," see *Babbitt v. Sweet Home Chapter, Communities for Great Ore.*, 515 U.S. 687, 698, 115 S.Ct. 2407, 132 L.Ed.2d 597 (1995). The mere fact that a threat-to-self defense reasonably falls within the general "job related" and "business necessity" standard does not mean that Congress accomplished nothing with its explicit provision for a defense based on threats to others. The provision made a conclusion clear that might otherwise have been fought over in litigation or administrative rulemaking. It did not lack a job to do merely because the EEOC might have adopted the same rule later in applying the general defense provisions, nor was its job any less responsible simply because the agency was left with the option to go a step further. A provision can be useful even without congressional attention being indispensable.

Accordingly, we reverse the judgment of the Court of Appeals and remand the case for proceedings consistent with this opinion.

It is so ordered.

ENDNOTES

1. 42 Fed. Reg. 22686 (May 4, 1977); S. Rep. 101-116; H.Rep. 101-485, Part 2, 51.
2. 28 CFR Section 41.31.
3. See, *Jasany v. U.S. Postal Service*, 755 F.2d 1244 (6th Cir. 1985).
4. S. Rep. 101-116,23; H.Rep. 101-485, Part 2, 52-53.
5. ADA, title I, Section 101(8).
6. EEOC Interpretive Guidelines, EEOC, 1994.
7. Id.
8. 56 Fed. Reg. 35,745 (July 26, 1991).
9. EEOC Interpretive Guidelines, 56 Fed. Reg. 35,751 (July 26, 1991).

6 Proactive Protections to Prevent Discrimination under the ADA

The Code of the West—Ten Principles to Live by:

1. Live Each Day with Courage.
2. Take Pride in Your Work.
3. Always Finish What You Start.
4. Do What Has to be Done.
5. Be Tough but Fair.
6. When You Make a Promise—Keep It.
7. Ride for The Brand.
8. Talk Less and Say More.
9. Remember Some Things Aren't for Sale.
10. Know When to Draw the Line.

Jim Owen in *Cowboy Ethics: What Wall Street Can Learn From The Code of the West*

America is another name for opportunity.

Ralph Waldo Emerson

LEARNING OBJECTIVES

1. Acquire an understanding of the issues impacting the safety function under the ADA and ADAAA.
2. Identify the methods of achieving and maintaining compliance with the ADA.
3. Acquire an understanding of the proactive measures to protect against potential liability under the ADA.

HYPOTHETICAL SITUATION

The human resource manager comes into your office and asks you to sign a document identifying that Employee XX is a safety and health risk to himself as well as fellow employees. You ask what this document is for and the human resource manager states, "It's so we don't have to let Employee XX back in the plan ... We don't want him hurting himself again or hurting anyone else." What are you going to do?

OVERVIEW

Safety professionals should investigate and ensure that their company or organization possess an up-to-date written policy which prohibits harassment and discrimination in the workplace. An effective and enforced written policy sets the tone for the workplace environment, assists employees in understanding the expectation of management, permits employees to understand what is appropriate behavior in the workplace and, above all, exhibits to employees that the company takes harassment and discrimination seriously. Additionally, most written antidiscrimination and harassment policies provide an avenue or method through which employees can raise complaints or questions in a confidential manner and be assured that the company will appropriate investigate and address the issue(s) or complaint in-house before a formal charge is filed with the EEOC or other legal action taken by the employee.

Generally, the company's antidiscrimination policy should be clear and easily understood, treat all employees fairly and consistently, provide a guideline as to what behavior is expected by management, and offer a clear method through which complaints and allegations are handled in a confidential manner. All levels of exempt and nonexempt personnel should receive training as to each and every aspect of the company policy, and this training should be conducted on at least an annual basis. Managerial employees should be provided additional training and education as to the specific expectations of the company in handling employee complaints of discrimination or harassment, as well as training in identifying and stopping possible discrimination and harassment in the workplace.

For safety professionals, it is important to have a firm grasp on the company's policy prohibiting discrimination and harassment in the workplace. Of particular importance is a working knowledge of the complaint and investigative procedures usually established in the company policy. Safety professionals should be aware that it is often far more cost-effective in terms of dollars as well as potential efficacy losses to address any complaint of discrimination in-house before a formal charge of discrimination is filed by the employee or applicant.

As part and parcel of most antidiscrimination policies is a method through which employees can provide a complaint of discrimination in a confidential manner, and the company can investigate and address the situation prior to the complainant walking out the doors of the company. The complaint portion of the company antidiscrimination policy usually identified where or how an employee or applicant can report the violation in a confidential manner. Some policies direct the employee or applicant to the human resource manager or other high-ranking managerial official within the operation while other companies provide an 800 telephone number and contact is made with a third party not directly linked to the company. Under any company antidiscrimination policy, the complaint must be handled in a confidential manner and the policy, as well as the reporting procedure, should identify that there will be no retaliation of any type in any manner for reporting.

Under many antidiscrimination policies, a proscribed procedure for investigating all complaints is often developed. Some antidiscrimination policies describe the methods utilized in this investigation while other polices simply address the reporting procedure. However, virtually all antidiscrimination policies possess a methodology through which the complaint filed by the employee or applicant is addressed, usually in-house, and the results of the investigation reported to the employee or applicant.

Safety professionals should be aware that the investigative procedures for most policies are conducted by a neutral in-house manager, attorney, or team of managers who investigate the complaint in a confidential manner. The investigative procedure utilized by many companies and organizations can include—but is not limited by, depending on the circumstances—an interview with the complaining employee or applicant, interview of any witnesses, analysis of the alleged site of the discrimination, review of e-mail or telephone messages, and other possible evidence. Safety professionals should be aware that most interviews are conducted separately and privately, and the interview is usually documented with the interviewer's notes. Safety professionals who may be on the investigative team must ensure that all information is maintained in a confidential manner.

Upon the completion of the investigation, the complaining employee or applicant is usually informed as to the findings of the investigation and any corrective action which has or will take place to rectify the situation. Safety professionals should be aware that this may or may not be the end of the situation. If the investigative process achieved the purposes of the applicant or employee, the matter may be dropped by the employee or applicant, based upon the corrective actions of the company or organization. However, the employee or applicant can still file a charge with the EEOC or pursue other legal action in the courts based upon the alleged discrimination or harassment. Safety professionals who serve in the capacity of an investigator should maintain strict confidentiality after the investigation is complete unless compelled by the court.

It is important that the company or organization's antidiscrimination policy is appropriately and effectively communicated with all employees. Many companies or organizations require each employee to acknowledge they have received, read, and understood the policy as part of the new employee orientation program or on an annual basis with current employees. Safety professionals should be aware that antidiscrimination and harassment training, as well as diversity training, can be conducted in conjunction with annual safety training opportunities. Training in the requirements of any antidiscrimination and harassment policy should be scheduled and conducted on at least an annual basis.

Safety professionals should be aware that written job descriptions are important when addressing the specific requirements of a particular job function. Safety professionals often possess a substantial amount of the information required on a written job description through their efforts in conducting and documenting job hazard or job safety analyses. Written job descriptions that have been addressed and implemented by the company or organization are often utilized to support the company's or organization's position regarding the requirements of a specific job function.

In the area of workers' compensation, safety professionals with responsibilities in the management of the workers' compensation function should be aware that many of the employees who incurred a work-related injury or illness may also qualify for protections under the ADA. Utilizing the three-prong test for an individual with a disability to "qualify" under the ADA, employees who incur a work-related injury or illness often meet the initial prong of a permanent mental or physical disability.

With workers' compensation claims where a permanent partial or permanent total disability rating has been provided by the attending physician or other consulting physician, the employee most often meets the criteria for this initial step in the qualifying analysis for a qualified individual with a disability (QID). The second prong of the QID analysis required the individual with a permanent mental or physical disability to possess a record of such disability. In many circumstances, the workers' compensation claim documentation would suffice in meeting this second prong in the QID analysis. With the achievement of these first two prongs of the QID analysis, the employee who incurred the work-related injury or illness possesses a substantial likelihood of also qualifying for protection under the ADA.

Given this qualification for protection under the ADA, safety professionals should pay careful attention to any requests for accommodation by employees, especially when an employee is returning to work after being off with the work-related injury or illness, or on restricted or light duty programs, training functions, and other activities where the qualified individual with a disability may be requesting a reasonable accommodation. As identified earlier in this chapter, safety professionals should listen to their employee and ask "How can I help you?" anytime an employee potentially could be requesting a reasonable accommodation.

In the area of defenses, two (2) possible defenses that usually impact the safety function are the undue hardship defense, and the safety and health risk defense. When a qualified individual with a disability makes a request for reasonable accommodation, many employers assess the request for reasonable accommodation and identify the costs and impact of the reasonable accommodation on their operations. Although the vast majority of the requests for reasonable accommodations are easily addressed and cost very little, circumstances may arise where the accommodation is denied based upon the undue hardship this accommodation would have on the operations or when the accommodation would create a safety or health risk to the qualified individual with a disability or to others in the workplace.

Under the undue hardship defense, safety professionals should be aware that undue hardship means a significant difficulty, disruption of the operation, or expense for the employer. Although safety professionals may hear from management that the requested accommodation costs too much, safety professionals should remind management that utilization of the undue hardship defense in not providing the requested accommodation to the qualified individual with a disability is not a simple shoot-from–the-hip decision. Utilizing the undue hardship defense requires the employer to prove several factors, including how much the cost of the accommodation will impact the workplace, the inability of the employer to absorb the cost of the accommodation, the detrimental impact of the accommodation on the workplace, and other potential factors. Safety professionals should be aware that the size of the company, the type of business, and other factors are also taken into account.

The safety and health defense can be a minefield for safety professionals. If the employer utilizes this defense, safety professionals may be placed in a difficult position in determining whether or not qualified individuals with a disability create a direct threat to the safety and health of themselves or others in the workplace. This issue may require the safety professional to evaluate and render a decision that will not only impact the qualified individual with a disability but also the company or organization. Safety professionals should be aware that the ADA does recognize that any individual who poses a direct threat to the health and safety of others that cannot be eliminated by reasonable accommodation may be disqualified from the particular job.* The term "direct threat" to others is defined by the EEOC as creating "a significant risk of substantial harm to the health and safety of the individual or others that cannot be eliminated by reasonable accommodation."† The determining factors that safety professionals should consider in making this determination include the duration of the risk, the nature and severity of the potential harm, and the likelihood that the potential harm will occur.‡ Safety professionals confronted with this issue should also consider the EEOC's Interpretive Guidelines, which state:

> *[If] an individual poses a direct threat as a result of a disability, the employer must determine whether a reasonable accommodation would either eliminate the risk or reduce it to an acceptable level. If no accommodation exists that would either eliminate the risk or reduce the risk, the employer may refuse to hire an applicant or may discharge an employee who poses a direct threat.§*

The area of testing can also be particularly problematic for safety professionals. After ADA, most preemployment testing, including physical examinations, psychological testing, and other testing often conducted during the prehiring phase of the application process is unlawful. Most employers have moved physical and psychological testing to the post-offer phase of the employment process. Although under the ADA, alcohol and controlled substance testing is permitted during the preemployment phase, safety professionals should carefully analyze this testing due to the potentially unlawful acquisition of health-related information during the medical history acquired prior to the actual alcohol and controlled substance testing. Although alcohol and controlled substance testing is legal in the private sector in most states, safety professionals should carefully review their individual state laws with the human resource department or legal counsel before implementing any type of employment-related alcohol and controlled substance testing.

Safety professionals should also provide special attention to any promotion-related testing being conducted within the company or organization. Testing has become more sophisticated and is often outsourced to testing companies. A careful analysis of any type of testing program, as well as the impact of individual state laws and the ADA's, should be at the top of any safety professional's list prior to implementing any type of employment-related testing.

It is important for safety professionals to slow down the decision-making process and ask "How can I help you?" and *listen* to their employees to avoid difficulties with the ADA. For any request for reasonable accommodation, the initial step is to ensure that the applicant or employee requesting the accommodation is qualified for the

---

* ADA, Section 103(b).
† EEOC Interpretive Guidelines, EEOC website (1994).
‡ Id.
§ 56 Fed.Reg.35,745 (July 26, 1991).

protections afforded under the ADA. If the individual requesting the accommodation is not qualified, there is often no need to go further with the accommodation. If the individual is qualified, often simply listening to the employee or applicant and discussing the accommodation can identify a workable and cost-effective method of achieving the accommodation.

Safety professionals should be aware that unlike many other laws, the qualified individual with a disability must request a reasonable accommodation. When the request is made, prudent safety professionals should stop, ask how can I help you, and listen before making any decisions which they may regret in the long run.

CHAPTER QUESTIONS

1. "Undue hardship" as a defense requires:
   a. Proof of financial hardship
   b. Depends on the size of the employer
   c. Employer's ability to absorb the costs
   d. All of the above
2. A qualified individual with a disability possesses:
   a. A permanent mental or physical disability
   b. A record of the disability
   c. None of the above
   d. A & B only
3. Preemployment physical testing is
   a. Unlawful under the ADA
   b. Lawful under the ADA
   c. Depends on state disability laws
   d. None of the above
4. A reasonable accommodation can be denied if
   a. The individual creates a safety and health risk to himself or herself.
   b. The individual creates a safety risk to contractors working on the site.
   c. The individual creates a safety and health risk to other employees.
   d. All of the above
5. In the hypothetical situation provided at the beginning of the chapter, what are you going to do?

Answers: 1—d; 2—d; 3—a; 4—d; 5—varied responses possible.

7 Defending a Claim of Discrimination under the ADA

The only limits to our realization of tomorrow will be our doubts of today.

Franklin D. Roosevelt

Torts are lawyers' happy hours ... Like double gins and fizz;

Spelled backward, *tort* is *trot* ... straight to the bank, that is.

Art Buck

LEARNING OBJECTIVES

1. Acquiring an understanding of the charge procedure under the ADA and ADAAA.
2. Identifying the methods to address reasonable accommodation requests under the ADA.
3. Acquiring an understanding of the defenses to protect against potential liability under the ADA.

HYPOTHETICAL SITUATION

An employee returns from a 3-month medical leave of absence after an automobile accident. Upon her return, the employee is assigned to return to her previous job on the assembly line. Before starting work, the employee comes to the safety professional and states, "You know I have been off after the accident ... I really don't think I can go back to the assembly line job because of my medical condition." What are you going to do?

OVERVIEW

The best way to defend a claim of discrimination under the ADA is to prevent the claim from being filed by an employee or applicant in the first place. However, safety professionals should be aware that it is relatively easy for employees or applicants to bring a charge of discrimination under the ADA with the EEOC or in state court. In fiscal year 2010, the EEOC received 25,265 claims of discrimination under the

ADA.* From this number of charges, 24,401 were resolved or approximately 96.5%. In resolving these charges, 10.6% were resolved through settlement, 6% of the charges were withdrawn by the person filing the charge, 16.4% were administratively closed, 62.2% were found to possess no reasonable cause, and 4.9% were found to possess reasonable cause.† Of the charges found to possess reasonable cause, 1.8% or 439 charges were successful, and 3.1% or 747 charges were unsuccessful. ‡ Additionally, 21.5% achieved a merit resolution and $76.1 million in monetary benefits were achieved for individuals filing the charge of discrimination under the ADA.§

The procedure for filing a charge of discrimination under the ADA is the same as with Title VII of the Civil Rights Act (see Chapter 4), and the agency tasked with enforcement of the ADA is the EEOC. As identified in the information below, the EEOC addresses the protections provided under the ADA, as well as resources to assist individuals who may have incurred discrimination.

Disability Discrimination

Disability discrimination occurs when an employer or other entity covered by the Americans with Disabilities Act, as amended, or the Rehabilitation Act, as amended, treats a qualified individual with a disability who is an employee or applicant unfavorably because she has a disability.

Disability discrimination also occurs when a *covered employer or other entity* treats an applicant or employee less favorably because she has a history of a disability (such as cancer that is controlled or in remission) or because she is believed to have a physical or mental impairment that is not transitory (lasting or expected to last six months or less) and minor (even if she does not have such an impairment).

The law requires an employer to provide reasonable accommodation to an employee or job applicant with a disability, unless doing so would cause significant difficulty or expense for the employer ("undue hardship").

The law also protects people from discrimination based on their relationship with a person with a disability (even if they do not themselves have a disability). For example, it is illegal to discriminate against an employee because her husband has a disability.

Note: Federal employees and applicants are covered by the Rehabilitation Act of 1973, instead of the Americans with Disabilities Act. The protections are mostly the same.

DISABILITY DISCRIMINATION AND WORK SITUATIONS

The law forbids discrimination when it comes to any aspect of employment, including hiring, firing, pay, job assignments, promotions, layoff, training, fringe benefits, and any other term or condition of employment.

---

* ADA charges for FY 97-10, EEOC website (2011).
† Id.
‡ Id.
§ Id.

DISABILITY DISCRIMINATION AND HARASSMENT

It is illegal to harass an applicant or employee because he has a disability, had a disability in the past, or is believed to have a physical or mental impairment that is not transitory (lasting or expected to last six months or less) and minor (even if he does not have such an impairment).

Harassment can include, for example, offensive remarks about a person's disability. Although the law doesn't prohibit simple teasing, offhand comments, or isolated incidents that aren't very serious, harassment is illegal when it is so frequent or severe that it creates a hostile or offensive work environment or when it results in an adverse employment decision (such as the victim's being fired or demoted).

The harasser can be the victim's supervisor, a supervisor in another area, a coworker, or someone who is not an employee of the employer, such as a client or customer.

DISABILITY DISCRIMINATION AND REASONABLE ACCOMMODATION

The law requires an employer to provide reasonable accommodation to an employee or job applicant with a disability, unless doing so would cause significant difficulty or expense for the employer.

A reasonable accommodation is any change in the work environment (or in the way things are usually done) to help a person with a disability apply for a job, perform the duties of a job, or enjoy the benefits and privileges of employment.

Reasonable accommodation might include, for example, making the workplace accessible for wheelchair users or providing a reader or interpreter for someone who is blind or hearing impaired.

While the federal antidiscrimination laws don't require an employer to accommodate an employee who must care for a disabled family member, the Family and Medical Leave Act (FMLA) may require an employer to take such steps. The Department of Labor enforces the FMLA. For more information, call 1-866-487-9243.

DISABILITY DISCRIMINATION AND REASONABLE ACCOMMODATION AND UNDUE HARDSHIP

An employer doesn't have to provide an accommodation if doing so would cause undue hardship to the employer.

Undue hardship means that the accommodation would be too difficult or too expensive to provide, in light of the employer's size, financial resources, and the needs of the business. An employer may not refuse to provide an accommodation just because it involves some cost. An employer does not have to provide the exact accommodation the employee or job applicant wants. If more than one accommodation works, the employer may choose which one to provide.

DEFINITION OF DISABILITY

Not everyone with a medical condition is protected by the law. In order to be protected, a person must be qualified for the job and have a disability as defined by the law.

A person can show that he or she has a disability in one of three ways:

- A person may be disabled if he or she has a physical or mental condition that substantially limits a major life activity (such as walking, talking, seeing, hearing, or learning).

- A person may be disabled if he or she has a history of a disability (such as cancer that is in remission).
- A person may be disabled if he is believed to have a physical or mental impairment that is not transitory (lasting or expected to last six months or less) and minor (even if he does not have such an impairment).

DISABILITY AND MEDICAL EXAMS DURING THE EMPLOYMENT APPLICATION AND INTERVIEW STAGE

The law places strict limits on employers when it comes to asking job applicants to answer medical questions, take a medical exam, or identify a disability.

For example, an employer may not ask a job applicant to answer medical questions or take a medical exam before extending a job offer. An employer also may not ask job applicants if they have a disability (or about the nature of an obvious disability). An employer may ask job applicants whether they can perform the job and how they would perform the job, with or without a reasonable accommodation.

DISABILITY AND MEDICAL EXAMS AFTER A JOB OFFER FOR EMPLOYMENT

After a job is offered to an applicant, the law allows an employer to condition the job offer on the applicant answering certain medical questions or successfully passing a medical exam, but only if all new employees in the same type of job have to answer the questions or take the exam.

DISABILITY AND MEDICAL EXAMS FOR PERSONS WHO HAVE STARTED WORKING AS EMPLOYEES

Once a person is hired and has started work, an employer generally can only ask medical questions or require a medical exam if the employer needs medical documentation to support an employee's request for an accommodation or if the employer believes that an employee is not able to perform a job successfully or safely because of a medical condition.

The law also requires that employers keep all medical records and information confidential and in separate medical files.

AVAILABLE RESOURCES

In addition to a variety of formal guidance documents, EEOC has developed a wide range of fact sheets, question-and-answer documents, and other publications to help employees and employers understand the complex issues surrounding disability discrimination. These titles include:

- Your Employment Rights as an Individual with a Disability
- Job Applicants and the ADA
- Veterans with Service-Connected Disabilities in the Workplace and the ADA
- Questions and Answers: Promoting Employment of Individuals with Disabilities in the Federal Workforce
- The Family and Medical Leave Act, the ADA, and Title VII of the Civil Rights Act of 1964
- The ADA: A Primer for Small Business
- Your Responsibilities as an Employer

- Small Employers and Reasonable Accommodation
- Work At Home/Telework as a Reasonable Accommodation
- Applying Performance and Conduct Standards To Employees With Disabilities
- Obtaining and Using Employee Medical Information as Part of Emergency Evacuation Procedures
- Veterans with Service-Connected Disabilities in the Workplace and the ADA—A Guide for Employers
- Pandemic Preparedness in the Workplace and the Americans with Disabilities Act
- Employer Best Practices for Workers with Caregiving Responsibilities
- Reasonable Accommodations for Attorneys with Disabilities
- How to Comply with the Americans with Disabilities Act: A Guide for Restaurants and Other Food Service Employers
- Final Report on Best Practices for the Employment of People with Disabilities In State Government
- *ABCs* of Schedule A Documents

THE ADA AMENDMENTS ACT

- Final Regulations Implementing the ADAAA
- Questions and Answers on the Final Rule Implementing the ADA Amendments Act of 2008
- Questions and Answers for Small Businesses: The Final Rule Implementing the ADA Amendments Act of 2008
- Fact Sheet on the EEOC's Final Regulations Implementing the ADAAA

QUESTIONS AND ANSWERS SERIES

- Health Care Workers and the Americans with Disabilities Act
- Deafness and Hearing Impairments in the Workplace and the Americans with Disabilities Act
- Blindness and Vision Impairments in the Workplace and the ADA
- The Americans with Disabilities Act's Association Provision
- Diabetes in the Workplace and the ADA
- Epilepsy in the Workplace and the ADA
- Persons with Intellectual Disabilities in the Workplace and the ADA
- Cancer in the Workplace and the ADA

MEDIATION AND THE ADA

- Questions and Answers for Mediation Providers: Mediation and the Americans with Disabilities Act (ADA)
- Questions and Answers for Parties to Mediation: Mediation and the Americans with Disabilities Act (ADA)*

Safety professionals should be aware that employees or applicants who believe the protections afforded to them under the ADA have been violated may file with the

---

* Disability Discrimination, EEOC website (2011).

EEOC, usually within 180 days from the date of the alleged violation, or with the individual state's fair employment practices agency or often state's human rights commission, often within 300 days from the date of the alleged discrimination.

Employees or applicants can file the complaint in person with the EEOC or state agency, online, or by telephone. The EEOC or state agency would determine whether to accept or reject the complaint and, if accepted, proceed with the investigation. The initial step in the investigation is usually written notification to the company or organization. If the EEOC or state agency rejects the complaint, the employee or applicant usually is usually notified in writing that the complaint is rejected and a "right to sue" is provided to the employee or applicant permitting them the ability to acquire legal counsel and pursue the complaint in civil court.

With the EEOC and correlating state agency's procedures discussed in detail in Chapters 3 and 4, the steps involved with an employee or applicant pursuing a civil action for an alleged violation of the ADA are as follows:

1. **Right to Sue Letter**—Employee or applicant received the "right to sue" letter from the EEOC or state agency and acquires legal counsel.
2. **Arbitration Clause**—If the employee agreed to arbitration through an employment related agreement (known as an "arbitration clause), the employee may be required to pursue the ADA complaint through the arbitration process.
3. **Complaint**—The civil action usually starts with legal counsel for the employee or applicant filing a complaint in the appropriate court with jurisdiction over the matter. A complaint document usually includes a list of the employee or applicant's causes of action, the alleged illegal activities by the company or organization, and facts to support each identified cause of action. The complaint also identified the damages the employee or applicant is seeking, such as money damages and punitive damages, although many complaints do not identify a specific dollar amount unless required by the court. Complaints are usually filed with the court, and a copy is served on the company or organization by certified mail or hand-delivery by the local sheriff or process server.
4. **Answer**—The company or organization can respond to the complaint in many different ways. The most common response is to provide a written answer where the company or organization addresses each allegation identified in the complaint with either an admission or a denial of each allegation. The company or organization can, in addition to the Answer, file a Motion to Dismiss with the court. A Motion to Dismiss is asking the court to reject the employee or applicant's complaint and stop the proceedings. A third response can be a Motion to Strike. Similar to a Motion to Dismiss, this motion asks the court to remove certain causes of action from the Complaint.
5. **Discovery**—If the case is not dismissed or otherwise resolved, the case usually moves to the discovery phase where the parties exchange information, documents, and other relevant evidence. Safety professionals should be aware that the discovery phase may last a month or two or even a year

or more, depending on the case, and the safety professional may be actively involved as a witness, as the company representative during site inspections, and in other activities. Some of the methods employed by the company and the employee or applicant, through their legal counsel to exchange information regarding the case include interrogatories (written questions requiring a written response), Demand for Production of Documents, Requests for Admissions, site inspections, physical examinations, psychological evaluations, subpoenas, and depositions.

6. **Pretrial**—When discovery is completed, the judge usually sets a date and establishes the guidelines for trial. At this point, the safety professional may be involved in assisting legal counsel in preparing the physical evidence or reviewing deposition testimony from potential witnesses. Legal counsel begins their preparation and strategy for trial, establishes their witness list, prepares exhibits, and otherwise prepares to present your case to the jury at trial.

7. **Settlement**—Safety professionals should be aware that often settlement negotiations are ongoing, and a substantial number of cases settle prior to an actual trial taking place. As each side begins to evaluate the evidence the other side possesses as well as the value of the case is estimated, each side acquires a clearer picture as to their chances of success, value of the case, and costs to be incurred.

8. **Jury Selection**—If the trial is to be by jury (rather than *de novo* or a bench trial which mean the case is heard and decided by a judge only), the jury must be selected. Usually local citizens are summoned to the courthouse and the attorneys for both sides questions the individuals as to their background, bias, and other factors. The attorneys for both sides, as well as the judge, can eliminate potential jurors from the pool until the requisite numbers of jurors are seated to hear the case.

9. **Trial**—At trial, the person or entity bringing the action is usually called "the plaintiff: and the entity or company in which the action is brought against is called "the defendant." The trial starts with opening statements from both sides, the direct and cross examination of witnesses, and a closing statement. Safety professionals should be aware that this may take a few days or several weeks. At the conclusion of the closing statements by both sides, the judge will provide the jury with instructions, and they will deliberate to provide a decision in the case.

Safety professionals should be aware that with many ADA cases, alternative methods of reaching a settlement or agreement are often utilized due to the cost and time involved in a trial. Some of the alternative methodologies include arbitration, where the parties agree to have the case heard and decided by one or a panel of impartial third party arbitrators. Arbitration can be binding (the decision is final), or the matter can go to trial if the arbitration fails. Mediation, which is often preferred by the EEOC, is a person (known as a mediator) who attempts to get the parties to resolve the conflict and promotes settlement. For private mediation in a civil action, an impartial mediator is often selected and agreed to by the parties and paid on an hourly basis in an attempt to settlement the matter in a cost-effective manner rather than having to incur the expense of a trial.

Safety professional possess a high probability of being called upon to serve as a witness in ADA cases. Safety professionals are often involved in taking the complaint by the employee involved in the investigation, or in other aspects of the situation before, during, and after the complaint is filed. To this end, safety professionals are often called upon as witnesses to provide the company's side of the case through their testimony.

When serving as a witness, it is essential that safety professionals properly prepare for their testimony. Testimony at trial, where the safety professional is under oath to tell the truth, and being cross examined by the opposition can be a bit intimidating, to say the least. If possible, safety professionals should request the opportunity to review all pertinent documents and request that legal counsel provide a mock direct and cross examination in preparation for trial. Safety professionals, whether in deposition or trial, should always be prepared to answer truthfully to all questions.

When in deposition or trial, safety professionals should always maintain their professionalism, no matter how heated the verbal exchange. Safety professionals should always think before responding and always listen very carefully to the questions posed by the opposition. If the question is not clear, safety professionals should politely ask for clarification and continue to ask for clarification until it is clear exactly what the question is requesting. And, on cross examination, only the question should be answered; do not volunteer information. Answer only the specific question asked by the opposition. If the question asks for a "yes" or "no" answer, only provide a "yes" or "no" answer. Do not explain your answer unless asked. And if the safety professional does not know the answer to the question posed, there is nothing wrong with simply saying, "I don't know." Always be consistent and do not speculate in your answer. Again, testifying at deposition or at trial can be very stressful. Always tell the truth—keep your emotions in check—and always be professional!

Safety professionals should be aware that they may be personally named as a defendant in an ADA lawsuit. This often happens in situation such as where the safety professional is allegedly the harasser or other actions, such as defamation, is alleged by the employee or applicant. Often, the safety professional is named in the action along with the company or organization. Safety professionals should be aware that this is often done by the employee or applicant's counsel for legal or strategic reasons (such as naming the safety professional to avoid having the case transferred from state court to federal court). Often, after the safety professional provides his or her deposition and other requested information, the safety professional is dismissed from the action and the suit continues against the company or organization.

In situations where the safety professional is a named defendant, it is not unusually for the company attorney to represent both the safety professional as well as the company at the onset of the litigation, so long as there is no conflict of interest between the safety professional and the company.

However, as the situation unfolds, safety professionals should be aware that if a conflict of interest arises or the interests of the company and the safety professional deviate, safety professionals should consider acquiring their own legal counsel to protect their interest. Safety professionals should be aware that if they acquire their own legal counsel, the safety professional is often responsible for the cost of this representation.

Safety professionals should be aware that in our litigious society, the potential of being involved in an ADA charge or in litigation involving the ADA is present as a result of the duties and responsibilities of many safety professionals. Safety professionals should be prepared for such actions and maintain their professionalism throughout the legal process. The ADA provides protections to qualified individuals with a disability as well as individuals associated with the qualified individual with a disability. Safety professionals who recognize and respect these protections in the workplace can minimize or eliminate the chances of an ADA charge or litigation involving the ADA.

CHAPTER QUESTIONS

1. The components of a civil trial include:
   a. Opening statements by the parties
   b. Direct and cross examination
   c. Closing statements by the parties
   d. All of the above
2. Safety professionals can be:
   a. A named defendant in an ADA civil action
   b. A witness in an ADA action
   c. An investigator in an ADA charge
   d. All of the above
3. The EEOC can
   a. Accept the charge
   b. Reject the charge
   c. Provide a right to sue letter
   d. All of the above
4. Safety professionals serving as a witness should:
   a. Argue with legal counsel
   b. Invoke the Fifth Amendment not to speak
   c. Always be professional
   d. None of the above
5. In the hypothetical situation identified above, what are you going to do?

Answers: 1—d; 2—d; 3—d; 4—c; 5—varied responses possible

8 Age Discrimination in Employment Act

If you wish to live long, you must be willing to grow old.

George Lawton

I credit my youthfulness at 80 to the fact of a cheerful disposition and contentment in every period of my life with what I was.

Oliver Wendell Holmes

LEARNING OBJECTIVES

1. Acquire an understanding of the requirements and protections afforded under the ADEA.
2. Acquire an understanding of the prohibited practices under the ADEA.
3. Acquire an understanding of the exclusions, exceptions and defenses under the ADEA.
4. Acquire an understanding of the Older Workers Benefits Protection Act.

HYPOTHETICAL SITUATION

Your company has advertised a position opening for an entry-level safety professional to work at plant level. Within the final applicant pool is a 54-year-old applicant with a bachelor's degree in safety and 2 year's experience, a 35-year-old applicant with a high school diploma and 15 year's experience in safety within your specific industry, and a 60-year-old applicant with a master's degree in safety with 30 year's experience in safety. What considerations should the safety professional consider regarding the protection afforded to any of these candidates under the ADEA?

OVERVIEW

With the current economic conditions, talk of reductions in Social Security and Medicare benefits in the future, insurance cost increases, losses in 401K plans, and many other factors, employees are generally remaining in the workplace longer or returning to the workforce, either by choice or economic necessity. Safety professionals are encountering new and different challenges in managing the safety and health of aging workers in the workplace. The physical skills set as well as cognitive skill level may diminish with time, however, the experience and expertise the older worker brings to the workplace is often invaluable. With this changing workplace, safety professionals should be aware of the various federal and state laws that offer protection to the older employee against discrimination in the workplace based upon their age.

The primary federal law offering protections to older workers is the Age Discrimination in Employment Act of 1967 (ADEA).* The purpose of the ADEA is to promote employment of older workers between the age of 40 and 70 years of age based on their ability rather than age and to prohibit arbitrary age discrimination in employment.† Similar to Title VII, the ADEA prohibits discrimination by companies or organizations, labor unions, and employment agencies against employees, members, referrals, and applicants. ‡

Safety professionals should be aware that the EEOC, the federal agency charged with enforcement of the ADEA identifies age discrimination as the following:

Age Discrimination

Age discrimination involves treating someone (an applicant or employee) less favorably because of his age.

The Age Discrimination in Employment Act (ADEA) only forbids age discrimination against people who are age 40 or older. It does not protect workers under the age of 40, although some states do have laws that protect younger workers from age discrimination.

It is not illegal for an *employer or other covered entity* to favor an older worker over a younger one, even if both workers are age 40 or older.

Discrimination can occur when the victim and the person who inflicted the discrimination are both over 40.

AGE DISCRIMINATION AND WORK SITUATIONS

The law forbids discrimination when it comes to any aspect of employment, including hiring, firing, pay, job assignments, promotions, layoff, training, fringe benefits, and any other term or condition of employment.

AGE DISCRIMINATION AND HARASSMENT

It is unlawful to harass a person because of his or her age.

Harassment can include, for example, offensive remarks about a person's age. Although the law doesn't prohibit simple teasing, offhand comments, or isolated incidents that aren't very serious, harassment is illegal when it is so frequent or severe that it creates a hostile or offensive work environment or when it results in an adverse employment decision (such as the victim being fired or demoted).

The harasser can be the victim's supervisor, a supervisor in another area, a coworker, or someone who is not an employee of the employer, such as a client or customer.

---

* 29 U.S.C.A. Section 623.
† Id.
‡ 29 U.S.C.A. Section 623(d).

AGE DISCRIMINATION AND EMPLOYMENT POLICIES/PRACTICES

An employment policy or practice that applies to everyone, regardless of age, can be illegal if it has a negative impact on applicants or employees age 40 or older and is not based on a reasonable factor other than age.

Safety professionals should be aware that the ADEA identified the following practices by companies or organizations as being unlawful:

- To fail or refuse to hire or to discharge any individual or otherwise discriminate against any individual with respect to his or her compensation, terms, conditions, or privileges of employment because of such individual's age;
- To limit, segregate, or classify its membership, or to classify or fail or refuse to refer for employment any individual, in a way which would tend to deprive that individual or employment opportunities, or which would limit such employment opportunities or otherwise adversely affect his or her status as an employee or as an applicant for employment, because of such individual's age;
- To cause or attempt to cause an employer to discriminate against an individual in violation of the Act.*

As noted above, the administration and enforcement of the ADEA is currently housed with the EEOC on the federal level. This has not always been the case. The ADEA was originally administered and enforced by the Wage and Hour Division of the U.S. Department of Labor and the Fair Labor Standards Act structure and mechanisms were utilized. With the Reorganization Plan No. 1 in 1978, the Equal Employment Opportunity Commission was provided authority over the ADEA and established it own guidelines; however, the ADEA must still be enforced through the procedures established under the Fair Labor Standards Act rather than those of Title VII with other similar federal antidiscrimination laws.†

Safety professionals should be aware that the ADEA was amended in 1990 by the Older Workers Benefit Protection Act (OWBPA).‡ Although the primary purpose of the OWBPA was to provide additional protections in the area of employee benefits, the OWBPA also addresses the issue of waivers of rights by protected older individuals. The OWBPA amended the ADEA to specifically prohibit companies or organizations from denying benefits to older employees. However, the OWBPA does, in specific situations, permit companies or organizations to reduce benefits based on the employee's age as long as the cost of providing the reduced benefits is the same as the cost of furnishing benefits to younger employees. In the area of waiver, safety professionals should be aware that the OWBPA contains very specific provisions through which older workers may waiver their rights to

---

* ADEA, EEOC website (2011

† Allen v. Marshall Fields & Co., 93 FRD 438 (ND Ill., 1982).

‡ Older Workers Benefit Protection Act (2008). Public Law 101-433. Also see, EEOC website located at www.eeoc.gov.

sue and/or waive their rights to benefits. Specifically, the OWBPA requires that any waiver must be:

1. *in writing and be understandable*
2. *specifically refer to ADEA rights or claims*
3. *not waive future rights or claim;*
4. *be in exchange for valuable consideration;*
5. *advise the person in writing to consult with an attorney before signing the waiver*
6. *provide the person with at least 21 days to consider the waiver or agreement and at least 7 days to revoke the agreement or waiver after signing* it.*

Safety professionals should be aware that the ADEA originally offered protection for individuals between the ages of 40 years of age and 65 years of age, and this was changed to move the upper age limit to 70 years of age. In 1986, Congress removed the upper age limit of 70 almost entirely from the ADEA, thus providing further protection for individuals age 40 and older. The lower age limit of 40 years of age has not been addressed by Congress and remains the same in the ADEA however prudent safety professionals should investigate the protected ages in their state statutes, which can be significantly different than the ADEA.

As described in detail in Chapter 9, safety professionals should be aware that an individual alleging a violation of the ADEA can file a charge with the EEOC or can bring a civil action in an appropriate court of law. Similar to the EEOC requirements for Title VII, the EEOC requires the charge to be filed within 180 days from the date of the alleged unlawful act unless the unlawful act occurred in a state that possesses a similar state agency that addresses state age discrimination laws where the time limit is usually 300 days. The ADEA charge, again like Title VII, must be in writing, identify the company or organization, and describe the alleged conduct in violation of the ADEA.† Again, similarly to the requirements of Title VII, companies or organizations are required to maintain records regarding benefits plans, merit systems, and preemployment applicants.‡

Under the ADEA, an individual is entitled to file a civil action in federal district court after waiting 60 days after filing with the EEOC if the following general conditions are met:

- *Sixty days has elapsed since the filing of the charge with the EEOC or state agency*
- *The age discrimination charge was filed within the 180 day time limitation (or 300 days with a state agency)*
- *The civil action is filed within the 2 or 3 year statute of limitations*§

---

* Id.
† 29 U.S.C.A. Section 626(d).
‡ 29 U.S.C.A. Section 626(d).
§ 29 U.S.C.A. Section 255(a).

In a civil action, the *prima facie* case that the individual alleging age discrimination must establish generally includes the following:

- The individual is in an age protected class under the ADEA
- The individual applied for and was qualified for the position or promotion for which the company or organization was seeking applicants
- The individual suffered an adverse employment action (e.g., discharge, demotion, etc.)
- After the individual's rejection, discharge, or demotion, the position remained open and the company or organization continued to seek applicants from people with the same qualifications as the individual.*

With ADEA charges or lawsuits, safety professionals should be aware that, in addition to direct or "smoking gun" evidence of age discrimination, individuals alleging age discrimination may utilize the disparate treatment theory, disparate impact theory, pretext, and circumstantial evidence. Defenses, although varied depending on the circumstances, can include the Bona Fide Occupational Qualification Exception (commonly referred to as "BFOQ"),† Reasonable Factors Other than Age (commonly referred to as "RFOA") defense, and Bona Fide seniority system defense.‡ Although safety professionals will seldom be involved in the actual charge or litigation, it is important for safety professionals to understand the underlying basis for any alleged charge of discrimination based on an employee or applicant's age.

With an expanding population of "baby boomer" workers over the age of 40 years of age entering or currently employed in the American workplace, safety professionals should be cognizant of the protections afforded by the ADEA and OWCPA, as well as individual state laws. Although the scope of the ADEA is substantially narrow than Title VII, safety professionals should take note of their actions and inactions that may impact, directly or indirectly, the older workers and applicants over the age of 40 years of age. Safety professionals should strive to ensure the reasonableness and practicability of their decision-making in light of the protections afforded to individuals under the ADEA, OWCPA, and state statutes. Safety professionals should always strive to maintain fair employment practices for all individuals derived through a fair, equitable, and ethical manner that affords legal protections to the company or organization in which they are employed.

It is important for safety professionals to understand who is protected under the ADEA, as well as the protections afforded to applicants and employees who may be within this protected class. Proactive measures to create and maintain a workplace free of discrimination and harassment should be the ultimate goal of the safety professional and any incident of age-based discrimination identified by the safety professional in the management of the safety functions should be reported to management in accordance with company policy.

---

* Id.
† 29 U.S.C.A. Section 623(f).
‡ 29 U.S.C.A. Section 623(f)(2).

CHAPTER QUESTIONS

1. ADEA is
   a. Age Delegation of Employment Amendment
   b. Age Discrimination in Employment Act
   c. Age Discrimination and Workplace Evaluation Act
   d. None of the above
2. The federal agency with jurisdiction over the ADEA today is
   a. U.S. Department of Labor
   b. Wage and Hour Division
   c. EEOC
   d. None of the above
3. The OWBPA is
   a. Older Workers Benefit Protection Act
   b. Occupational Warning & Benefactor Protection Act
   c. Older Worker Benevolent Protection Act
   d. None of the above
4. Under the ADEA and OWBPA, an employee may waive his/her rights only
   a. In oral form before a witness
   b. In written form only
   c. In written form and a waiting period
   d. None of the above
5. In the hypothetical situation described in the opening of the chapter, what are you going to do?

Answers: 1—b; 2—c; 3—a; 4—c; 5—varied responses possible.

Supreme Court of the United States

WESTERN AIR LINES, INC., Petitioner, v. Charles G. CRISWELL et al.*

No. 83–1545

Argued January 14, 1985 Decided June 17, 1985

Flight engineers who had been forced to retire at age 60 and pilots who had been denied reassignment as flight engineers upon reaching age 60 brought age discrimination action against the airline. The District Court, 514 F.Supp. 384, entered judgment on jury verdict in favor of employees, and airline appealed. The Court of Appeals for the Ninth Circuit, 709 F.2d 544, affirmed. The Supreme Court, Justice Stevens, held that: (1) to establish bona fide occupational qualification defense, employer must establish that age is a legitimate proxy for safety-

---

* From Westlaw and modified for the purpose of this text.

related job qualifications in that it is impossible or highly impractical to deal with older employees on an individualized basis; (2) jury was not required to defer to airline's selection of job qualifications which were reasonable in light of the safety risks; (3) standard adopted in the Age Discrimination in Employment Act is one of reasonable necessity, not reasonableness; (4) employer must establish more than a rational basis in fact for believing that identification of unqualified persons cannot occur in an individualized basis; and (5) even in cases involving public safety, the Act does not permit the trier of fact to give complete deference to the employer's decision.

Affirmed.

**2744 SYLLABUS [FN*]

[FN*] The syllabus constitutes no part of the opinion of the Court but has been prepared by the Reporter of Decisions for the convenience of the reader. See *United States v. Detroit Lumber Co.*, 200 U.S. 321, 337, 26 S.Ct. 282, 287, 50 L.Ed. 499.

*400 The Age Discrimination in Employment Act of 1967 (ADEA) generally prohibits mandatory retirement before age 70, but § 4(f)(1) of the Act provides an exception "where age is a bona fide occupational qualification [BFOQ] reasonably necessary to the normal operation of the particular business." Petitioner airline company requires that its flight engineers, who are members of the cockpit crews of petitioners' aircraft but do not operate flight controls unless both the pilot and the copilot become incapacitated, retire at age 60. A Federal Aviation Administration regulation prohibits any person from serving as a pilot or copilot after reaching his 60th birthday. Certain of the respondents, who include flight engineers forced to retire at age 60 and pilots who, upon reaching 60, were denied reassignment as flight engineers, brought suit in Federal District Court against petitioner, contending that the age 60 retirement requirement for flight engineers violated the ADEA. Petitioner defended, in part, on the theory that the requirement is a BFOQ "reasonably necessary" to the safe operation of the airline. The physiological and psychological capabilities of persons over age 60, and the ability to detect disease or a precipitous decline in such capabilities on the basis of individual medical examinations, were the subject of conflicting expert testimony presented by the parties. The jury instructions included statements that the "BFOQ defense is available only if it is reasonably necessary to the normal operation or essence **2745 of [petitioner's] business"; "the essence of [petitioner's] business is the safe transportation of [its] passengers"; and petitioner could establish a BFOQ by proving both that "it was highly impractical for [petitioner] to deal with each [flight engineer] over age 60 on an individualized basis to determine his particular ability to perform his job safely" and that some flight engineers "over age 60 possess traits of a physiological, psychological or other nature which preclude safe and efficient job performance that cannot be ascertained by means other than knowing their age." The District Court entered judgment based on the jury's verdict for the plaintiffs, and the Court of Appeals affirmed, rejecting petitioner's contention that the BFOQ instruction was insufficiently deferential to petitioner's legitimate concern for the safety of its passengers.

*401 *Held:*

1. The ADEA's restrictive language, its legislative history, and the consistent interpretation of the administrative agencies charged with enforcing the statute establish that the BFOQ exception was meant to be an extremely narrow exception to the general prohibition of age discrimination contained in the ADEA. Pp. 2749–2751.

2. The relevant considerations for resolving a BFOQ defense to an age-based qualification purportedly justified by safety interests are whether the job qualification is "reasonably necessary" to the overriding interest in public safety, and whether the employer is compelled to rely on age as a proxy for the safety-related job qualification validated in the first inquiry. The latter showing may be made by the employer's establishing either (a) that it had reasonable cause to believe that all or substantially all persons over the age qualification would be unable to perform safely the duties of the job, or (b) that it is highly impractical to deal with the older employees on an individualized basis. Pp. 2751–2753.

3. The jury here was properly instructed on the elements of the BFOQ defense under the above standard, and the instructions were sufficiently protective of public safety. Pp. 2753–2756.

   (a) Petitioner's contention that the jury should have been instructed to defer to petitioner's selection of job qualifications for flight engineers "that are reasonable in light of the safety risks" is at odds with Congress' decision, in adopting the ADEA, to subject such decisions to a test of objective justification in a court of law. The BFOQ standard adopted in the statute is one of "reasonable necessity," not reasonableness. The public interest in safety is adequately reflected in instructions that track the statute's language. Pp. 2753–2755.

   (b) The instructions were not defective for failing to inform the jury that an airline must conduct its operations "with the highest possible degree of safety." Viewing the record as a whole, the jury's attention was adequately focused on the importance of safety to the operation of petitioner's business. P. 2755.

   (c) There is no merit to petitioner's contention that the jury should have been instructed under the standard that the ADEA only requires that the employer establish "a rational basis in fact" for believing that identification of those persons lacking suitable qualifications cannot be made on an individualized basis. Such standard conveys a meaning that is significantly different from that conveyed by the statutory phrase "reasonably necessary," and is inconsistent with the preference for individual evaluation expressed in the language and legislative history of the ADEA. Nor can such standard be justified on the ground that an employer must be allowed to resolve the controversy in a conservative *402 manner when qualified experts disagree as to whether persons over a certain age can be dealt with on an individual basis. Such argument incorrectly assumes that all expert opinion is entitled to equal weight, and virtually ignores the function of the trier of fact in evaluating conflicting testimony. Pp. 2755–2756.

709 F.2d 544 (CA9 1983), affirmed.

JUSTICE **STEVENS** DELIVERED THE OPINION OF THE COURT.

The petitioner, Western Air Lines, Inc., requires that its flight engineers retire at age 60. Although the Age Discrimination in Employment Act of 1967 (ADEA), *403 29 U.S.C. §§ 621-634, generally prohibits mandatory retirement before age 70, the Act provides an exception "where age is a bona fide occupational qualification [BFOQ] reasonably necessary to the normal operation of the particular business." [FN1] A jury concluded that Western's mandatory retirement rule did not qualify as a BFOQ even though it purportedly was adopted for safety reasons. The question here is whether the jury was properly instructed on the elements of the BFOQ defense.[FN2]

FN1. Section 4(f)(1) of the ADEA provides:

"It shall not be unlawful for an employer ...

"(1) to take any action otherwise prohibited ... where age is a bona fide occupational qualification reasonably necessary to the normal operation of the particular business...." 81 Stat. 603, 29 U.S.C. § 623(f)(1).

FN2. In *Trans World Airlines, Inc. v. Thurston,* 469 U.S. 111, 105 S.Ct. 613, 83 L.Ed.2d 523 (1985), decided earlier this Term, TWA allowed flight engineers to continue working past age 60, and allowed pilots to downbid to flight engineer positions provided that they were able to find an open position prior to their 60th birthdays. See *id.,* at 115-116, 105 S.Ct., at 618-619. Pilots who were displaced for any reason besides the Federal Aviation Administration's age 60 rule, however, were permitted to "bump" less senior persons occupying flight engineer positions without waiting for vacancies to occur. We held that this transfer policy discriminated among pilots on the basis of age, and violated the ADEA. Since TWA did not impose an under-age-60 qualification for flight engineers, however, it had no occasion to rely on the same BFOQ theory presented here by Western.

I

In its commercial airline operations, Western operates a variety of aircraft, including the Boeing 727 and the McDonnell-Douglas DC-10. These aircraft require three crew members in the cockpit: a captain, a first officer, and a flight engineer. "The 'captain' is the pilot and controls the aircraft. He is responsible for all phases of its operation. The "first officer" is the copilot and assists the captain. The 'flight engineer' usually monitors a side-facing instrument panel. He does not operate the flight controls unless the captain and the first officer become incapacitated." *Trans World Airlines, Inc. v. Thurston,* 469 U.S. 111, 114, 105 S.Ct. 613, 618, 83 L.Ed.2d 523 (1985).

*404 A regulation of the Federal Aviation Administration (FAA) prohibits any person from serving as a pilot or first officer on a commercial flight "if that person has reached his 60th birthday." 14 CFR § 121.383(c) (1985). The FAA has justified the retention of mandatory retirement for pilots on the theory that "incapacitating medical events" and "adverse psychological, emotional, and physical changes" occur as a consequence of aging. "The inability to detect or predict with precision an individual's risk of sudden or subtle incapacitation, in the face of known age-related risks, counsels against relaxation of the rule." 49 Fed.Reg. 14695 (1984). See also 24 Fed.Reg. 9776 (1959).

At the same time, the FAA has refused to establish a mandatory retirement age for flight engineers. "While a flight engineer has important duties which contribute to the safe operation of the airplane, he or she may not assume the responsibilities of the pilot in command." 49 Fed.Reg., at 14694. Moreover, available statistics establish that flight engineers have rarely been a contributing cause or factor in commercial aircraft accidents or "incidents." *Ibid.*

In 1978, respondents Criswell and Starley were captains operating DC-10s for **2747 Western. Both men celebrated their 60th birthdays in July 1978. Under the collective-bargaining agreement in effect between Western and the union, cockpit crew members could obtain open positions by bidding in order of seniority.[FN3] In order to avoid mandatory retirement *405 under the FAA's under-age-60 rule for pilots, Criswell and Starley applied for reassignment as flight engineers. Western denied both requests, ostensibly on the ground that both employees were members of the company's retirement plan which required all crew members to retire at age 60.[FN4] For the same reason, respondent Ron, a career flight engineer, was also retired in 1978 after his 60th birthday.

FN3. While this lawsuit was proceeding to trial, Criswell and Starley also pursued their remedies under the collective-bargaining agreement. The System Wide Board of Adjustment, over a dissent, ultimately ruled that the contract provision that appeared to authorize the pilots' downbidding was only intended to allow senior pilots operating narrow-body equipment to bid for first officer or flight engineer positions on wide-body aircraft. App. to Pet. for Cert. A84-A90. Since Criswell and Starley were already serving on wide-body aircraft, the provision did not apply to them. The Board also concluded that the provision would not support a transfer "for the obvious purpose of evading the application of [the] agreed retirement plan." *Id.,* at A89. Western relied on this ground in its motion for summary judgment, but the District Court concluded that material questions of fact remained on the question of whether age was a substantial and determinative factor in the denial of the downbids. *Id.,* at A81.

FN4. The Western official who was responsible for the decision to retire the plaintiffs conceded that "the sole basis" for the denial of the applications of Criswell, Starley, and Ron was the same: "the provision in the pension plan regarding retirement at age 60." Tr. 1163. In addition, he admitted that he had "no personal knowledge" of any safety rationale for the under-age-60 rule for flight engineers, *id.,* at 2059, nor had it played any significant role in his decision to retire them. See *id.,* at 61, 2027-2033, 2056-2057. The airline sent Starley and Ron form letters informing them of its "considered judgment after examining all of the applicable statutory law that since you have been a member of our Pilot retirement plan, that we cannot continue your employment beyond the normal retirement date of age 60." See App. 89, 91.

Mandatory retirement provisions similar to those contained in Western's pension plan had previously been upheld under the ADEA. *United Air Lines, Inc. v. McMann,* 434 U.S. 192, 98 S.Ct. 444, 54 L.Ed.2d 402 (1977). As originally enacted in 1967, the Act provided an exception to its general proscription of age discrimination for any actions undertaken "to observe the terms of a ... bona fide employee benefit plan such as a retirement, pension, or insurance plan, which is not a subterfuge to evade the purposes of this Act."[FN5] In April 1978, however, Congress amended the statute to prohibit employee benefit plans from requiring the involuntary retirement of any employee because of age.[FN6]

FN5. § 4(f)(2), 81 Stat. 603, 29 U.S.C. § 623(f)(2).

FN6. 92 Stat. 189, 29 U.S.C. § 623(f)(2).

Criswell, Starley, and Ron brought this action against Western contending that the under-age-60 qualification for *406 the position of flight engineer violated the ADEA. In the District Court, Western defended, in part, on the theory that the age-60 rule is a BFOQ "reasonably necessary" to the safe operation of the airline.[FN7] All parties submitted evidence concerning the nature of the flight engineer's tasks, the physiological and psychological traits required to perform them, and the availability of those traits among persons over age 60.

FN7. Western also contended that its denials of the downbids by pilots Starley and Criswell were based on "reasonable factors other than age." 29 U.S.C. § 623(f)(1); see n. 10, *infra.*

As the District Court summarized, the evidence at trial established that the flight engineer's "normal duties are less critical to the safety of flight than those of a pilot." 514 F.Supp. 384, 390 (CD Cal.1981). The flight engineer, however, does have critical functions in emergency situations and, of course, might cause considerable disruption in the event of his own medical emergency.

The actual capabilities of persons over age 60, and the ability to detect disease or **2748 a precipitous decline in their faculties, were the subject of conflicting medical testimony. Western's expert witness, a former FAA Deputy Federal Air Surgeon,[FN8] was especially concerned about the possibility of a "cardiovascular event" such as a heart attack. He testified that "with advancing age the likelihood of onset of disease increases and that in persons over age 60 it could not be predicted whether and when such diseases would occur." *Id.,* at 389.

FN8. Although the witness had served with the FAA for seven years ending in 1979, he conceded that throughout his tenure at the FAA he never had advocated that the agency extend the age-60 rule to flight engineers. Tr. 1521

The plaintiffs' experts, on the other hand, testified that physiological deterioration is caused by disease, not aging, and that "it was feasible to determine on the basis of individual medical examinations whether flight deck crew members, including those over age 60, were physically qualified to continue *407 to fly." *Ibid.* These conclusions were corroborated by the nonmedical evidence:

"The record also reveals that both the FAA and the airlines have been able to deal with the health problems of pilots on an individualized basis. Pilots who have been grounded because of alcoholism or cardiovascular disease have been recertified by the FAA and allowed to resume flying. Pilots who were unable to pass the necessary examination to maintain their FAA first class medical certificates, but who continued to qualify for second class medical certificates were allowed to 'down-grade' from pilot to [flight engineer]. There is nothing in the record to indicate that these flight deck crew members are physically better able to perform their duties than flight engineers over age 60 who have not experienced such events or that they are less likely to become incapacitated." *Id.,* at 390.

Moreover, several large commercial airlines have flight engineers over age 60 "flying the line" without any reduction in their safety record. *Ibid.*

The jury was instructed that the "BFOQ defense is available only if it is reasonably necessary to the normal operation or essence of defendant's business." Tr. 2626. The jury was informed that "the essence of Western's business is the safe transportation of their passengers." *Ibid.* The jury was also instructed:

"One method by which defendant Western may establish a BFOQ in this case is to prove:

"(1) That in 1978, when these plaintiffs were retired, it was highly impractical for Western to deal with each second officer over age 60 on an individualized basis to determine his particular ability to perform his job safely; and

"(2) That some second officers over age 60 possess traits of a physiological, psychological or other nature *408 which preclude safe and efficient job performance that cannot be ascertained by means other than knowing their age.

"In evaluating the practicability to defendant Western of dealing with second officers over age 60 on an individualized basis, with respect to the medical testimony, you should consider the state of the medical art as it existed in July 1978." *Id.,* at 2627.

The jury rendered a verdict for the plaintiffs, and awarded damages. After trial, the District Court granted equitable relief, explaining in a written opinion why it found no merit in Western's BFOQ defense to the mandatory retirement rule. 514 F.Supp., at 389–391.[FN9]

FN9. After the judgment in the *Criswell* action, eight other pilots and one career flight officer filed a separate action seeking similar relief. A preliminary injunction was granted on behalf of the flight engineer, and Western appealed. The Court of Appeals consolidated the appeal with Western's appeal in *Criswell,* and affirmed the preliminary injunction. 709 F.2d 544, 558-559 (CA9 1983). The plaintiffs in the collateral action are respondents here.

On appeal, Western made various arguments attacking the verdict and judgment below, but the Court of Appeals affirmed in all respects. 709 F.2d 544 (CA9 1983). **2749 In particular, the Court of Appeals rejected Western's contention that the instruction on the BFOQ defense was insufficiently deferential to the airline's legitimate concern for the safety of its passengers. *Id.,* at 549-551. We granted certiorari to consider the merits of this question. 469 U.S. 815, 105 S.Ct. 80, 83 L.Ed.2d 28 (1984).[FN10]

FN10. One of Western's claims in the trial court was that its refusal to allow pilots to serve as flight engineers after they reached age 60 was based on "reasonable factors other than age" (RFOA), namely, a facially neutral policy embodied in its collective-bargaining agreement which prohibited downbidding. See nn. 3 and 7, *supra*. The jury rejected this defense in its verdict. On appeal, Western claimed that the instructions had improperly required it to bear the burden of proof on the RFOA issue inasmuch as the burden of persuasion on the issue of age discrimination is at all times on the plaintiff. Cf. *Texas Dept. of Community Affairs v. Burdine*, 450 U.S. 248, 101 S.Ct. 1089, 67 L.Ed.2d 207 (1981); *Furnco Construction Co. v. Waters*, 438 U.S. 567, 98 S.Ct. 2943, 57 L.Ed.2d 957 (1978). The Court of Appeals rejected this claim on the merits. 709 F.2d, at 552-553. We granted certiorari to consider the merits of this question, 469 U.S. 815, 105 S.Ct. 80, 83 L.Ed.2d 28 (1984), but as we read the instructions the burden *was* placed on the plaintiffs on the RFOA issue. The general instruction on the question of discrimination provided that the "burden of proof is on the plaintiffs to show discriminatory treatment on the basis of age." App. 58. The instructions expressly informed the jury when the burden shifted to the defendant to prove various issues, e.g., *id.*, at 60 (business necessity); *id.*, at 61 (BFOQ), but did not so inform the jury in the RFOA instruction, *id.*, at 62–63. Because the plaintiffs were assigned the burden of proof, we need not consider whether it would have been error to assign it to the defendant.

*409 II

Throughout the legislative history of the ADEA, one empirical fact is repeatedly emphasized: the process of psychological and physiological degeneration caused by aging varies with each individual. "The basic research in the field of aging has established that there is a wide range of individual physical ability regardless of age." [FN11] As a result, many older American workers perform at levels equal or superior to their younger colleagues.

FN11. Report of the Secretary of Labor, The Older American Worker: Age Discrimination in Employment 9 (1965) (hereinafter Report), EEOC, Legislative History of the Age Discrimination in Employment Act 26 (1981) (hereinafter Legislative History). See also S.Rep. No. 95–493, p. 2 (1977), U.S.Code Cong. & Admin.News 1978, pp. 504, 505, Legislative History 435 ("Scientific research ... indicates that chronological age alone is a poor indicator of ability to perform a job").

In 1965, the Secretary of Labor reported to Congress that despite these well-established medical facts there "is persistent and widespread use of age limits in hiring that in a great many cases can be attributed only to arbitrary discrimination against older workers on the basis of age and regardless of ability." [FN12] Two years later, the President recommended that Congress enact legislation to abolish arbitrary age limits on *410 hiring. Such limits, the President declared, have a devastating effect on the dignity of the individual and result in a staggering loss of human resources vital to the national economy. [FN13]

FN12. Report, at 21, Legislative History 37.

FN13. "Hundreds of thousands not yet old, not yet voluntarily retired, find themselves jobless because of arbitrary age discrimination. Despite our present low rate of unemployment, there has been a persistent average of 850,000 people age 45 and over who are unemployed.

* * *

"In economic terms, this is a serious and senseless loss to a nation on the move. But the greater loss is the cruel sacrifice in happiness and well-being which joblessness imposes on these citizens and their families." H.R.Doc. No. 40, 90th Cong., 1st Sess., 7 (1967), Legislative History 61.

After further study,[FN14] Congress responded with the enactment of the ADEA. The preamble declares that the purpose of the ADEA is "to promote employment of older persons based on their ability rather than age [and] to prohibit arbitrary age discrimination in employment." 81 Stat. 602, 29 U.S.C. § 621(b). Section 4(a)(1) makes it "unlawful for an employer ... to fail or refuse to hire or to discharge any individual or otherwise discriminate against any **2750 individual with respect to his compensation, terms, conditions, or privileges of employment, because of such individual's age." 81 Stat. 603, 29 U.S.C. § 623(a) (1). This proscription presently applies to all persons between the ages of 40 and 70. 29 U.S.C. § 631(a).

FN14. See *EEOC v. Wyoming,* 460 U.S. 226, 230, 103 S.Ct. 1054, 1057, 75 L.Ed.2d 18 (1983).

The legislative history of the 1978 Amendments to the ADEA makes quite clear that the policies and substantive provisions of the Act apply with especial force in the case of mandatory retirement provisions. The House Committee on Education and Labor reported:

"Increasingly, it is being recognized that mandatory retirement based solely upon age is arbitrary and that chronological age alone is a poor indicator of ability to perform a job. Mandatory retirement does not take *411 into consideration actual differing abilities and capacities. Such forced retirement can cause hardships for older persons through loss of roles and loss of income. Those older persons who wish to be re-employed have a much more difficult time finding a new job than younger persons.

"Society, as a whole, suffers from mandatory retirement as well. As a result of mandatory retirement, skills and experience are lost from the work force resulting in reduced GNP. Such practices also add a burden to Government income maintenance programs such as social security." [FN15]

FN15. H.R.Rep. No. 95-527, pt. 1, p. 2 (1977), Legislative History 362. Cf. S.Rep. No. 95–493, p. 4 (1977), Legislative History 437 ("The committee believes that the arguments for retaining existing mandatory retirement policies are largely based on misconceptions rather than upon a careful analysis of the facts"). In the 1978 Amendments, Congress narrowed an exception to the ADEA which had previously authorized involuntary retirement under limited circumstances. See *supra,* at ----.

In both 1967 and 1978, however, Congress recognized that classifications based on age, like classifications based on religion, sex, or national origin, may sometimes serve as a necessary proxy for neutral employment qualifications essential to the employer's business. The diverse employment situations in various industries, however, forced Congress to adopt a "case-by-case basis ... as the underlying rule in the administration of the legislation." H.R.Rep. No. 805, 90th

Cong., 1st Sess., 7 (1967), U.S.Code Cong. & Admin.News 1967, pp. 2213, 2220, Legislative History 80.[FN16] Congress offered only general guidance on when an age classification *412 might be permissible by borrowing a concept and statutory language from Title VII of the Civil Rights Act of 1964[FN17] and providing that such a classification is lawful "where age is a bona fide occupational qualification reasonably necessary to the normal operation of the particular business." 29 U.S.C. § 623(f)(1).

FN16. "Many different types of employment situations prevail. Administration of this law must place emphasis on case-by-case basis, with unusual working conditions weighed on their own merits. The purpose of this legislation, simply stated, is to insure that age, within the limits prescribed herein, is not a determining factor in a refusal to hire." S.Rep. No. 723, 90th Cong., 1st Sess., 7 (1967), Legislative History 111.

FN17. Section 703(e) of Title VII permits classifications based on religion, sex or national origin in those certain instances "where religion, sex, or national origin is a bona fide occupational qualification reasonably necessary to the normal operation of that particular business or enterprise." 42 U.S.C. § 2000e-2(e)(1).

Shortly after the passage of the Act, the Secretary of Labor, who was at that time charged with its enforcement, adopted regulations declaring that the BFOQ exception to the ADEA has only "limited scope and application" and "must be construed narrowly." 33 Fed. Reg. 9172 (1968), 29 CFR § 860.102(b) (1984). The Equal Employment Opportunity Commission (EEOC) adopted the same narrow construction of the BFOQ exception after it was assigned authority for enforcing the statute. 46 Fed. Reg. 47727 (1981), 29 CFR § 1625.6 (1984). The restrictive language of the statute and the consistent interpretation of **2751 the administrative agencies charged with enforcing the statute convince us that, like its Title VII counterpart, the BFOQ exception "was in fact meant to be an extremely narrow exception to the general prohibition" of age discrimination contained in the ADEA. *Dothard v. Rawlinson,* 433 U.S. 321, 334, 97 S.Ct. 2720, 2729, 53 L.Ed.2d 786 (1977).

III

In *Usery v. Tamiami Trail Tours, Inc.,* 531 F.2d 224 (1976), the Court of Appeals for the Fifth Circuit was called upon to evaluate the merits of a BFOQ defense to a claim of age discrimination. Tamiami Trail Tours, Inc., had a policy of refusing to hire persons over age 40 as intercity bus drivers. At trial, the bus company introduced testimony supporting its theory that the hiring policy was a BFOQ based *413 upon safety considerations—the need to employ persons who have a low risk of accidents. In evaluating this contention, the Court of Appeals drew on its Title VII precedents, and concluded that two inquiries were relevant.

First, the court recognized that some job qualifications may be so peripheral to the central mission of the employer's business that *no* age discrimination can be "reasonably *necessary* to the normal operation of the particular business."[FN18] 29 U.S.C. § 623(f)(1). The bus company justified the age qualification for hiring its drivers on safety considerations, but the court concluded that this claim was to be evaluated under an objective standard:

FN18. *Diaz v. Pan American World Airways, Inc.,* 442 F.2d 385 (CA5), cert. denied, 404 U.S. 950, 92 S.Ct. 275, 30 L.Ed.2d 267 (1971), provided authority for this proposition. In *Diaz* the court had rejected Pan American's claim that a female-only qualification for the position of in-flight cabin attendant was a BFOQ under Title VII. The District Court had upheld the qualification as a BFOQ finding that the airline's passengers preferred the "pleasant environment" and the "cosmetic effect" provided by female attendants, and that most men were unable to perform effectively the "non-mechanical functions" of the job. The Court of Appeals rejected the BFOQ defense concluding that these considerations "are tangential to the essence of the business involved." 442 F.2d, at 388.

"[T]he job qualifications which the employer invokes to justify his discrimination must be *reasonably necessary* to the essence of his business—here, the *safe* transportation of bus passengers from one point to another. The greater the safety factor, measured by the likelihood of harm and the probable severity of that harm in case of an accident, the more stringent may be the job qualifications designed to insure safe driving." 531 F.2d, at 236. This inquiry "adjusts to the safety factor" by ensuring that the employer's restrictive job qualifications are "reasonably necessary" to further the overriding interest in public safety. *Ibid.* In *Tamiami,* the court noted that no one had seriously *414 challenged the bus company's safety justification for hiring drivers with a low risk of having accidents.

Second, the court recognized that the ADEA requires that age qualifications be something more than "convenient" or "reasonable"; they must be "reasonably necessary ... to the particular business," and this is only so when the employer is compelled to rely on age as a proxy for the safety-related job qualifications validated in the first inquiry.[FN19] This showing could be made in two ways. The employer could establish that it "... 'had reasonable cause to believe, that is, a factual basis for believing, that all or substantially all [persons **2752 over the age qualifications] would be unable to perform safely and efficiently the duties of the job involved.'" [FN20] In *Tamiami,* the employer did not seek to justify its hiring qualification under this standard.

FN19. *Weeks v. Southern Bell Telephone & Telegraph Co.,* 408 F.2d 228 (CA5 1969), provided authority for this proposition. In *Weeks* the court rejected Southern Bell's claim that a male-only qualification for the position of switchman was a BFOQ under Title VII. Southern Bell argued, and the District Court had found, that the job was "strenuous," but the court observed that that "finding is extremely vague." *Id.,* at 234. The court rejected the BFOQ defense concluding that "using these class stereotypes denies desirable positions to a great many women perfectly capable of performing the duties involved." *Id.,* at 236. Moreover, the employer had made no showing that it was "impossible or highly impractical to deal with women on an individualized basis." *Id.,* at 235, n. 5.

FN20. 531 F.2d, at 235 (quoting *Weeks v. Southern Bell Telephone & Telegraph Co.,* 408 F.2d, at 235).

Alternatively, the employer could establish that age was a legitimate proxy for the safety-related job qualifications by proving that it is "... impossible or highly impractical" to deal with the older employees on an individualized basis.[FN21] "One method by which the employer can carry this burden is to establish that some members of the discriminated-against class possess a trait precluding safe and efficient job performance *415 that cannot be ascertained by means other than knowledge of the applicant's membership in the class." *Id.*, at 235. In *Tamiami,* the medical evidence on this point was conflicting, but the District Court had found that individual examinations could not determine which individuals over the age of 40 would be unable to operate the buses safely. The Court of Appeals found that this finding of fact was not "clearly erroneous," and affirmed the District Court's judgment for the bus company on the BFOQ defense. *Id.*, at 238.

FN21. 531 F.2d, at 235 (quoting *Weeks v. Southern Bell Telephone & Telegraph Co.,* 408 F.2d, at 235, n. 5).

Congress, in considering the 1978 Amendments, implicitly endorsed the two-part inquiry identified by the Fifth Circuit in the *Tamiami* case. The Senate Committee Report expressed concern that the amendment prohibiting manda-tory retirement in accordance with pension plans might imply that mandatory retirement could not be a BFOQ:

"For example, in certain types of particularly arduous law enforcement activ-ity, there may be a factual basis for believing that substantially all employees above a specified age would be unable to continue to perform safely and effi-ciently the duties of their particular jobs, and it may be impossible or impractical to determine through medical examinations, periodic reviews of current job per-formance and other objective tests the employees' capacity or ability to continue to perform the jobs safely and efficiently.

"Accordingly, the committee adopted an amendment to make it clear that where these two conditions are satisfied and where such a bona fide occupa-tional qualification has therefore been established, an employer may lawfully require mandatory retirement at that specified age." S.Rep. No. 95-493, pp. 10–11 (1977), U.S.Code Cong. & Admin.News 1978, pp. 513, 514, Legislative History 443–444.

The amendment was adopted by the Senate, but deleted by the Conference Committee because it "neither added to nor *416 worked any change upon pres-ent law." FN22 H.R.Conf.Rep. No. 95-950, p. 7 (1978), *U.S.Code Cong. & Admin. News* 1978, p. 528, Legislative History 518.

FN22. Senator Javits, an active proponent of the legislation, obviously viewed the BFOQ defense as a narrow one when he explained that it could be proved when "the employer can demonstrate that there is an objective, factual basis for believing that virtually all employees above a certain age are unable to safely perform the duties of their jobs and where, in addition, there is no practical medi-cal or performance test to determine capacity." 123 Cong.Rec. 34319 (1977), Legislative History 506. See also H.R.Rep. No. 95-527, pt. 1, p. 12, Legislative History 372.

[1] ☑ Every Court of Appeals that has confronted a BFOQ defense based on safety considerations has analyzed the problem consistently with the *Tamiami* stan-dard.FN23 **2753 An EEOC regulation embraces the same criteria.FN24 Considering the narrow language of the BFOQ exception, the parallel treatment of such ques-tions under Title VII, and the uniform application of the standard by the federal courts, the EEOC, and Congress, we conclude that this two-part inquiry properly

*417 identifies the relevant considerations for resolving a BFOQ defense to an age-based qualification purportedly justified by considerations of safety.

FN23. See, e.g., *Monroe v. United Air Lines, Inc.*, 736 F.2d 394 (CA7 1984), cert. denied, 470 U.S. 1004, 105 S.Ct. 1356, 84 L.Ed.2d 378 (1985); *Johnson v. American Airlines, Inc.*, 745 F.2d 988, 993-994 (CA5 1984), cert. pending, No. 84-1271; 709 F.2d, at 550 (case below); *Orzel v. City of Wauwatosa Fire Dept.*, 697 F.2d 743, 752-753 (CA7), cert. denied, 464 U.S. 992, 104 S.Ct. 484, 78 L.Ed.2d 680 (1983); *Tuohy v. Ford Motor Co.*, 675 F.2d 842, 844-845 (CA6 1982); *Smallwood v. United Air Lines, Inc.*, 661 F.2d 303, 307 (CA4 1981), cert. denied, 456 U.S. 1007, 102 S.Ct. 2299, 73 L.Ed.2d 1302 (1982); *Arritt v. Grisell*, 567 F.2d 1267, 1271 (CA4 1977). Cf. *Harriss v. Pan American World Airways, Inc.*, 649 F.2d 670, 676-677 (CA9 1980) (Title VII).

FN24. 46 Fed.Reg. 47727 (1981), 29 CFR § 1625.6(b) (1984):

"An employer asserting a BFOQ defense has the burden of proving that (1) the age limit is reasonably necessary to the essence of the business, and either (2) that all or substantially all individuals excluded from the job involved are in fact disqualified, or (3) that some of the individuals so excluded possess a disqualifying trait that cannot be ascertained except by reference to age. If the employer's objective in asserting a BFOQ is the goal of public safety, the employer must prove that the challenged practice does indeed effectuate that goal and that there is no acceptable alternative which would better advance it or equally advance it with less discriminatory impact."

IV

In the trial court, Western preserved an objection to any instruction in the *Tamiami* mold, claiming that "any instruction pertaining to the statutory phrase 'reasonably necessary to the normal operation of [defendant's] business' ... is irrelevant to and confusing for the deliberations of the jury." [FN25] Western proposed an instruction that would have allowed it to succeed on the BFOQ defense by proving that "in 1978, when these plaintiffs were retired, there existed a *rational basis in fact* for defendant to believe that use of [flight engineers] over age 60 on its DC-10 airliners would increase the likelihood of risk to its passengers." [FN26] The proposed instruction went on to note that the jury might rely on the FAA's age 60 rule for pilots to establish a BFOQ under this standard "without considering any other evidence."[FN27] It also noted that the medical evidence submitted by the parties might provide a "rational basis in fact."

FN25. Record, Doc. No. 164 (objections to plaintiffs proposed BFOQ instruction).

FN26. *Ibid.* (Defendant's Proposed Instruction No. 19) (emphasis added). In support of the "rational basis in fact" language in the proposed instruction Western cited language in the Seventh Circuit's opinion in *Hodgson v. Greyhound Lines, Inc.*, 499 F.2d 859 (1974), cert. denied, 419 U.S. 1122, 95 S.Ct. 805, 42 L.Ed.2d 822 (1975), which had been criticized by the Fifth Circuit panel in *Tamiami* and which the Seventh Circuit later repudiated. *Orzel v. City of Wauwatosa Fire Dept.*, 697 F.2d, at 752-753. Western also relied on the District Court's opinion in *Tuohy v. Ford Motor Co.*, 490 F.Supp. 258 (ED Mich.1980), which was reversed on appeal, 675 F.2d 842 (CA6 1982).

FN27. Record, Doc. No. 164 (Defendant's Proposed Instruction No. 19.1).

On appeal, Western defended its proposed instruction, and the Court of Appeals soundly rejected it. 709 F.2d, at 549-551. In this Court, Western slightly changes its course. *418 The airline now acknowledges that the *Tamiami* standard identifies the relevant general inquiries that must be made in evaluating the BFOQ defense. However, Western claims that in several respects the instructions given below were insufficiently protective of public safety. Western urges that we interpret or modify the *Tamiami* standard to weigh these concerns in the balance.

REASONABLY NECESSARY JOB QUALIFICATIONS

Western relied on two different kinds of job qualifications to justify its mandatory retirement policy. First, it argued that flight engineers should have a low risk of incapacitation or psychological and physiological deterioration. At this vague level of analysis respondents have not seriously disputed—nor could they—that the qualification of good health for a vital crew member is reasonably necessary to the essence of the airline's operations. Instead, they have argued that age is not a necessary proxy for that qualification.

[2] ☑ On a more specific level, Western argues that flight engineers must meet the same stringent qualifications as pilots, and **2754 that it was therefore quite logical to extend to flight engineers the FAA's age-60 retirement rule for pilots. Although the FAA's rule for pilots, adopted for safety reasons, is relevant evidence in the airline's BFOQ defense, it is not to be accorded conclusive weight. *Johnson v. Mayor and City Council of Baltimore,* 472 U.S. 353, 370-371, 105 S.Ct. 2717, 2726-2727, 86 L.Ed.2d 286. The extent to which the rule is probative varies with the weight of the evidence supporting its safety rationale and "the congruity between the ... occupations at issue." At 371, 105 S.Ct., at 2727. In this case, the evidence clearly established that the FAA, Western, and other airlines all recognized that the qualifications for a flight engineer were less rigorous than those required for a pilot.[FN28]

FN28. As the Court of Appeals noted, the "jury heard testimony that Western itself allows a captain under the age of sixty who cannot, for health reasons, continue to fly as a captain or co-pilot to downbid to a position as second officer. [In addition,] half the pilots flying in the United States are flying for major airlines which do not require second officers to retire at the age of sixty, and ... there are over 200 such second officers currently flying on wide-bodied aircraft." 709 F.2d, at 552. See also *supra,* at 2748.

[3] ☑ [4] ☑ *419 In the absence of persuasive evidence supporting its position, Western nevertheless argues that the jury should have been instructed to defer to "Western's selection of job qualifications for the position of [flight engineer] that are reasonable in light of the safety risks." Brief for Petitioner 30. This proposal is plainly at odds with Congress' decision, in adopting the ADEA, to subject such management decisions to a test of objective justification in a court of law. The BFOQ standard adopted in the statute is one of "reasonable necessity," not reasonableness.

In adopting that standard, Congress did not ignore the public interest in safety. That interest is adequately reflected in instructions that track the language of the statute. When an employer establishes that a job qualification has been carefully formulated to respond to documented concerns for public safety, it will not be overly burdensome to persuade a trier of fact that the qualification is "reasonably necessary" to safe operation of the business. The uncertainty implicit in the concept of managing safety risks always makes it "reasonably necessary" to err on the side of caution in a close case.[FN29] The employer cannot be expected to establish the risk of an airline accident "to a certainty, for certainty would require running the risk until a tragic accident would *420 prove that the judgment was sound." *Usery v. Tamiami Trail Tours, Inc.,* 531 F.2d, at 238. When the employer's argument has a credible basis in the record, it is difficult to believe that a jury of laypersons—many of whom no doubt have flown or could expect to fly on commercial air carriers—would not defer in a close case to the airline's judgment. Since the instructions in this case would not have prevented the airline from raising this contention to the jury in closing argument, we are satisfied that the verdict is a consequence of a defect in Western's proof rather than a defect in the trial court's instructions.[FN30]

FN29. Several Courts of Appeals have recognized that safety considerations are relevant in making or reviewing findings of fact. See, e.g., *Levin v. Delta Air Lines, Inc.,* 730 F.2d 994, 998 (CA5 1984); *Orzel v. City of Wauwatosa Fire Dept.,* 697 F.2d, at 755; *Tuohy v. Ford Motor Co.,* 675 F.2d, at 845; *Murnane v. American Airlines, Inc.,* 215 U.S.App.D.C. 55, 58, 667 F.2d 98, 101 (1981), cert. denied, 456 U.S. 915, 102 S.Ct. 1770, 72 L.Ed.2d 174 (1982); *Hodgson v. Greyhound Lines, Inc.,* 499 F.2d, at 863. Such considerations, of course, are only relevant at the margin of a close case, and do not relieve the employer from its burden of establishing the BFOQ by the preponderance of credible evidence.

FN30. Moreover, we do not find that petitioner's proposed instructions made any reference to the notion of deference to the expertise of the employer, except insofar as that concept was implicit in the "rational basis in fact" standard reflected in its proposed instructions. As we reject that standard as inconsistent with the statute, *infra,* at 2755-2756, we are somewhat reluctant to fault the trial judge for not giving an instruction that was not requested.

WESTERN'S STATUTORY SAFETY OBLIGATION

[5] ☑ The instructions defined the essence of Western's business as "the safe transportation **2755 of their passengers." Tr. 2626. Western complains that this instruction was defective because it failed to inform the jury that an airline must conduct its operations "with the highest possible degree of safety." [FN31]

FN31. This standard is set forth in the Federal Aviation Act, which provides, in part:

"In prescribing standards, rules, and regulations, and in issuing certificates under this subchapter, the Secretary of Transportation shall give full consideration to the duty resting upon air carriers to perform their services with *the highest possible degree of safety* in the public interest ... " 49 U.S.C.App. § 1421(b) (emphasis added).

Jury instructions, of course, "may not be judged in artificial isolation," but must be judged in the "context of the overall charge" and the circumstances of the case. See *Cupp v. Naughten,* 414 U.S. 141, 147, 94 S.Ct. 396, 400, 38 L.Ed.2d 368 (1973). In this case, the instructions characterized safe transportation as the "essence" *421 of Western's business and specifically referred to the importance of "safe and efficient job performance" by flight engineers. Tr. 2627. Moreover, in closing argument counsel pointed out that because "safety is the essence of Western's business," the airline strives for "the highest degree possible of safety." [FN32] Viewing the record as a whole, we are satisfied that the jury's attention was adequately focused on the importance of safety to the operation of Western's business. Cf. *United States v. Park,* 421 U.S. 658, 674, 95 S.Ct. 1903, 1912, 44 L.Ed.2d 489 (1975).

FN32. "We have tried to present, throughout the case, our view that safety is the essence of Western's business. It is the core, it is what the air passenger service business is all about. We have a duty to our passengers, which we consider to be the most important duty of all the business operations that we engage in, including making money. Our first duty is that the passengers and the crews on all our aircraft are safe. And we attempt to render to them the highest degree possible of safety." Tr. 2514.

AGE AS A PROXY FOR JOB QUALIFICATIONS

[6] ☑ Western contended below that the ADEA only requires that the employer establish "a rational basis in fact" for believing that identification of those persons lacking suitable qualifications cannot occur on an individualized basis.[FN33] This "rational basis in fact" standard would have been tantamount to an instruction to return a verdict in the defendant's favor. Because that standard conveys a meaning that is significantly different from that conveyed by the statutory phrase "reasonably necessary," it was correctly rejected by the trial court.[FN34]

FN33. In this Court Western proposes a "factual basis" standard. We do not perceive any substantial difference between this standard and the instruction that it sought below, and we discuss the question as it was raised in the proposed instructions, and discussed in the Court of Appeals.

FN34. This standard has been rejected by nearly every court to consider it. 709 F.2d, at 550-551 (case below); *Orzel v. City of Wauwatosa Fire Dept.,* 697 F.2d, at 755-756; *Tuohy v. Ford Motor Co.,* 675 F.2d, at 845; *Harriss v. Pan American World Airways, Inc.,* 649 F.2d, at 677; *Arritt v. Grisell,* 567 F.2d, at 1271; *Usery v. Tamiami Trail Tours, Inc.,* 531 F.2d, at 235-236.

[7] ☑ *422 Western argues that a "rational basis" standard should be adopted because medical disputes can never be proved "to a certainty" and because juries should not be permitted "to resolve bona fide conflicts among medical experts respecting the adequacy of individualized testing." Reply Brief for Petitioner 9, n. 10. The jury, however, need not be convinced beyond all doubt that medical testing is impossible, but only that the proposition is true "on a preponderance of the evidence." Moreover, Western's attack on the wisdom of assigning the resolution of complex questions to 12 laypersons is inconsistent with the structure of the ADEA. Congress expressly decided that problems involving age discrimination in employment should be resolved on a "case-by-case basis" by **2756 proof to a jury.[FN35]

FN35. *Supra,* at 2750, and n. 16; 29 U.S.C. § 626(c)(2); *Lorillard v. Pons,* 434 U.S. 575, 98 S.Ct. 866, 55 L.Ed.2d 40 (1978).

[8] ☑ [9] ☑ The "rational basis" standard is also inconsistent with the preference for individual evaluation expressed in the language and legislative history of the ADEA.[FN36] Under the Act, employers are to evaluate employees between the ages of 40 and 70 on their merits and not their age. In the BFOQ defense, Congress provided a limited exception to this general principle, but required that employers validate any discrimination as "reasonably necessary to the normal operation of the particular business." It might well be "rational" to require mandatory retirement at *any* age less than 70, but that result would not comply with Congress' direction that employers must justify the rationale for the age chosen. Unless an employer can establish a substantial basis for believing that all or nearly all employees above an age lack the qualifications required for the position, the age selected for mandatory retirement less than 70 must be an age at which it *423 is highly impractical for the employer to insure by individual testing that its employees will have the necessary qualifications for the job.

FN36. Indeed, under a "rational basis" standard a jury might well consider that its "inquiry is at an end" with an expert witness' articulation of any "plausible reaso[n]" for the employer's decision. Cf. *United States Railroad Retirement Board v. Fritz,* 449 U.S. 166, 179, 101 S.Ct. 453, 461, 66 L.Ed.2d 368 (1980).

[10] ☑ Western argues that its lenient standard is necessary because "where qualified experts disagree as to whether persons over a certain age can be dealt with on an individual basis, an employer must be allowed to resolve that controversy in a conservative manner." Reply Brief for Petitioner 8-9. This argument incorrectly assumes that all expert opinion is entitled to equal weight, and virtually ignores the function of the trier of fact in evaluating conflicting testimony. In this case, the jury may well have attached little weight to the testimony of Western's expert witness. See *supra,* at 2747-2748, and n. 8. A rule that would require the jury to defer to the judgment of any expert witness testifying for the employer, no matter how unpersuasive, would allow some employers to give free reign to the stereotype of older workers that Congress decried in the legislative history of the ADEA.

[11] ☑ When an employee covered by the Act is able to point to reputable businesses in the same industry that choose to eschew reliance on mandatory retirement earlier than age 70, when the employer itself relies on individualized testing in similar circumstances, and when the administrative agency with primary responsibility for maintaining airline safety has determined that individualized testing is not impractical for the relevant position, the employer's attempt to justify its decision on the basis of the contrary opinion of experts—solicited for the purposes of litigation—is hardly convincing on any objective standard short of complete deference. Even in cases involving public safety, the ADEA plainly does not permit the trier of fact to give complete deference to the employer's decision.

The judgment of the Court of Appeals is
Affirmed.

Justice POWELL took no part in the decision of this case.

9 Proactive Protections to Prevent Discrimination under the ADEA

The secret of staying young is to live honestly, eat slowly, and lie about your age."

Lucille Ball

If one would understand older people, one should first forget age. Oldness is not so much passing a certain birthday as it is the rearrangement of a complicated set of physical, mental, social, and economic circumstances. One must not label a man who has lived a lot of years as an old person. For an individual who has early formed good habits of living, picked up the important techniques of adjustment, and acquired a good attitude or philosophy, life continues to be an ever-increasing adventure in development. Development can continue at sixty, seventy, and eighty as surely as it did in youth.

William W. Turk

LEARNING OBJECTIVES

1. Acquire an understanding of the potential pitfalls for safety professionals under the ADEA.
2. Acquire an understanding of the proactive practices to prevent discrimination under the ADEA.
3. Acquire an understanding of the exclusions, exceptions, and defenses under the ADEA and OWBPA.

HYPOTHETICAL SITUATION

The safety professionals are interviewing potential applicants for a new start-up operation that will require the selected candidate to work 10–12 hours per day, every Saturday and be on call 24 hours a day, 7 days a week for at least the initial year of operations. The current candidate is a male with 30 years of safety experience identified on his resume. During the interview, the candidate asks, "I see that this job requires a lot of hours ….. Do you think I am too old for this job?" How are you going to respond?

OVERVIEW

With the aging workforce, the potential of a safety professional confronting situations where age discrimination may be present has increased, and will continue to increase, in the workplace. Safety professionals should be cognizant of the protections afforded to applicants and employees over the age of 40 under the ADEA and identify situations where the potential of discrimination may be present. As identified in the EEOC posting below, ADEA provided protections not only in hiring but in a broad spectrum of activities in which safety professionals are involved or responsible for in the workplace.

Starts Here Age Discrimination

Age discrimination involves treating someone (an applicant or employee) less favorably because of his age.

The Age Discrimination in Employment Act (ADEA) only forbids age discrimination against people who are age 40 or older. It does not protect workers under the age of 40, although some states do have laws that protect younger workers from age discrimination.

It is not illegal for an *employer or other covered entity* to favor an older worker over a younger one, even if both workers are age 40 or older.

Discrimination can occur when the victim and the person who inflicted the discrimination are both over 40.

AGE DISCRIMINATION AND WORK SITUATIONS

The law forbids discrimination when it comes to any aspect of employment, including hiring, firing, pay, job assignments, promotions, layoff, training, fringe benefits, and any other term or condition of employment.

AGE DISCRIMINATION AND HARASSMENT

It is unlawful to harass a person because of his or her age.

Harassment can include, for example, offensive remarks about a person's age. Although the law doesn't prohibit simple teasing, offhand comments, or isolated incidents that aren't very serious, harassment is illegal when it is so frequent or severe that it creates a hostile or offensive work environment or when it results in an adverse employment decision (such as the victim's being fired or demoted).

The harasser can be the victim's supervisor, a supervisor in another area, a coworker, or someone who is not an employee of the employer, such as a client or customer.

AGE DISCRIMINATION AND EMPLOYMENT POLICIES/PRACTICES

An employment policy or practice that applies to everyone, regardless of age, can be illegal if it has a negative impact on applicants or employees age 40 or older and is not based on a reasonable factor other than age.*

Safety professionals should identify and become familiar with your company's or organization's ADEA policy and procedures. In many companies, ADEA may be incorporated into one policy that includes Title VII, ADA, and other federal laws, and the same reporting and investigative procedures utilized for ADEA. As with

---

* EEOC website (2011).

other federal laws, safety professionals should be aware that there is a posting requirement for ADEA, and the EEOC is the governing federal agency. Again, as with the other federal antidiscrimination laws, safety professionals should be aware that many states also possess similar laws protecting older workers and many of these laws may offer more protection than the ADEA.

Three areas of protection under the ADEA that often intersect with the safety function and where safety professionals should pay special attention include the areas of training, harassment, and job assignments. As safety professionals are aware, the safety function requires a substantial amount of safety training with regard to a wide variety of different OSHA standards and is often ongoing throughout the year. This training can be in-house or external in nature, depending on the subject matter. Safety professionals should be cognizant of the protections afforded to employees over the age of 40 and ensure that all selections for participation in training are based on reasonable factors and not on the age of the prospective participant. For example, an employee is 1 year from retirement. The safety professional is making a selection to attend a specific training course that requires travel and other expenses. Would the safety professional be discriminating against the ADEA-protected employee if the employee was not selected to participate in the training because he/she will be retiring in 1 year?

Although in-house safety training is usually open to all employees and often required participation, safety professionals should be cognizant to ensure that selection for participation is based upon reasonable factors other than age. Safety training, especially specialized training that can impact such employment areas as promotion, could be considered to be a benefit and thus become an issue in any ADEA action if the training is denied.

In the area of workplace harassment or the prevention of a hostile work environment, safety professionals should be aware that what may be considered to be horseplay by some may be considered workplace harassment to others. The harassment is viewed through the eyes of the employee being harassed. Although most safety professionals do not permit any type of horseplay in their operations, it is important for safety professionals to be aware of other activities, such as name calling, practical jokes, and other similar activities that may also create a hostile work environment for an individual employee who is protected under the ADEA as well as other federal antidiscrimination laws.

As with other antidiscrimination laws, safety professionals should be cognizant of the protection provided to employees over the age of 40 in the area of job assignments. Although safety professionals may consider reasonable factors other than age when assigning specific work assignments, safety professionals should be aware that using age as the primary reason for selecting a specific job assignment for an employee, especially if there is a wage differential between job assignments, may create a discriminatory situation. Although situations involving termination, demotion, failure to promote, and similar disciplinary or human resource situations may be the primary issue involved in any ADEA charge or litigation, safety professionals should be aware that any situation where the employee believes he or she has been discriminated against as a result of their age may become part of the employee's case against the company or organization.

Of particular importance for safety professionals with workers' compensation responsibilities is the recognition of the protections provided to employees over the age of 40 who may have incurred a work-related injury or illness, and the requirements of the ADEA and OWBPA with regards to any waiver and release of ADEA rights that may be part of a severance agreement, which may include a workers' compensation settlement. Safety professionals can find that the employer, in order to minimize the risk of potential litigation, may include a workers' compensation settlement as part of a severance agreement where the employer offers the employee money, extended benefits, or other compensation in exchange for a release or waiver of all claims or liability connected to the relationship between the employee and employer. In essence, the workers' compensation claim may be included within an agreement for the employee to terminate the relationship with the employer, with the employer providing money or other benefits in exchange for the employee's promise not to sue the employer.

Safety professionals should be aware that this agreement is often called a *severance agreement* or *termination agreement*. Additionally, safety professionals should be aware that this agreement often contains very specific terms and agreements between the employer and employee which is included within a written contract or document.

Generally, the waiver and/or release within the severance agreement is valid when the employee knowingly and voluntarily signed the document. Again, generally, the severance agreement identifies the employee and employer, specifies the particular details of the relationship, contains the waiver and/or release, and notes the consideration (money or benefits) the employee will receive in exchange for the employee's waiver of the right to sue the employer. The severance agreement usually also addresses any future rights or responsibilities of the employee and employer, and addresses compliance with applicable state or federal laws. However, safety professionals should be aware that under the requirements of the ADEA and OWBPA, seven additional requirements must be met in order for a waiver and/or release of a potential age discrimination claim to be met. It must:

1. Be in writing and written in a manner clearly understood.
2. Specifically refer to rights or claims arising under the ADEA.
3. Advise the employee to consult an attorney before accepting the agreement.
4. Provide the employee with a minimum of 21 days to consider the offer.
5. Provide the employee seven days to revoke his or her acceptance (or signature).
6. Not include rights and claims that arise after the date when the agreement is executed.
7. Be supported by consideration (money or benefits) in addition to that which the employee is already entitled.*

Safety professionals should also be aware that the ADEA and OWBPA requirements for waiver and/or release apply to RIFS, downsizing, layoffs, involuntary

---

* "Understanding Waiver of Discrimination Claims in Employee Severance Agreements," EEOC website (2011)

terminations and other situations where an employee over the age of 40 is separated from the employer. Although safety professionals usually are not directly involved in these types of employment separations, it is important for safety professionals to keep these requirements in mind in every situation in which employees over the age of 40 are terminated or separated from the company. Following are a sample waiver and general release:

Sample Waiver and General Release: Group Layoffs of Employees Age 40 and Over

The following example illustrates one way in which the required OWBPA information could be presented to employees as part of a waiver agreement and is not intended to suggest that employers must follow this format. Rather, each waiver agreement should be individualized based on an employer's particular organizational structure and the average comprehension and education of the employees in the decisional unit subject to termination. For another example of how the required information might be presented, see 29 C.F.R. § 1625.22(f)(vii).

Although this sample addresses only OWBPA issues, most severance agreements also ask employees to waive all claims against the employer, including claims arising under any federal, state, and local laws. See paragraph 6 below.

Dear [Employee]:

This letter will constitute the agreement between you and [your employer]("the Company") on the terms of your separation from the Company (hereinafter the "Agreement"). **The Agreement will be effective on the date specified below.**

1. Your employment will terminate on _____ [*date*].
   OR
   You have agreed to resign on _____ [*date*]. Your last day of work will be _____ [*date.*]

2. In consideration of your acceptance of this Agreement, the Company will pay you an extra _____ [*week's/month's*] salary at your current rate of $_____ per [*week/month*], less customary payroll deductions, to be paid within five (5) business days after the effective date of this Agreement as defined in paragraph 7 below. This severance pay will be in addition to your earned salary and accrued vacation pay or leave to which you are entitled.

3. Except as to claims that cannot be released under applicable law, you waive and release any and all claims you have or might have against the Company.... These claims include, but are not limited to claims for discrimination arising under federal, state, and local statutory or common law, including Title VII of the Civil Rights Act, the Americans with Disabilities Act, the Age Discrimination in Employment Act, the Genetic Information and Discrimination Act, and [state law].

4. The following information is required by OWBPA:
   You acknowledge that on _____, you were given 45 days to consider and accept the terms of this Agreement and that you were advised to consult with an attorney about the Agreement before signing it. To accept the Agreement, please date and sign this letter and return

it to me. Once you do so, you will still have seven (7) additional days from the date you sign to revoke your acceptance ("revocation period"). If you decide to revoke this Agreement after signing and returning it, you must give me a written statement of revocation or send it to me by fax, electronic mail, or registered mail. If you do not revoke during the seven-day revocation period, this Agreement will take effect on the eighth (8th) day after the date you the sign the Agreement.

[Paragraphs 3, 4, and 5 may address benefits, unemployment compensation, references, return of property, confidentiality, etc.]

The class, unit, or group of individuals covered by the program includes all employees in the _____ [plant, location, area, etc.] whose employment is being terminated in the reduction in force during the following period :_____). All employees in _____[plant, location, area, etc.] whose employment is being terminated are eligible for the program.

The following is a listing of the ages and job titles of employees who were and were not selected for layoff [or termination] and offered consideration for signing the waiver. Except for those employees selected for layoff [or termination], no other employee is eligible or offered consideration in exchange for signing the waiver:

Job title	Age	# Selected	# Not selected
(1) Bookkeepers	25	2	4
	28	1	7
	45	6	2
(2) Accountants	63	1	0
	24	3	5
(3) Retail sales clerks	29	1	7
	40	2	1
(4) Wholesale clerks	33	0	3
	51	2	1

Sincerely,

_____

On behalf of _____*[the company name]*
By signing this letter, I acknowledge that I have had the opportunity to consult with an attorney of my choice; that I have carefully reviewed and considered this Agreement; that I understand the terms of the Agreement; and that I voluntarily agree to them.

Date: _____

_____

Employee*

* Appendix B—Understanding Waivers of Discrimination Claims in Employee Severance Agreements, EEOC website (2011).

CHAPTER QUESTIONS

1. Harassment, under the ADEA, can include:
   a. Offensive remarks about a person's age
   b. Teasing about a person's clothing
   c. Offhand comments about an employee's personality
   d. None of the above
2. Three areas of importance for safety professionals to prevent discrimination based on age include:
   a. Training
   b. Harassment
   c. Job assignments
   d. All of the above
3. Horseplay can be:
   a. Discrimination under the ADEA
   b. Viewed through the eyes of the victim
   c. Create a hostile work environment
   d. All of the above
4. ADEA and OWBPA required waiver and/or release to:
   a. Be in writing
   b. Provide a 21-day waiting period.
   c. Advise review by legal counsel
   d. All of the above
5. In the hypothetical situation provided at the start of the chapter, how would you respond?

Answers: 1—a; 2—d; 3—d; 4—d; 5—varied responses possible.

10 Defending a Claim of Discrimination under the ADEA

If you have ten thousand regulations you destroy all respect for the law.

Winston Churchill

Morality cannot be legislated, but behavior can be regulated.
Judicial decrees may not change the heart, but they can restrain the heartless.

Martin Luther King Jr.

LEARNING OBJECTIVES

1. Acquire an understanding of the defenses to an ADEA action.
2. Acquire an understanding of a bona fide occupational qualification.
3. Acquire an understanding of the potential procedural and technical deficiencies that could be utilized as a defense in an ADEA action.

HYPOTHETICAL SITUATION

Your safety staff member files a charge of age discrimination under the ADEA based upon her performance evaluation and the fact that she did not receive a promotion to senior staff member. Another female staff member, age 39, received the promotion. In her charge, the safety staff member believes she did not receive the promotion because she was 50 years of age. In a meeting with legal counsel, you are asked to identify your defenses to this charge. What is your response?

OVERVIEW

Safety professionals should be aware that defending an ADEA charge or an ADEA action in a court of law is not usually within the safety function responsibilities; however, it is important for safety professionals to understand the complexities of the possible defenses as well as the safety professional's possible role as a witness or support for the defense team. Safety professionals should be aware that as a result of several recent U.S. Supreme Court cases, including but not limited to *Gross v.*

FBL Financial Services, Inc.,* *Smith v. City of Jackson,*† *Kentucky Retirement Systems v. EEOC,*‡ and other related cases, the burden of proof and the landscape of ADEA actions have been modified.

As with Title VII and other federal antidiscrimination laws, the agency responsible for ADEA charges is the Equal Employment Opportunities Commission (EEOC), and the same 180 days from the date of the alleged unlawful act deadline is applicable (300 days for most state antidiscrimination laws). One of the first possible defenses that safety professionals should check is to ensure that the ADEA charge has been filed in a timely way with the appropriate federal or state governmental agency. In the event that the charge was not filed in writing within the specified time period, safety professionals should be aware that this may be grounds to challenge the charge. Safety professionals should also be aware that under Section 6 of the Portal-To-Portal Act,§ ADEA civil lawsuits must be filed within 2 years from the date of the alleged misconduct, unless the actions constitute a willful violation wherein the time limit for filing is 3 years.

Safety professionals should also be aware that there are several conditions that an employee must meet prior to bringing an ADEA civil action in a court of law. First, the employee must wait 60 days after filing the charge with the EEOC or state agency before initiating the lawsuit in a court of law.¶

Safety professionals should also be aware that the EEOC can attempt to resolve the issue through conciliation during this 60-day time period. Second, the ADEA charge with the EEOC of state agency must be filed within the federal time limit of 180 days or state time limit of 300 days, respectively. And third, the ADEA civil lawsuit must be filed within 2–3 years for cases alleging willful misconduct. Safety professionals should also be aware that not only can an employee bring a civil action but the EEOC may bring a lawsuit against the company or organization on behalf of the employee in any court of competent jurisdiction seeking damages.**

Safety professionals should be aware that the two primary defenses in an ADEA action include Bona Fide Occupational Qualifications (BFOQ) and Reasonable Factors Other than Age (RFOA). Starting with BFOQ, this defense is similar to the BFOQ defense under Title VII (Chapter 4). Under the BFOQ defense, the ADEA permits employers to take certain actions that may be discriminatory if the actions are necessary to the normal operation of the company or organization.††

However, the ADEA narrowly construes and limits the applications of the BFOQ defense. Safety professionals should be aware that many of the ADEA cases where the BFOQ defense is utilized involve safety-related factors or precautions taken by the company or organization.‡‡ Safety professionals should be aware that safety considerations or even public interest may justify the use of the BFOQ exception, how-

---

* 129, S.Ct.2343 (2009).
† 544 U.S. 228 (2005).
‡ 128 S.Ct. 2361 (2008).
§ 29 U.S.C.A. Section 255.
¶ 29 U.S.C.A. 626(d).
** 29 U.S.C.A. Section 216(e).
†† 29 U.S.C.A. Section 623(f).
‡‡ See, for example, Hodgson v. Greyhound Lines, Inc., 499 F.2d 859 (1974).

ever, many companies who assume that a certain employee of a protected age would not qualify for a specific job is usually not sufficient to justify a disqualification from the specific job.*

Safety professionals should be aware that in order to justify an exclusion of employees or applicants of a certain protected age group under the BFOQ exception, the company or organization must show the following:

- The company "either has a factual basis for believing that all, or substantially all, members of the affected age group would be unable to perform the duties (of the job) safely and efficiently; or
- That, because of physiological or psychological degeneration associated with some members of the age group, but not necessarily all, there is no practical way to deal with such affected members of the age group on an individual basis."†

As can be seen, the BFOQ defense does place a burden upon the company, as well as the safety professional, to meet the requirements of this possible defense. Safety professionals should also be aware that bona fide seniority systems are permitted under the ADEA so long as the seniority system is based on the length of service as the primary criteria, effectively communicated to all employees and uniformly applied to all employees irrespective of the employee's age.‡

Safety professionals should be aware that one of the most common defenses to ADEA actions is the RFOA defense. In essence, this defense involves the company's or organization's position that their decision which resulted in the alleged discrimination was based on other reasonable factors other than age. Safety professionals should be aware that the reason the RFOA defense is available under the ADEA is due to Congress's intent to not require employers to hire because of the applicant's age but to ensure that age was not a determining factor in the decision-making process.§ Additionally, safety professionals should be aware that ADEA does not prohibit a company or organization from terminating an employee over the age of 40 when the involuntary termination is based on good cause.¶ Safety professionals should also be aware that other factors, such as production standards and educational requirements can be utilized as a permissible method of differentiation where these factors are job related and applied uniformly for all applicants or employees.**

Of particular importance for safety professionals with medical responsibilities is the area of employment-related testing and screening. Safety professionals should take careful note of the various federal antidiscrimination laws, including the ADEA, which impact the post-offer employment and screening processes utilized by many companies and organizations to evaluate applicants and for internal promotion and evaluation purposes. Safety professionals should carefully analyze and assess

---

* See, for example, Usery v. Tamiami Trail Tours, Inc., 531 F.2d 224 (1976).
† Houghton v. McDonnell Douglas Corp., 553 F.2d 561 (1977).
‡ 29 CFR Section 860.105(c).
§ 29 CFR Section 860.103(c).
¶ 29 U.S.C.A. Section 623(f)(3).
** 29 CFR Section 860.103(f)(2).

their testing and screening processes and procedures to ensure that compliance is achieved and maintained with all federal and state laws and requirements.

Employment Tests and Selection Procedures

Employers often use tests and other selection procedures to screen applicants for hire and employees for promotion. There are many different types of tests and selection procedures, including cognitive tests, personality tests, medical examinations, credit checks, and criminal background checks.

The use of tests and other selection procedures can be a very effective means of determining which applicants or employees are most qualified for a particular job. However, use of these tools can violate the federal antidiscrimination laws if an employer intentionally uses them to discriminate based on race, color, sex, national origin, religion, disability, or age (40 or older). Use of tests and other selection procedures can also violate the federal antidiscrimination laws if they disproportionately exclude people in a particular group by race, sex, or another covered basis, unless the employer can justify the test or procedure under the law.

On May 16, 2007, the EEOC held a public meeting on Employment Testing and Screening. Witnesses addressed legal issues related to the use of employment tests and other selection procedures. (To read the testimony of these witnesses, see the EEOC's website at http://eeoc.gov/eeoc/meetings/archive/5-16-07/index.html.)

This fact sheet provides technical assistance on some common issues relating to the federal antidiscrimination laws, and the use of tests and other selection procedures in the employment process.

BACKGROUND

- Title VII of the Civil Rights Act of 1964 (Title VII), the Americans with Disabilities Act of 1990 (ADA), and the Age Discrimination in Employment Act of 1967 (ADEA) prohibit the use of discriminatory employment tests and selection procedures.
- There has been an increase in employment testing due in part to post 9/11 security concerns, as well as concerns about workplace violence, safety, and liability. In addition, the large-scale adoption of online job applications has motivated employers to seek efficient ways to screen large numbers of online applicants in a nonsubjective way.
- The number of discrimination charges raising issues of employment testing, and exclusions based on criminal background checks, credit reports, and other selection procedures, reached a high point in FY 2007 at 304 charges.

TYPES OF EMPLOYMENT TESTS AND SELECTION PROCEDURES

Examples of employment tests and other selection procedures, many of which can be administered online, include the following:

- Cognitive tests assess reasoning, memory, perceptual speed and accuracy, and skills in arithmetic and reading comprehension, as well as knowledge of a particular function or job;

- Physical ability tests measure the physical ability to perform a particular task or the strength of specific muscle groups, as well as strength and stamina in general;
- Sample job tasks (e.g., performance tests, simulations, work samples, and realistic job previews) assess performance and aptitude on particular tasks;
- Medical inquiries and physical examinations, including psychological tests, assess physical or mental health;
- Personality tests and integrity tests assess the degree to which a person has certain traits or dispositions (e.g., dependability, cooperativeness, safety) or aim to predict the likelihood that a person will engage in certain conduct (e.g., theft, absenteeism);
- Criminal background checks provide information on arrest and conviction history;
- Credit checks provide information on credit and financial history;
- Performance appraisals reflect a supervisor's assessment of an individual's performance; and
- English proficiency tests determine English fluency.

GOVERNING EEO LAWS

- Title VII of the Civil Rights Act of 1964
  - Title VII prohibits employment discrimination based on race, color, religion, sex, or national origin.
  - With respect to tests in particular, Title VII permits employment tests as long as they are not "designed, intended or used to discriminate because of race, color, religion, sex, or national origin." 42 U.S.C. § 2000e-2(h). Title VII also imposes restrictions on how to score tests. Employers are not permitted to (1) adjust the scores of, (2) use different cutoff scores for, or (3) otherwise alter the results of employment-related tests on the basis of race, color, religion, sex, or national origin. *Id.* at §2000e-2(l).
  - Title VII prohibits both "disparate treatment" and "disparate impact" discrimination.
    - Title VII prohibits *intentional* discrimination based on race, color, religion, sex, or national origin. For example, Title VII forbids a covered employer from testing the reading ability of African-American applicants or employees but not testing the reading ability of their white counterparts. This is called **"disparate treatment"** discrimination. Disparate treatment cases typically involve the following issues:
      - Were people of a different race, color, religion, sex, or national origin treated differently?
      - Is there any evidence of bias, such as discriminatory statements?
      - What is the employer's reason for the difference in treatment?
      - Does the evidence show that the employer's reason for the difference in treatment is untrue, and that the real reason for the different treatment is race, color, religion, sex, or national origin?
- Title VII also prohibits employers from using neutral tests or selection procedures that have the *effect* of disproportionately excluding persons based on race, color, religion, sex, or national origin, where the tests or selection procedures are not "job-related and consistent with business necessity." This is called **"disparate impact"** discrimination.

Disparate impact cases typically involve the following issues:

- – Does the employer use a particular employment practice that has a disparate impact on the basis of race, color, religion, sex, or national origin? For example, if an employer requires that all applicants pass a physical agility test, does the test disproportionately screen out women? Determining whether a test or other selection procedure has a disparate impact on a particular group ordinarily requires a statistical analysis.
- – If the selection procedure has a disparate impact based on race, color, religion, sex, or national origin, can the employer show that the selection procedure is job-related and consistent with business necessity? An employer can meet this standard by showing that it is necessary to the safe and efficient performance of the job. The challenged policy or practice should therefore be associated with the skills needed to perform the job successfully. In contrast to a general measurement of applicants' or employees' skills, the challenged policy or practice must evaluate an individual's skills as related to the particular job in question.
- – If the employer shows that the selection procedure is job-related and consistent with business necessity, can the person challenging the selection procedure demonstrate that there is a less discriminatory alternative available? For example, is another test available that would be equally effective in predicting job performance but would not disproportionately exclude the protected group?

  See 42 U.S.C. § 2000e-2 (k). This method of analysis is consistent with the seminal Supreme Court decision about disparate impact discrimination, *Griggs v. Duke Power Co.*, 401 U.S. 424 (1971).
- In 1978, the EEOC adopted the Uniform Guidelines on Employee Selection Procedures or "UGESP" under Title VII. *See* 29 C.F.R. Part 1607.*[15] UGESP provided uniform guidance for employers about how to determine if their tests and selection procedures were lawful for purposes of Title VII disparate impact theory.
  - – UGESP outlines three different ways employers can show that their employment tests and other selection criteria are job-related and consistent with business necessity. These methods of demonstrating job-relatedness are called "test validation." UGESP provides detailed guidance about each method of test validation.
- Title I of the Americans with Disabilities Act (ADA)
  - Title I of the ADA prohibits private employers and state and local governments from discriminating against qualified individuals with disabilities on the basis of their disabilities.
  - The ADA specifies when an employer may require an applicant or employee to undergo a medical examination, that is, a procedure or test that seeks information about an individual's physical or mental impairments or health. The ADA also specifies when an employer may make "disability-related inquiries," that is, inquiries that are likely to elicit information about a disability.

---

* [15]The Departments of Labor and Justice and the Office of Personnel Management (then called the Civil Service Commission) issued UGESP along with the EEOC. (Employers' Testing and Selection Procedures Fact Sheet, EEOC website [2011].)

- When hiring, an employer may not ask questions about disability or require medical examinations until *after* it makes a conditional job offer to the applicant. 42 U.S.C. §12112 (d)(2);
- After making a job offer (but before the person starts working), an employer may ask disability-related questions and conduct medical examinations as long as it does so for *all individuals entering the same job category*. *Id.* at § 12112(d)(3); and
- With respect to *employees*, an employer may ask questions about disability or require medical examinations only if doing so is *job-related and consistent with business necessity*. Thus, for example, an employer could request medical information when it has a *reasonable belief*, based on *objective evidence*, that a particular employee will be unable to perform essential job functions or will pose a direct threat because of a medical condition, or when an employer receives a request for a *reasonable accommodation* and the person's disability and/or need for accommodation is not obvious. *Id.* at § 12112(d)(4).

- The ADA also makes it unlawful to:
  - Use employment tests that screen out or tend to screen out an individual with a disability or a class of individuals with disabilities unless the test, as used by the employer, is shown to be job-related and consistent with business necessity. 42 U.S.C. § 12112(b)(6);
  - Fail to select and administer employment tests in the most effective manner to ensure that test results accurately reflect the skills, aptitude, or whatever other factor that such test purports to measure, rather than reflecting an applicant's or employee's impairment. *Id.* at § 12112(b)(7); and
  - Fail to make reasonable accommodations, including in the administration of tests, to the known physical or mental limitations of an otherwise qualified individual with a disability who is an applicant or employee, unless such accommodation would impose an undue hardship. *Id.* at § 12112(b)(5).

- The Age Discrimination in Employment Act (ADEA)
  - The ADEA prohibits discrimination based on age (40 and over) with respect to any term, condition, or privilege of employment. Under the ADEA, covered employers may not select individuals for hiring, promotion, or reductions in force in a way that unlawfully discriminates on the basis of age.
  - The ADEA prohibits *disparate treatment* discrimination, that is, intentional discrimination based on age. For example, the ADEA forbids an employer from giving a physical agility test only to applicants over age 50, based on a belief that they are less physically able to perform a particular job, but not testing younger applicants.

- The ADEA also prohibits employers from using neutral tests or selection procedures that have a *discriminatory impact* on persons based on age (40 or older), unless the challenged employment action is based on a *reasonable factor other than age*. *Smith v. City of Jackson*, 544 U.S. 228 (2005). Thus, if a test or other selection procedure has a disparate impact based on age, the employer must show that the test or device chosen was a reasonable one.

RECENT EEOC LITIGATION AND SETTLEMENTS

A number of recent EEOC enforcement actions illustrating basic EEO principles focus on testing.

- **Title VII and Cognitive Tests**: *Less discriminatory alternative for cognitive test with disparate impact. EEOC v. Ford Motor Co. and United Automobile Workers of America*, involved a court-approved settlement agreement on behalf of a nationwide class of African-Americans who were rejected for an apprenticeship program after taking a cognitive test known as the Apprenticeship Training Selection System (ATSS). The ATSS was a written cognitive test that measured verbal, numerical, and spatial reasoning in order to evaluate mechanical aptitude. Although it had been validated in 1991, the ATSS continued to have a statistically significant disparate impact by excluding African-American applicants. Less discriminatory selection procedures were subsequently developed that would have served Ford's needs, but Ford did not modify its procedures. In the settlement agreement, Ford agreed to replace the ATSS with a selection procedure, to be designed by a jointly-selected industrial psychologist, that would predict job success and reduce adverse impact. Additionally, Ford paid $8.55 million in monetary relief.
- **Title VII and Physical Strength Tests**: *Strength test must be job-related and consistent with business necessity if it disproportionately excludes women.* In *EEOC v. Dial Corp.*, women were disproportionately rejected for entry-level production jobs because of a strength test. The test had a significant adverse impact on women; prior to the use of the test, 46% of hires were women; after use of the test, only 15% of hires were women. Dial defended the test by noting that it looked like the job and use of the test had resulted in fewer injuries to hired workers. The EEOC established through expert testimony, however, that the test was considerably more difficult than the job and that the reduction in injuries occurred 2 years before the test was implemented, most likely due to improved training and better job rotation procedures. On appeal, the Eighth Circuit upheld the trial court's finding that Dial's use of the test violated Title VII under the disparate impact theory of discrimination. *See* http://www.eeoc.gov/press/11-20-06.html
- **ADA and Test Accommodation**: *Employer must provide reasonable accommodation on pre-employment test for hourly, unskilled manufacturing jobs.* The EEOC settled *EEOC v. Daimler Chrysler Corp.*, a case brought on behalf of applicants with learning disabilities who needed reading accommodations during a pre-employment test given for hourly unskilled manufacturing jobs. The resulting settlement agreement provided monetary relief for 12 identified individuals and the opportunity to take the hiring test with the assistance of a reader. The settlement agreement also required that the employer provide a reasonable accommodation on this particular test to each applicant who requested a reader and provided documentation establishing an ADA disability. The accommodation consisted of either a reader for all instructions and all written parts of the test, or an audiotape providing the same information.

EMPLOYER BEST PRACTICES FOR TESTING AND SELECTION

- Employers should administer tests and other selection procedures without regard to race, color, national origin, sex, religion, age (40 or older), or disability.

- Employers should ensure that employment tests and other selection procedures are properly validated for the positions and purposes for which they are used. The test or selection procedure must be job-related and its results appropriate for the employer's purpose. While a test vendor's documentation supporting the validity of a test may be helpful, the employer is still responsible for ensuring that its tests are valid under UGESP.
- If a selection procedure screens out a protected group, the employer should determine whether there is an equally effective alternative selection procedure that has less adverse impact and, if so, adopt the alternative procedure. For example, if the selection procedure is a test, the employer should determine whether another test would predict job performance but not disproportionately exclude the protected group.
- To ensure that a test or selection procedure remains predictive of success in a job, employers should keep abreast of changes in job requirements and should update the test specifications or selection procedures accordingly.
- Employers should ensure that tests and selection procedures are not adopted casually by managers who know little about these processes. A test or selection procedure can be an effective management tool, but no test or selection procedure should be implemented without an understanding of its effectiveness and limitations for the organization, its appropriateness for a specific job, and whether it can be appropriately administered and scored.
- For further background on experiences and challenges encountered by employers, employees, and job seekers in testing, see the testimony from the Commission's meeting on testing, located on the EEOC's public website at: http://eeoc.gov/eeoc/meetings/archive/5-16-07/index.html.
- For general information on discrimination Title VII, the ADA and the ADEA see EEOC's website at http://www.eeoc.gov/laws/statutes/index.cfm

Safety professionals should also be aware that there are a number of other technical and procedural defenses which involve deficiencies within the legal process including such deficiencies as failure to show the company employed more 20 employees, failure to meet the time limitations and failure to show the applicant or employee was over the age of 40. Other defenses to an ADEA action that safety professionals should be aware of can include such defenses as reliance on an official written administrative regulation, reliance on an EEOC interpretation, or other related documents.

CHAPTER QUESTIONS

1. The BFOQ defense involves:
   a. Medical examinations
   b. PPE adjustments
   c. Good faith reliance on EEOC interpretation
   d. None of the above
2. The RFOA defense involves:
   a. Factors other than age.
   b. Factors addressing the specific age of the employee

    c. Factors addressing the age of the entire workforce
    d. None of the above

3. A seniority system:
    a. Is automatically a violation of the ADEA
    b. Is permitted under specific circumstances
    c. Is only permitted in a unionized operation
    d. None of the above

4. A procedural deficiency in the ADEA action:
    a. Can be utilized as a defense
    b. Cannot be used as a defense
    c. Can only be used by the EEOC
    d. None of the above

5. In the hypothetical situation identified at the start of the chapter, how would you respond?

Answers: 1—d; 2—a; 3—b; 4—a; 5—various responses possible.

11 The Genetic Information Nondiscrimination Act of 2008 (GINA)

I believe in human dignity as the source of national purpose, human liberty as the source of national action, the human heart as the source of national compassion, and in the human mind as the source of our invention and our ideas.

John F. Kennedy

1. Do more than exist, live.
2. Do more that touch, feel.
3. Do more than look, observe.
4. Do more than read, absorb.
5. Do more than hear, listen.
6. Do more than listen, understand.
7. Do more than think, ponder.
8. Do more than talk, say something.

John H. Rhoades

LEARNING OBJECTIVES

1. Acquire an understanding of the requirements of GINA.
2. Acquire an understanding of the interplay of GINA with other federal antidiscrimination laws.

HYPOTHETICAL SITUATION

An employee comes to the safety professional and states, "I hear that the company is now using our medical information to test our DNA ….. Can they fire me if I my test is bad?" How are you going to respond?

OVERVIEW

Safety professionals should be aware of the Genetic Information Nondiscrimination Act (GINA), which took effect on November 21, 2009. Although not directly linked to the safety function, safety professionals should be aware that GINA prohibits discrimination in the workplace based upon an individual employee's genetic tests, genetic tests of family members, or a disease or disorder which is genetically related in employee's family members. For safety professionals, the primary areas where

GINA may impact the safety function in many companies and organizations is in workers' compensation, health insurance, medical examinations, medical documentation, and related fields. Safety professionals should pay special attention at any time the issue of genetic information surfaces and ensure that all requirements under GINA are strictly enforced.

Press Release

November 9, 2010

EEOC ISSUES GENETIC INFORMATION
NONDISCRIMINATION ACT FINAL REGULATIONS

Law Prohibits Using Genetic Information to Make Employment Decisions

The U.S. Equal Employment Opportunity Commission (EEOC) today issued *final regulations* implementing the employment provisions (Title II) of the Genetic Information Nondiscrimination Act of 2008 (GINA). GINA prohibits use of genetic information to make decisions about health insurance and employment, and restricts the acquisition and disclosure of genetic information. Title II of GINA represents the first legislative expansion of the EEOC's jurisdiction since the Americans with Disabilities Act of 1990.

The regulations were approved by a unanimous vote of the Commission, and include clarifications and refinements made in response to comments received during the notice and comment period. Said EEOC Chair Jacqueline A. Berrien: "The final regulations implementing GINA reflect the concerted effort by all Commissioners to ensure that workers, job applicants, and employers will have clear guidance concerning the implementation of this new law. These regulations are also a testimony to the tireless work of the late Paul Steven Miller, who was a commissioner of the EEOC and a leader in the movement to protect individuals against discrimination based on family medical history or genetic information for many years." The GINA regulations are the first issued by the EEOC since Chair Berrien and Commissioners Chai R. Feldblum and Victoria A. Lipnic joined Commissioners Stuart J. Ishimaru and Constance Barker on the Commission in April, 2010.

Congress enacted GINA with strong bipartisan support in 2008, in response to concerns that patients would decline to take advantage of the increasing availability of genetic testing out of concern that they could lose their jobs or health insurance if such tests revealed adverse information. Title II of GINA prohibits employment discrimination based on genetic information, and restricts the acquisition and disclosure of genetic information. Genetic information includes information about individuals' genetic tests and the tests of their family members; family medical history; requests for and receipt of genetic services by an individual or a family member; and genetic information about a fetus carried by an individual or family member or of an embryo legally held by the individual or family member using assisted reproductive technology.

The final regulations provide examples of genetic tests; more fully explain GINA's prohibition against requesting, requiring, or purchasing genetic information; provide model language employers can use when requesting medical information from employees to avoid acquiring genetic information; and describe how

GINA applies to genetic information obtained via electronic media, including web sites and social networking sites.

"I am pleased that the new Commission was able to complete the GINA regulations and make some common-sense changes based on the public record," said EEOC Commissioner Victoria A. Lipnic. "While fulfilling the law's purpose to protect individuals from genetic discrimination, I believe these final regulations properly balance and reflect the needs and realities of the workplace and preserve the appropriate means for employers to offer health and wellness plans."

The Commission has also issued two question-and-answer documents on the final GINA regulations,*

Safety professionals should be aware that GINA consists of Title I, which addressed genetic nondiscrimination in health insurance, and Title II, which prohibits employment-related discrimination on the basis of genetic information. Safety professionals should be aware that the regulations and guidance for Title I, which is primarily focused on health insurance, is provided by the Department of Labor, Health, and Human Services and the Treasury. The regulations and guidance for Title II, which primarily addressed employment-related discrimination based on genetic information, is provided by the EEOC.

Safety professionals should be aware that genetic information involves the following:

(A) In general, the term "genetic information" means, with respect to any individual, information about—
  (i) such individual's genetic tests,
  (ii) the genetic tests of family members of such individual, and
  (iii) the manifestation of a disease or disorder in family members of such an individual.
(B) Inclusion of genetic services and participation in genetic research. Such a term includes, with respect to any individual, any request for, or receipt of, genetic services, or participation in clinical research which includes genetic services, by such an individual or any family member of such individual.
(C) Exclusions. The term "genetic information" shall not include information about the sex or age of any individual.†

Additionally, of particular importance to safety professional is the fact that genetic monitoring includes employee evaluations of potential exposures to toxic or other substances in the workplace, which possess the potential of causing "chromosomal damage or evidence of increased occurrence of mutations, that may have developed in the course of employment due to exposure to toxic substances in the workplace, in order to identify, evaluate, and respond to the effects of or control adverse environmental exposures in the workplace".‡ For safety professionals contracting certain monitoring activities, GINA defined genetic testing as "an analysis of human DNA, RNA, chromosomes, proteins, or metabolites, that detects genotypes, mutations, or chromosomal changes."§ Safety professionals should be aware that identified genetic services to include:

---

* Press Release, EEOC website (2010).
† The Genetic Information Nondiscrimination Act or 2008, EEOC website (2011).
‡ Id.
§ Id.

(A) a genetic test;
(B) genetic counseling (including obtaining, interpreting, or assessing genetic infor-
     mation); or
(C) genetic education.*

Safety professionals should be aware that unlawful employment practice under
GINA can include:

(1) to fail or refuse to hire, or to discharge, any employee, or otherwise to discrimi-
    nate against any employee with respect to the compensation, terms, conditions,
    or privileges of employment of the employee, because of genetic information
    with respect to the employee; or
(2) to limit, segregate, or classify the employees of the employer in any way that
    would deprive or tend to deprive any employee of employment opportunities or
    otherwise adversely affect the status of the employee as an employee, because
    of genetic information with respect to the employee.
(b) Acquisition of genetic Information. It shall be an unlawful employment
     practice for an employer to request, require, or purchase genetic information
     with respect to an employee or a family member of the employee ... ."†

For safety professionals who may be conduction genetic-related monitoring or testing as required by OSHA, state plan, MSHA, or the Atomic Energy Commission to
identify the biological effects of toxic substances in the workplace, this information
can only be utilized if :

(A) the employer provides written notice of the genetic monitoring to the employee;
(B)
     (i)   the employee provides prior, knowing, voluntary, and written authorization; or
     (ii) the genetic monitoring is required by Federal or State law;
(C) the employee is informed of individual monitoring results;
(D) the monitoring is in compliance with ... regulations.‡

Safety professionals should further be aware that the company or organization,
"excluding any licensed health care professional or board certified genetic counselor that is involved in the genetic monitoring program, receives the results
of the monitoring only in aggregate terms that do not disclose the identity of
specific employees"§ and a company or organization "conducts DNA analysis for law enforcement purposes as a forensic laboratory or for purposes of
human remains identification, and requests or requires genetic information of
such employer's employees, but only to the extent that such genetic information
is used for analysis of DNA identification markers for quality control to detect
sample contamination."¶

Similarly, safety professionals should be aware that GINA can impact the training function for not only companies but also for labor organizations. "It shall be an
unlawful employment practice for any employer, labor organization, or joint labor–

---

* Id.
† Id
‡ Id.
§ Id
¶ Id.

management committee controlling apprenticeship or other training or retraining, including on-the-job training programs—

(1) to discriminate against any individual because of genetic information with respect to the individual in admission to, or employment in, any program established to provide apprenticeship or other training or retraining;

(2) to limit, segregate, or classify the applicants for or participants in such apprenticeship or other training or retraining, or fail or refuse to refer for employment any individual, in any way that would deprive or tend to deprive any individual of employment opportunities, or otherwise adversely affect the status of the individual as an employee, because of genetic information with respect to the individual; or

(3) to cause or attempt to cause an employer to discriminate against an applicant for or a participant in such apprenticeship or other training or retraining in violation of this title.

(b) Acquisition of Genetic Information. It shall be an unlawful employment practice for an employer, labor organization, or joint labor–management committee described in subsection (a) to request, require, or purchase genetic information with respect to an individual or a family member of the individual except—

(1) where the employer, labor organization, or joint labor-management committee inadvertently requests or requires family medical history of the individual or family member of the individual;

(2) where—

(A) health or genetic services are offered by the employer, labor organization, or joint labor-management committee, including such services offered as part of a wellness program;

(B) the individual provides prior, knowing, voluntary, and written authorization;

(C) only the individual (or family member if the family member is receiving genetic services) and the licensed health care professional or board certified genetic counselor involved in providing such services receive individually identifiable information concerning the results of such services; and

(D) any individually identifiable genetic information provided under subparagraph (C) in connection with the services provided under subparagraph (A) is only available for purposes of such services and shall not be disclosed to the employer, labor organization, or joint labor–management committee except in aggregate terms that do not disclose the identity of specific individuals;

(3) where the employer, labor organization, or joint labor–management committee requests or requires family medical history from the individual to comply with the certification provisions of section 103 of the Family and Medical Leave Act of 1993 (29 U.S.C. 2613) or such requirements under State family and medical leave laws;

(4) where the employer, labor organization, or joint labor–management committee purchases documents that are commercially and publicly available (including newspapers, magazines, periodicals, and books, but not including medical databases or court records) that include family medical history;

(5) where the information involved is to be used for genetic monitoring of the biological effects of toxic substances in the workplace, but only if—

(A)   the employer, labor organization, or joint labor–management commit-
      tee provides written notice of the genetic monitoring to the individual;
(B)
      (i)   the individual provides prior, knowing, voluntary, and written
             authorization; or
      (ii)   the genetic monitoring is required by Federal or State law;
(C)   the individual is informed of individual monitoring results;
(D)   the monitoring is in compliance with—
      (i)   any Federal genetic monitoring regulations, including any such
             regulations that may be promulgated by the Secretary of Labor
             pursuant to the Occupational Safety and Health Act of 1970 (29
             U.S.C. 651 et seq.), the Federal Mine Safety and Health Act of
             1977 (30 U.S.C. 801 et seq.), or the Atomic Energy Act of 1954 (42
             U.S.C. 2011 et seq.); or
      (ii)   State genetic monitoring regulations, in the case of a State that is
             implementing genetic monitoring regulations under the authority
             of the Occupational Safety and Health Act of 1970 (29 U.S.C. 651
             et seq.); and
(E)   the employer, labor organization, or joint labor–management commit-
      tee, excluding any licensed health care professional or board certified
      genetic counselor that is involved in the genetic monitoring program,
      receives the results of the monitoring only in aggregate terms that do
      not disclose the identity of specific individuals; or
(6) where the employer conducts DNA analysis for law enforcement purposes as a
    forensic laboratory or for purposes of human remains identification, and requests
    or requires genetic information of such employer's apprentices or trainees, but
    only to the extent that such genetic information is used for analysis of DNA
    identification markers for quality control to detect sample contamination.*

Safety professionals should become familiar with the requirements under Title II,
Sections 206 and 210, which address the confidentiality of genetic information,
medical information that is not considered genetic information, and the interaction with the current HPPA and other medical information requirements. Under
Section 206,

> *"If an employer, employment agency, labor organization, or joint labor–man-*
> *agement committee possesses genetic information about an employee or mem-*
> *ber, such information shall be maintained on separate forms and in separate*
> *medical files and be treated as a confidential medical record of the employee or*
> *member. An employer, employment agency, labor organization, or joint labor–*
> *management committee shall be considered to be in compliance with the main-*
> *tenance of information requirements of this subsection with respect to genetic*
> *information subject to this subsection that is maintained with and treated as a*
> *confidential medical record under section 102(d)(3)(B) of the Americans With*
> *Disabilities Act (42 U.S.C. 12112(d)(3)(B)).*
>
> *(b)   Limitation on Disclosure. An employer, employment agency, labor orga-*
> *nization, or joint labor–management committee shall not disclose genetic*
> *information concerning an employee or member except—*

---

* Id.

> (1) to the employee or member of a labor organization (or family member if the family member is receiving the genetic services) at the written request of the employee or member of such organization;
>
> (2) to an occupational or other health researcher if the research is conducted in compliance with the regulations and protections provided for under part 46 of title 45, Code of Federal Regulations ...."*

Safety professionals should be aware that Section 206 also addresses the release of genetic information by an order of the court and in conjunction with the HPPA regulations. Additionally, safety professionals should be aware that under Section 210, "An employer, employment agency, labor organization, or joint labor–management committee shall not be considered to be in violation of this title based on the use, acquisition, or disclosure of medical information that is not genetic information about a manifested disease, disorder, or pathological condition of an employee or member, including a manifested disease, disorder, or pathological condition that has or may have a genetic basis."†

With our fast-changing technology, genetic testing has become more affordable and is being utilized on a broader scale than in the past. Safety professionals who are involved in any aspect of any type of genetic testing should become familiar with the requirements of GINA and ensure strict compliance with the requirements of the relatively new law. As with virtually all federal antidiscrimination laws, safety professionals should be aware that GINA prohibits any type of discrimination, harassment, or retaliation and the primary governmental agency tasked with enforcement is the EEOC.

CHAPTER QUESTIONS

1. GINA is:
   a. The Genetic Information Nondiscrimination Act of 2008
   b. The Genetic Information Discrimination Prevention Act of 2009
   c. The Genetic Information Nondisclosure Act of 2010
   d. None of the above
2. GINA prohibits:
   a. Discrimination based on genetic information
   b. Harassment based on genetic information
   c. Retaliation based upon genetic information
   d. All of the above
3. Genetic information must be:
   a. Maintained on separate forms
   b. Maintained in separate files
   c. Maintained in a confidential manner
   d. All of the above
4. The primary governmental agency or agencies responsible for Title I and title II of GINA is/are:
   a. Title I—HHS and Title II—EEOC

---

* Id. at Section 206.
† Id at Section 210.

   b. Title I—EEOC and Title II—HHS
   c. EEOC for both Title I and Title II
   d. None of the above
5. How would you respond to the hypothetical question posed above.

Answers: 1—a; 2—d; 3—d; 4—a; 5—varied.

The Supreme Court of the United States

INTERNATIONAL UNION, UNITED AUTOMOBILE, AEROSPACE AND AGRICULTURAL IMPLEMENT WORKERS OF AMERICA, UAW, et al., Petitioners v. JOHNSON CONTROLS, INC.*

No. 89–1215.

Argued Oct. 10, 1990.
Decided March 20, 1991.

Class action was brought challenging employer's policy barring all women, except those whose infertility was medically documented, from jobs involving actual or potential lead exposure exceeding Occupational Safety and Health Administration (OSHA) standard. The United States District Court for the Eastern District of Wisconsin, 680 F.Supp. 309, granted summary judgment for employer. On appeal, the Court of Appeals for the Seventh Circuit, 886 F.2d 871, affirmed, and certiorari was granted. The Supreme Court, Justice Blackmun, held that: (1) employer's policy was facially discriminatory, and (2) employer did not establish that sex was a bona fide occupational qualification (BFOQ).

Reversed and remanded.

Justice White concurred in part and concurred in the judgment and filed opinion in which Chief Justice Rehnquist and Justice Kennedy joined.

Justice Scalia filed opinion concurring in the judgment.

**1197 SYLLABUS FN*

FN* The syllabus constitutes no part of the opinion of the Court but has been prepared by the Reporter of Decisions for the convenience of the reader. See *United States v. Detroit Lumber Co.,* 200 U.S. 321, 337, 26 S.Ct. 282, 287, 50 L.Ed. 499.

A primary ingredient in respondent's battery manufacturing process is lead, occupational exposure to which entails health risks, including the risk of harm to any fetus carried by a female employee. After eight of its employees became pregnant while maintaining blood lead levels exceeding that noted by the Occupational Safety and Health Administration (OSHA) as critical for a worker planning to have a family, respondent announced a policy barring all women,

---

* From Westlaw, and modified for the purposes of this text.

except those whose infertility was medically documented, from jobs involving actual or potential lead exposure exceeding the OSHA standard. Petitioners, a group including employees affected by respondent's fetal protection policy, filed a class action in the District Court, claiming that the policy constituted sex discrimination violative of Title VII of the Civil Rights Act of 1964, as amended. The court granted summary judgment for respondent, and the Court of Appeals affirmed. The latter court held that the proper standard for evaluating the policy was the business necessity inquiry applied by other Circuits; that respondent was entitled to summary judgment because petitioners had failed to satisfy their burden of persuasion as to each of the elements of the business necessity defense under *Wards Cove Packing Co. v. Atonio,* 490 U.S. 642, 109 S.Ct. 2115, 104 L.Ed.2d 733; and that even if the proper evaluative standard was bona fide occupational **1198 qualification (BFOQ) analysis, respondent still was entitled to summary judgment because its fetal protection policy is reasonably necessary to further the industrial safety concern that is part of the essence of respondent's business.

Held: Title VII, as amended by the Pregnancy Discrimination Act (PDA), forbids sex-specific fetal protection policies. Pp. 1202–1210.

(a) By excluding women with childbearing capacity from lead-exposed jobs, respondent's policy creates a facial classification based on gender and explicitly discriminates against women on the basis of their sex under § 703(a) of Title VII. Moreover, in using the words "capable of bearing children" as the criterion for exclusion, the policy explicitly classifies on the basis of potential for pregnancy, which classification must be *188 regarded, under the PDA, in the same light as explicit sex discrimination. The Court of Appeals erred in assuming that the policy was facially neutral because it had only a discriminatory effect on women's employment opportunities, and because its asserted purpose, protecting women's unconceived offspring, was ostensibly benign. The policy is not neutral because it does not apply to male employees in the same way as it applies to females, despite evidence about the debilitating effect of lead exposure on the male reproductive system. Also, the absence of a malevolent motive does not convert a facially discriminatory policy into a neutral policy with a discriminatory effect. Cf. *Phillips v. Martin Marietta Corp.,* 400 U.S. 542, 91 S.Ct. 496, 27 L.Ed.2d 613. Because respondent's policy involves disparate treatment through explicit facial discrimination, the business necessity defense and its burden shifting under *Wards Cove* are inapplicable here. Rather, as indicated by the Equal Employment Opportunity Commission's enforcement policy, respondent's policy may be defended only as a BFOQ, a more stringent standard than business necessity. Pp. 1202–1204.

(b) The language of both the BFOQ provision set forth in § 703(e)(1) of Title VII, which allows an employer to discriminate on the basis of sex "in those certain instances where ... sex ... is a [BFOQ] reasonably necessary to the normal operation of [the] particular business" and the PDA provision that amended Title VII, which specifies that, unless pregnant employees differ from others "in their ability or inability to work," they must be "treated the same" as other employees "for all employment-related purposes," as well as these provisions' legislative history and the case law, prohibit an employer from discriminating against a woman because of her capacity to become pregnant unless her reproductive potential prevents her from performing the duties of her job. The so-called safety exception to the BFOQ is limited to instances in which sex or pregnancy actually interferes with the employee's ability to perform, and the employer must direct its concerns in this regard to those aspects of the woman's job-related activities that fall within the "essence" of the particular business. *Dothard v. Rawlinson,* 433 U.S. 321, 333, 335, 97 S.Ct. 2720, 2728-29, 2729-30, 53 L.Ed.2d 786; *Western Air Lines, Inc. v. Criswell,* 472 U.S. 400, 413, 105 S.Ct. 2743, 2751, 86 L.Ed.2d 321. The unconceived fetuses of respondent's female employees are neither customers nor third parties whose safety is essential to the business of battery manufacturing. Pp. 1204–1207.

(c) Respondent cannot establish a BFOQ. Fertile women, as far as appears in the record, participate in the manufacture of batteries as efficiently as anyone else. Moreover, respondent's professed concerns about the welfare of the next generation do not suffice to establish a BFOQ of female sterility. Title VII, as amended by the PDA, mandates that decisions about the welfare of future children be left to the parents *189 who conceive, bear, support, and raise them rather than to the employers who hire those parents or the courts. Pp. 1207–1208.

**1199 (d) An employer's tort liability for potential fetal injuries and its increased costs due to fertile women in the workplace do not require a different result. If, under general tort principles, Title VII bans sex-specific fetal-protection policies, the employer fully informs the woman of the risk, and the employer has not acted negligently, the basis for holding an employer liable seems remote at best. Moreover, the incremental cost of employing members of one sex cannot justify a discriminatory refusal to hire members of that gender. See, e.g., *Los Angeles Dept. of Water and Power v. Manhart,* 435 U.S. 702, 716–718, and n. 32, 98 S.Ct. 1370, 1379–1380, and n. 32, 55 L.Ed.2d 657. Pp. 1208–1210.

886 F.2d 871 (CA7 1989), reversed and remanded.

BLACKMUN, J., delivered the opinion of the Court, in which MARSHALL, STEVENS, O'CONNOR, and SOUTER, JJ., joined. WHITE, J., filed an opinion concurring in part and concurring in the judgment, in which REHNQUIST, C.J., and KENNEDY, J., joined, *post,* p. 1210. SCALIA, J., filed an opinion concurring in the judgment, *post,* p. 1216.

Marsha S. Berzon argued the cause for petitioners. With her on the briefs were *Jordan Rossen, Ralph O. Jones,* and *Laurence Gold.*

Stanley S. Jaspan argued the cause for respondent. With him on the briefs were *Susan R. Maisa, Anita M. Sorensen, Charles G. Curtis, Jr.,* and *John P. Kennedy.**

*Briefs of *amici curiae* urging reversal were filed for the United States et al. by *Solicitor General Starr, Assistant Attorney General Dunne, Deputy Solicitor General Roberts, Deputy Assistant Attorney General Clegg, Clifford M. Sloan, David K. Flynn, Charles A. Shanor, Gwendolyn Young Reams, Lorraine C. Davis,* and *Carolyn L. Wheeler;* for the State of California et al. by *John K. Van de Kamp,* Attorney General, *Andrea Sheridan Ordin,* Chief Assistant Attorney General, *Marian M. Johnston,* Supervising Deputy Attorney General, and *Manuel M. Medeiros,* Deputy Attorney General; for the Commonwealth of Massachusetts et al. by *James M. Shannon,* Attorney General of Massachusetts, *Jennifer Wriggins, Marjorie Heins,* and *Judith E. Beals,* Assistant Attorneys General, and by the Attorneys General for their respective jurisdictions as follows: *Robert K. Corbin* of Arizona, *Clarine Nardi Riddle* of Connecticut, *Charles M. Oberly III* of Delaware, *Robert A. Butterworth* of Florida, *William J. Guste, Jr.,* of Louisiana, *James E. Tierney* of Maine, *Frank J. Kelley* of Michigan, *Hubert H. Humphrey III* of Minnesota, *Robert M. Spire* of Nebraska, *Robert J. Del Tufo* of New Jersey, *Robert Abrams* of New York, *Anthony J. Celebrezze, Jr.,* of Ohio, *Robert H. Henry* of Oklahoma, *Hector Rivera-Cruz* of Puerto Rico, *Jim Mattox* of Texas, *Jeffrey L. Amestoy* of Vermont, *Godfrey R. de Castro* of the Virgin Islands, and *Kenneth O. Eikenberry* of Washington; for the American Civil Liberties Union et al. by *Joan E. Bertin, Elisabeth A. Werby,* and *Isabelle Katz Pinzler;* for the American Public Health Association et al. by *Nadine Taub* and *Suzanne L. Mager;* for Equal Rights Advocates et al. by *Susan Deller Ross* and *Naomi R. Cahn;* for the NAACP Legal Defense and Educational Fund, Inc., et al., by *Julius LeVonne Chambers, Charles Stephen Ralston,* and *Ronald L. Ellis;* and for Trial Lawyers for Public Justice by *Arthur H. Bryant.*

Briefs of *amici curiae* urging affirmance were filed for the Chamber of Commerce of the United States of America by *Timothy B. Dyk, Willis J. Goldsmith, Stephen A. Bokat,* and *Robin S. Conrad;* for Concerned Women for America by *Jordan W. Lorence, Cimron Campbell,* and *Wendell R. Bird;* for the Equal Employment Advisory Council et al. by *Robert E. Williams, Douglas S. McDowell, Garen E. Dodge, Jan S. Amundson,* and *Quentin Riegel;* for the Industrial Hygiene Law Project by *Jack Levy* and *Ilise Levy Feitshans;* for the National Safe Workplace Institute by *James D. Holzhauer;* for the United States Catholic Conference by *Mark E. Chopko* and *John A. Liekweg;* and for the Washington Legal Foundation by *Daniel J. Popeo, Paul D. Kamenar,* and *John C. Scully.*

Briefs of *amici curiae* were filed for the Association of the Bar of the City of New York et al. by *Sidney S. Rosdeitcher, Evelyn Cohn, Janet Gallagher, Janice Goodman, Arthur Leonard,* and *Jim Williams;* for the Natural Resources Defense Council, Inc., by *Thomas O. McGarity* and *Albert H. Meyerhoff;* and for the Pacific Legal Foundation et al. by *Ronald A. Zumbrun* and *Anthony T. Caso.* *190 Justice BLACKMUN delivered the opinion of the Court.

In this case we are concerned with an employer's gender-based fetal protection policy. May an employer exclude a fertile female employee from certain jobs because of its concern for the health of the fetus the woman might conceive?

I

Respondent Johnson Controls, Inc., manufactures batteries. In the manufacturing process, the element lead is a primary ingredient. Occupational exposure to lead entails health risks, including the risk of harm to any fetus carried by a female employee.

*191 Before the Civil Rights Act of 1964, 78 Stat. 241, became law, Johnson Controls did not employ any woman in a battery-manufacturing job. In June 1977, however, it announced its first official policy concerning its employment of women in lead-exposure work:

> "[P]rotection of the health of the unborn child is the immediate and direct responsibility of the prospective parents. While the medical profession and the company can support them in the exercise of this responsibility, it cannot assume it for them without simultaneously infringing their rights as persons.
>
> "... Since not all women who can become mothers wish to become mothers (or will become mothers), it would appear to be illegal discrimination to treat all who are capable of pregnancy as though they will become pregnant." App. 140.

Consistent with that view, Johnson Controls "stopped short of excluding women capable of bearing children from lead exposure," *id.*, at 138, but emphasized that a woman who expected to have a child should not choose a job in which she would have such exposure. The company also required a woman who wished to be considered for employment to sign a statement that she had been advised of the risk of having a child while she was exposed to lead. The statement informed the woman that although there was evidence "that women exposed to lead have a higher rate of abortion," this evidence was "not as clear ... as the relationship between cigarette smoking and cancer," but that it was, "medically speaking, just good sense not to run that risk if you want children and do not want to expose the unborn child to risk, however small ...." *Id.*, at 142–143.

Five years later, in 1982, Johnson Controls shifted from a policy of warning to a policy of exclusion. Between 1979 and 1983, eight employees became pregnant while maintaining **1200 blood lead levels in excess of 30 micrograms per deciliter. Tr. of Oral Arg. 25, 34. This appeared to be the critical level *192 noted by the Occupational Safety and Health Administration (OSHA) for a worker who was planning to have a family. See 29 CFR § 1910.1025 (1990). The company responded by announcing a broad exclusion of women from jobs that exposed them to lead:

> "... [I]t is [Johnson Controls'] policy that women who are pregnant or who are capable of bearing children will not be placed into jobs involving lead exposure or which could expose them to lead through the exercise of job bidding, bumping, transfer or promotion rights." App. 85–86.

The policy defined "women ... capable of bearing children" as "[a]ll women except those whose inability to bear children is medically documented." *Id.,* at 81. It further stated that an unacceptable work station was one where, "over the past year," an employee had recorded a blood lead level of more than 30 micrograms per deciliter or the work site had yielded an air sample containing a lead level in excess of 30 micrograms per cubic meter. *Ibid.*

II

In April 1984, petitioners filed in the United States District Court for the Eastern District of Wisconsin a class action challenging Johnson Controls' fetal protection policy as sex discrimination that violated Title VII of the Civil Rights Act of 1964, as amended, 42 U.S.C. § 2000e *et seq.* Among the individual plaintiffs were petitioners Mary Craig, who had chosen to be sterilized in order to avoid losing her job, Elsie Nason, a 50-year-old divorcee, who had suffered a loss in compensation when she was transferred out of a job where she was exposed to lead, and Donald Penney, who had been denied a request for a leave of absence for the purpose of lowering his lead level because he intended to become a father. Upon stipulation of the parties, the District Court certified a class consisting of "all past, present and future production and maintenance employees" in United Auto Workers bargaining *193 units at nine of Johnson Controls' plants "who have been and continue to be affected by [the employer's] fetal-protection policy implemented in 1982." No. 84-C-0472 (Feb. 25, 1985), pp. 1, 2.

The District Court granted summary judgment for defendant-respondent Johnson Controls. 680 F.Supp. 309 (1988). Applying a three-part business necessity defense derived from fetal protection cases in the Courts of Appeals for the Fourth and Eleventh Circuits, the District Court concluded that while "there is a disagreement among the experts regarding the effect of lead on the fetus," the hazard to the fetus through exposure to lead was established by "a considerable body of opinion"; that although "[e]xpert opinion has been provided which holds that lead also affects the reproductive abilities of men and women ... [and] that these effects are as great as the effects of exposure of the fetus ... a great body of experts are of the opinion that the fetus is more vulnerable to levels of lead that would not affect adults"; and that petitioners had "failed to establish that there is an acceptable alternative policy which would protect the fetus." *Id.,* at 315–316. The court stated that, in view of this disposition of the business necessity defense, it did not "have to undertake a bona fide occupational qualification's [sic] (BFOQ) analysis." *Id.,* at 316, n. 5.

The Court of Appeals for the Seventh Circuit, sitting en banc, affirmed the summary judgment by a 7-to-4 vote. 886 F.2d 871 (1989). The majority held that the proper standard for evaluating the fetal protection policy was the defense of business necessity; that Johnson Controls was entitled to summary judgment under that defense; and that even if the proper standard was a BFOQ, Johnson Controls still was entitled to summary judgment.

The Court of Appeals, see *id.,* at 883–885, first reviewed fetal protection opinions from the Eleventh and Fourth Circuits. See **1201 *Hayes v. Shelby Memorial Hospital,* 726 F.2d 1543 (CA11 1984), and *194 *Wright v. Olin Corp.,*

697 F.2d 1172 CA4 1982). Those opinions established the three-step business necessity inquiry: whether there is a substantial health risk to the fetus; whether transmission of the hazard to the fetus occurs only through women; and whether there is a less discriminatory alternative equally capable of preventing the health hazard to the fetus. 886 F.2d, at 885. The Court of Appeals agreed with the Eleventh and Fourth Circuits that "the components of the business necessity defense the courts of appeals and the EEOC have utilized in fetal-protection cases balance the interests of the employer, the employee and the unborn child in a manner consistent with Title VII." *Id.*, at 886. The court further noted that, under *Wards Cove Packing Co. v. Atonio,* 490 U.S. 642, 109 S.Ct. 2115, 104 L.Ed.2d 733 (1989), the burden of persuasion remained on the plaintiff in challenging a business necessity defense, and—unlike the Fourth and Eleventh Circuits—it thus imposed the burden on the plaintiffs for all three steps. 886 F.2d, at 887–893. Cf. *Hayes,* 726 F.2d, at 1549, and *Wright,* 697 F.2d, at 1187.

Applying this business necessity defense, the Court of Appeals ruled that Johnson Controls should prevail. Specifically, the court concluded that there was no genuine issue of material fact about the substantial health-risk factor because the parties agreed that there was a substantial risk to a fetus from lead exposure. 886 F.2d, at 888–889. The Court of Appeals also concluded that, unlike the evidence of risk to the fetus from the mother's exposure, the evidence of risk from the father's exposure, which petitioners presented, "is, at best, speculative and unconvincing." *Id.,* at 889. Finally, the court found that petitioners had waived the issue of less discriminatory alternatives by not adequately presenting it. It said that, in any event, petitioners had not produced evidence of less discriminatory alternatives in the District Court. *Id.,* at 890–893.

Having concluded that the business necessity defense was the appropriate framework and that Johnson Controls satisfied *195 that standard, the court proceeded to discuss the BFOQ defense and concluded that Johnson Controls met that test, too. *Id.,* at 893–894. The en banc majority ruled that industrial safety is part of the essence of respondent's business, and that the fetal-protection policy is reasonably necessary to further that concern. Quoting *Dothard v. Rawlinson,* 433 U.S. 321, 335, 97 S.Ct. 2720, 2729–2730, 53 L.Ed.2d 786 (1977), the majority emphasized that, in view of the goal of protecting the unborn, "more is at stake" than simply an individual woman's decision to weigh and accept the risks of employment. 886 F.2d, at 898.

Judges Cudahy and Posner dissented and would have reversed the judgment and remanded the case for trial. Judge Cudahy explained: "It may (and should) be difficult to establish a BFOQ here but I would afford the defendant an opportunity to try." *Id.,* at 901. "[T]he BFOQ defense need not be narrowly limited to matters of worker productivity, product quality and occupational safety." *Id.,* at 902, n. 1. He concluded that this case's "painful complexities are manifestly unsuited for summary judgment." *Id.,* at 902.

Judge Posner stated: "I think it a mistake to suppose that we can decide this case once and for all on so meager a record." *Ibid.* He, too, emphasized that, under Title VII, a fetal protection policy which explicitly applied just to women could be defended only as a BFOQ. He observed that Title VII defines a BFOQ defense as a "bona fide occupational qualification reasonably necessary to the normal operation" of a business, and that "the 'normal operation' of a business encompasses ethical, legal, and business concerns about the effects of an employer's activities on third parties." *Id.,* at 902 and 904. He emphasized, however, that whether a particular policy is lawful is a question of fact that should ordinarily be resolved at trial. *Id.,* at 906. Like Judge Cudahy, he stressed that "it will be the rare case where the **1202 lawfulness of such a policy can be decided on the defendant's motion for summary judgment." *Ibid.*

*196 Judge Easterbrook, also in dissent and joined by Judge Flaum, agreed with Judges Cudahy and Posner that the only defense available to Johnson Controls was the BFOQ. He concluded, however, that the BFOQ defense would not prevail because respondent's stated concern for the health of the unborn was irrelevant to the operation of its business under the BFOQ. He also viewed the employer's concern as irrelevant to a woman's ability or inability to work under the Pregnancy Discrimination Act's amendment to Title VII, 92 Stat. 2076, 42 U.S.C. § 2000e(k). Judge Easterbrook also stressed what he considered the excessive breadth of Johnson Controls' policy. It applied to all women (except those with medical proof of incapacity to bear children). Although most women in an industrial labor force do not become pregnant, most of those who do become pregnant will have blood lead levels under 30 micrograms per deciliter, and most of those who become pregnant with levels exceeding that figure will bear normal children anyway. 886 F.2d, at 912–913. "Concerns about a tiny minority of women cannot set the standard by which all are judged." *Id.,* at 913.

With its ruling, the Seventh Circuit became the first Court of Appeals to hold that a fetal protection policy directed exclusively at women could qualify as a BFOQ. We granted certiorari, 494 U.S. 1055, 110 S.Ct. 1522, 108 L.Ed.2d 762 (1990), to resolve the obvious conflict between the Fourth, Seventh, and Eleventh Circuits on this issue, and to address the important and difficult question whether an employer, seeking to protect potential fetuses, may discriminate against women just because of their ability to become pregnant. [FN1]

FN1. Since our grant of certiorari, the Sixth Circuit has reversed a District Court's summary judgment for an employer that had excluded fertile female employees from foundry jobs involving exposure to specified concentrations of airborne lead. See *Grant v. General Motors Corp.,* 908 F.2d 1303 (1990). The court said: "We agree with the view of the dissenters in *Johnson Controls* that fetal-protection policies perforce amount to overt sex discrimination, which cannot logically be recast as disparate impact and cannot be countenanced without proof that infertility is a BFOQ ... [P]laintiff ... has alleged a claim of overt discrimination that her employer may justify only through the BFOQ defense." *Id.,* at 1310.

In *Johnson Controls, Inc. v. Fair Employment & Housing Commission,* 218 Cal.App.3d 517, 267 Cal.Rptr. 158 (1990), the court held respondent's fetal protection policy invalid under California's fair-employment law.

*197 III

The bias in Johnson Controls' policy is obvious. Fertile men, but not fertile women, are given a choice as to whether they wish to risk their reproductive health for a particular job. Section 703(a) of the Civil Rights Act of 1964, 78 Stat. 255, as amended, 42 U.S.C. § 2000e-2(a), prohibits sex-based classifications in terms and conditions of employment, in hiring and discharging decisions, and in other employment decisions that adversely affect an employee's status. FN2 Respondent's fetal protection policy explicitly discriminates against women on the basis of their sex. The policy excludes women with childbearing capacity from lead-exposed jobs and so creates a facial classification based on gender. Respondent assumes as much in its brief before this Court. Brief for Respondent 17, n. 24.

FN2. The statute reads:

"It shall be an unlawful employment practice for an employer—

"(1) to fail or refuse to hire or discharge any individual, or otherwise to discriminate against any individual with respect to his compensation, terms, conditions, or privileges of employment, because of such individual's race, color, religion, sex, or national origin; or

"(2) to limit, segregate, or classify his employees or applicants for employment in any way which would deprive or tend to deprive any individual of employment opportunities or otherwise adversely affect his status as an employee, because of such individual's race, color, religion, sex, or national origin."

**1203 [1] ☑ Nevertheless, the Court of Appeals assumed, as did the two appellate courts that already had confronted the issue, that sex-specific fetal protection policies do not involve facial discrimination. 886 F.2d, at 886–887; *Hayes,* 726 F.2d, at 1547; *Wright,* 697 F.2d, at 1190. These courts analyzed the policies as though they were facially neutral, and had only a *198 discriminatory effect upon the employment opportunities of women. Consequently, the courts looked to see if each employer in question had established that its policy was justified as a business necessity. The business necessity standard is more lenient for the employer than the statutory BFOQ defense. The Court of Appeals here went one step further and invoked the burden-shifting framework set forth in *Wards Cove Packing Co. v. Atonio,* 490 U.S. 642, 109 S.Ct. 2115, 104 L.Ed.2d 733 (1989), thus requiring petitioners to bear the burden of persuasion on all questions. 886 F.2d, at 887–888. The court assumed that because the asserted reason for the sex-based exclusion (protecting women's unconceived offspring) was ostensibly benign, the policy was not sex-based discrimination. That assumption, however, was incorrect.

First, Johnson Controls' policy classifies on the basis of gender and childbearing capacity, rather than fertility alone. Respondent does not seek to protect the unconceived children of all its employees. Despite evidence in the record about the debilitating effect of lead exposure on the male reproductive system, Johnson Controls is concerned only with the harms that may befall the unborn offspring of its female employees. Accordingly, it appears that Johnson Controls would have lost in the Eleventh Circuit under *Hayes* because its policy does not "effectively and equally protec[t] the offspring of all employees." 726 F.2d, at 1548. This Court faced a conceptually similar situation in *Phillips v. Martin Marietta Corp.,* 400 U.S. 542, 91 S.Ct. 496, 27 L.Ed.2d 613 (1971), and found sex discrimination because the policy established "one hiring policy for women and another for men—each having pre-school-age children." *Id.,* at 544, 91 S.Ct., at 498. Johnson Controls' policy is facially discriminatory because it requires only a female employee to produce proof that she is not capable of reproducing.

Our conclusion is bolstered by the Pregnancy Discrimination Act (PDA), 42 U.S.C. § 2000e(k), in which Congress explicitly provided that, for purposes of Title VII, discrimination "on the basis of sex" includes discrimination "because *199 of or on the basis of pregnancy, childbirth, or related medical conditions." FN3 "The Pregnancy Discrimination Act has now made clear that, for all Title VII purposes, discrimination based on a woman's pregnancy is, on its face, discrimination because of her sex." *Newport News Shipbuilding & Dry Dock Co. v. EEOC,* 462 U.S. 669, 684, 103 S.Ct. 2622, 2631, 77 L.Ed.2d 89 (1983). In its use of the words "capable of bearing children" in the 1982 policy statement as the criterion for exclusion, Johnson Controls explicitly classifies on the basis of potential for pregnancy. Under the PDA, such a classification must be regarded, for Title VII purposes, in the same light as explicit sex discrimination. Respondent has chosen to treat all its female employees as potentially pregnant; that choice evinces discrimination on the basis of sex.

FN3. The Act added subsection (k) to § 701 of the Civil Rights Act of 1964 and reads in pertinent part:

"The terms 'because of sex' or 'on the basis of sex' [in Title VII] include, but are not limited to, because of or on the basis of pregnancy, childbirth, or related medical conditions; and women affected by pregnancy, childbirth, or related medical conditions shall be treated the same for all employment-related purposes ... as other persons not so affected but similar in their ability or inability to work...."

[2] ☑ [3] ☑ We concluded above that Johnson Controls' policy is not neutral because it does not apply to the reproductive capacity of the company's male employees in the same way as it applies to that of the females. Moreover, the absence of a malevolent motive does **1204 not convert a facially discriminatory policy into a neutral policy with a discriminatory effect. Whether an employment practice involves disparate treatment through explicit facial discrimination does not depend on why the employer discriminates but rather on the explicit terms of the discrimination. In *Martin Marietta, supra,* the motives underlying the employers' express exclusion of women did not alter the intentionally discriminatory character of the policy. Nor did the arguably benign motives lead to consideration of a business necessity defense. The question *200 in that case was whether the discrimination in question could be justified under § 703(e) as a BFOQ. The beneficence of an employer's purpose does not undermine the conclusion that an explicit gender-based policy is sex discrimination under § 703(a) and thus may be defended only as a BFOQ.

The enforcement policy of the Equal Employment Opportunity Commission accords with this conclusion. On January 24, 1990, the EEOC issued a Policy Guidance in the light of the Seventh Circuit's decision in the present case. App. to Pet. for Cert. 127a. The document noted: "For the plaintiff to bear the burden of proof in a case in which there is direct evidence of a facially discriminatory policy is wholly inconsistent with settled Title VII law." *Id.,* at 133a. The Commission concluded: "[W]e now think BFOQ is the better approach." *Id.,* at 134a.

In sum, Johnson Controls' policy "does not pass the simple test of whether the evidence shows 'treatment of a person in a manner which but for that person's sex would be different.' " *Los Angeles Dept. of Water and Power v. Manhart,* 435 U.S. 702, 711, 98 S.Ct. 1370, 1377, 55 L.Ed.2d 657 (1978), quoting Developments in the Law, Employment Discrimination and Title VII of the Civil Rights Act of 1964, 84 Harv.L.Rev. 1109, 1170 (1971). We hold that Johnson Controls' fetal-protection policy is sex discrimination forbidden under Title VII unless respondent can establish that sex is a "bona fide occupational qualification."

IV

Under § 703(e)(1) of Title VII, an employer may discriminate on the basis of "religion, sex, or national origin in those certain instances where religion, sex, or national origin is a bona fide occupational qualification reasonably necessary to the normal operation of that particular business or enterprise." 42 U.S.C. § 2000e-2(e)(1). We therefore turn to the question whether Johnson Controls' fetal protection policy *201 is one of those "certain instances" that come within the BFOQ exception.

The BFOQ defense is written narrowly, and this Court has read it narrowly. See, e.g., *Dothard v. Rawlinson,* 433 U.S. 321, 332–337, 97 S.Ct. 2720, 2728–2731, 53 L.Ed.2d 786 (1977); *Trans World Airlines, Inc. v. Thurston,* 469 U.S. 111, 122–125, 105 S.Ct. 613, 622–624, 83 L.Ed.2d 523 (1985). We have read the BFOQ language of § 4(f) of the Age Discrimination in Employment Act of 1967 (ADEA), 81 Stat. 603, as amended, 29 U.S.C. § 623(f)(1), which tracks the BFOQ provision in Title VII, just as narrowly. See *Western Air Lines, Inc. v. Criswell,* 472 U.S. 400, 105 S.Ct. 2743, 86 L.Ed.2d 321 (1985). Our emphasis on the restrictive scope of the BFOQ defense is grounded on both the language and the legislative history of § 703.

[4] ☑ The wording of the BFOQ defense contains several terms of restriction that indicate that the exception reaches only special situations. The statute thus limits the situations in which discrimination is permissible to "certain instances" where sex discrimination is "reasonably necessary" to the "normal operation" of the "particular" business. Each one of these terms—*certain, normal, particular*—prevents the use of general subjective standards and favors an objective, verifiable requirement. But the most telling term is "occupational"; this indicates that these objective, verifiable requirements must concern job-related skills and aptitudes.

**1205 Justice WHITE defines "occupational" as meaning related to a job. *Post,* at 1210, n. 1. According to him, any discriminatory requirement imposed by an employer is "job-related" simply because the employer has chosen to make the requirement a condition of employment. In effect, he argues that sterility may be an occupational qualification for women because Johnson Controls has chosen to require it. This reading of "occupational" renders the word mere surplusage. "Qualification" by itself would encompass an employer's idiosyncratic requirements. By modifying "qualification" with "occupational," Congress narrowed the term to qualifications that affect an employee's ability to do the job.

*202 Johnson Controls argues that its fetal protection policy falls within the so-called safety exception to the BFOQ. Our cases have stressed that discrimination on the basis of sex because of safety concerns is allowed only in narrow circumstances. In *Dothard v. Rawlinson,* this Court indicated that danger to a woman herself does not justify discrimination. 433 U.S., at 335, 97 S.Ct. at 2729–2730. We there allowed the employer to hire only male guards in contact areas of maximum-security male penitentiaries only because more was at stake than the "individual woman's decision to weigh and accept the risks of employment." *Ibid.* We found sex to be a BFOQ inasmuch as the employment of a female guard would create real risks of safety to others if violence broke out because the guard was a woman. Sex discrimination was tolerated because sex was related to the guard's ability to do the job-maintaining prison security. We also required in *Dothard* a high correlation between sex and ability to perform job functions and refused to allow employers to use sex as a proxy for strength. although it might be a fairly accurate one.

Similarly, some courts have approved airlines' layoffs of pregnant flight attendants at different points during the first five months of pregnancy on the ground that the employer's policy was necessary to ensure the safety of passengers. See *Harriss v. Pan American World Airways, Inc.,* 649 F.2d 670 (CA9 1980); *Burwell v. Eastern Air Lines, Inc.,* 633 F.2d 361 (CA4 1980), cert. denied, 450 U.S. 965, 101 S.Ct. 1480, 67 L.Ed.2d 613 (1981); *Condit v. United Air Lines, Inc.,* 558 F.2d 1176 (CA4 1977), cert. denied, 435 U.S. 934, 98 S.Ct. 1510, 55 L.Ed.2d 531 (1978); *In re National Airlines, Inc.,* 434 F.Supp. 249 (S.D.Fla.1977). In two of these cases, the courts pointedly indicated that fetal, as opposed to passenger, safety was best left to the mother. *Burwell,* 633 F.2d, at 371; *National Airlines,* 434 F.Supp., at 259.

We considered safety to third parties in *Western Airlines, Inc. v. Criswell, supra,* in the context of the ADEA. We focused upon "the nature of the flight engineer's tasks," and the "actual capabilities of persons over age 60" in relation to *203 those tasks. 472 U.S., at 406, 105 S.Ct., at 2747. Our safety concerns were not independent of the individual's ability to perform the assigned tasks, but rather involved the possibility that, because of age-connected debility, a flight engineer might not properly assist the pilot, and might thereby cause a safety emergency. Furthermore, although we considered the safety of third parties in *Dothard* and *Criswell,* those third parties were indispensable to the particular business at issue. In *Dothard,* the third parties were the inmates; in *Criswell,* the third parties were the passengers on the plane. We stressed that in order to qualify as a BFOQ, a job qualification must relate to the "essence," *Dothard,* 433 U.S., at 333, 97 S.Ct., at 2751, (emphasis deleted), or to the "central mission of the employer's business," *Criswell,* 472 U.S., at 413, 105 S.Ct., at 2751.

Justice WHITE ignores the "essence of the business" test and so concludes that "protecting fetal safety while carrying out the duties of battery manufacturing is as much a legitimate concern as is safety to third parties in guarding prisons (*Dothard*) or flying airplanes (*Criswell*)." *Post,* at 1213. By limiting his discussion to cost and **1206 safety concerns and rejecting the "essence of the business" test that our case law has established, he seeks to expand what is now the narrow BFOQ defense. Third-party safety considerations properly entered into the BFOQ analysis in *Dothard* and *Criswell* because they went to the core of the employee's job performance. Moreover, that performance involved the central purpose of the enterprise. *Dothard,* 433 U.S., at 335, 97 S.Ct., at 2729–2730 ("The essence of a correctional counselor's job is to maintain prison security"); *Criswell,* 472 U.S., at 413, 105 S.Ct., at 2751 (the central mission of the airline's business was the safe transportation of its passengers). Justice WHITE attempts to transform this case into one of customer safety. The unconceived fetuses of Johnson Controls' female employees, however, are neither customers nor third parties whose safety is essential to the business of battery manufacturing. No one can disregard the possibility of injury to future children; the BFOQ, however, *204 is not so broad that it transforms this deep social concern into an essential aspect of battery making.

[5] ☑ [6] ☑ Our case law, therefore, makes clear that the safety exception is limited to instances in which sex or pregnancy actually interferes with the employee's ability to perform the job. This approach is consistent with the language of the BFOQ provision itself, for it suggests that permissible distinctions based on sex must relate to ability to perform the duties of the job. Johnson Controls suggests, however, that we expand the exception to allow fetal protection policies that mandate particular standards for pregnant or fertile women. We decline to do so. Such an expansion contradicts not only the language of the BFOQ and the narrowness of its exception, but also the plain language and history of the PDA.

The PDA's amendment to Title VII contains a BFOQ standard of its own: Unless pregnant employees differ from others "in their ability or inability to work," they must be "treated the same" as other employees "for all employment-related purposes." 42 U.S.C. § 2000e(k). This language clearly sets forth Congress' remedy for discrimination on the basis of pregnancy and potential pregnancy. Women who are either pregnant or potentially pregnant must be treated like others "similar in their ability ... to work." *Ibid.* In other words, women as capable of doing their jobs as their male counterparts may not be forced to choose between having a child and having a job.

Justice WHITE asserts that the PDA did not alter the BFOQ defense. *Post,* at 1213. He arrives at this conclusion by ignoring the second clause of the Act, which states that "women affected by pregnancy, childbirth, or related medical conditions shall be treated the same for all employment-related purposes ... as other persons not so affected but similar in their ability or inability to work." 42 U.S.C. § 2000e(k). Until this day, every Member of this Court had acknowledged that "[t]he second clause [of the PDA] could not be clearer: it mandates that pregnant employees 'shall be *205 treated the same for all employment-related purposes' as nonpregnant employees similarly situated with respect to their ability or inability to work." *California Federal Savings and Loan Assn. v. Guerra,* 479 U.S. 272, 297, 107 S.Ct. 683, 698, 93 L.Ed.2d 613 (1987) (WHITE, J., dissenting). Justice WHITE now seeks to read the second clause out of the Act.

The legislative history confirms what the language of the PDA compels. Both the House and Senate Reports accompanying the legislation indicate that this statutory standard was chosen to protect female workers from being treated differently from other employees simply because of their capacity to bear children. See Amending Title VII, Civil Rights Act of 1964, S.Rep. No. 95–331, pp. 4–6 (1977):

> "Under this bill, the treatment of pregnant women in covered employment must focus not on their condition alone but on the actual effects of that condition on their **1207 ability to work. Pregnant women who are able to work must be permitted to work on the same conditions as other employees ....
>
> "... [U]nder this bill, employers will no longer be permitted to force women who become pregnant to stop working regardless of their ability to continue."

See also Prohibition of Sex Discrimination Based on Pregnancy, H.R.Rep. No. 95–948, pp. 3–6 (1978), *U.S. Code Cong. & Admin. News* 1978, p. 4749.

This history counsels against expanding the BFOQ to allow fetal-protection policies. The Senate Report quoted above states that employers may not require a pregnant woman to stop working at any time during her pregnancy unless she is unable to do her work. Employment late in pregnancy often imposes risks on the unborn child, see "Walking a Tightrope: Pregnancy, Parenting, and Work," in *Double Exposure* 196, 196–202 (W. Chavkin ed. 1984), but Congress indicated that the employer may take into account only the woman's ability to get her job done. See Becker, *206 From *Muller v. Oregon* to Fetal Vulnerability Policies, 53 *U.Chi.Law.Rev.* 1219, 1255–1256 (1986). With the PDA, Congress made clear that the decision to become pregnant or to work while being either pregnant or capable of becoming pregnant was reserved for each individual woman to make for herself.

[7] ☑ We conclude that the language of both the BFOQ provision and the PDA which amended it, as well as the legislative history and the case law, prohibit an employer from discriminating against a woman because of her capacity to become pregnant unless her reproductive potential prevents her from performing the duties of her job. We reiterate our holdings in *Criswell* and *Dothard* that an employer must direct its concerns about a woman's ability to perform her job safely and efficiently to those aspects of the woman's job-related activities that fall within the "essence" of the particular business. [FN4]

FN4. Justice WHITE predicts that our reaffirmation of the narrowness of the BFOQ defense will preclude considerations of privacy as a basis for sex-based discrimination. *Post,* at 1214, n. 8. We have never addressed privacy-based sex discrimination and shall not do so here because the sex-based discrimination at issue today does not involve the privacy interests of Johnson Controls' customers. Nothing in our discussion of the "essence of the business test," however, suggests that sex could not constitute a BFOQ when privacy interests are implicated. See, e.g., *Backus v. Baptist Medical Center,* 510 F.Supp. 1191 (ED Ark.1981) (essence of obstetrics nurse's business is to provide sensitive care for patient's intimate and private concerns), vacated as moot, 671 F.2d 1100 (CA8 1982).

V

[8] ☑ We have no difficulty concluding that Johnson Controls cannot establish a BFOQ. Fertile women, as far as appears in the record, participate in the manufacture of batteries as efficiently as anyone else. Johnson Controls' professed moral and ethical concerns about the welfare of the next generation do not suffice to establish a BFOQ of female sterility. Decisions about the welfare of future children must be left to the parents who conceive, bear, support, and raise them rather than to the employers who hire those parents. Congress has mandated this choice through Title VII, as amended by the *207 PDA. Johnson Controls has attempted to exclude women because of their reproductive capacity. Title VII and the PDA simply do not allow a woman's dismissal because of her failure to submit to sterilization.

Nor can concerns about the welfare of the next generation be considered a part of the "essence" of Johnson Controls' business. Judge Easterbrook in this case pertinently observed: "It is word play to say that 'the job' at Johnson [Controls] is to make batteries without risk to fetuses in the same way 'the job' at Western Air Lines is to fly planes without crashing." 886 F.2d, at 913.

**1208 Johnson Controls argues that it must exclude all fertile women because it is impossible to tell which women will become pregnant while working with lead. This argument is somewhat academic in light of our conclusion that the company may not exclude fertile women at all; it perhaps is worth noting, however, that Johnson Controls has shown no "factual basis for believing that all or substantially all women would be unable to perform safely and efficiently the duties of the job involved." *Weeks v. Southern Bell Tel. & Tel. Co.,* 408 F.2d 228, 235 (CA5 1969), quoted with approval in *Dothard,* 433 U.S., at 333, 97 S.Ct., at 2751. Even on this sparse record, it is apparent that Johnson Controls is concerned about only a small minority of women. Of the eight pregnancies reported among the female employees, it has not been shown that any of the babies have birth defects or other abnormalities. The record does not reveal the birth rate for Johnson Controls' female workers, but national statistics show that approximately 9% of all fertile women become pregnant each year. The birthrate drops to 2% for blue collar workers over age 30. See Becker, 53 U.Chi.L.Rev., at 1233. Johnson Controls' fear of prenatal injury, no matter how sincere, does not begin to show that substantially all of its fertile women employees are incapable of doing their jobs.

*208 VI

A word about tort liability and the increased cost of fertile women in the workplace is perhaps necessary. One of the dissenting judges in this case expressed concern about an employer's tort liability and concluded that liability for a potential injury to a fetus is a social cost that Title VII does not require a company to ignore. 886 F.2d, at 904–905. It is correct to say that Title VII does not prevent the employer from having a conscience. The statute, however, does prevent sex-specific fetal protection policies. These two aspects of Title VII do not conflict.

More than 40 States currently recognize a right to recover for a prenatal injury based either on negligence or on wrongful death. See, e.g., *Wolfe v. Isbell,* 291 Ala. 327, 333–334, 280 So.2d 758, 763 (1973); *Simon v. Mullin,* 34 Conn.Sup. 139, 147, 380 A.2d 1353, 1357 (1977). See also Note, 22 Suffolk U.L.Rev. 747, 754–756, and nn. 54, 57, and 58 (1988) (listing cases). According to Johnson Controls, however, the company complies with the lead standard developed by OSHA and warns its female employees about the damaging effects of lead. It is worth noting that OSHA gave the problem of lead lengthy consideration and concluded that "there is no basis whatsoever for the claim that women of childbearing age should be excluded from the workplace in order to protect the fetus or the course of pregnancy." 43 Fed.Reg. 52952, 52966 (1978). See also *id.,* at 54354, 54398. Instead, OSHA established a series of mandatory protections which, taken together, "should effectively minimize any risk to the fetus and newborn child." *Id.,* at 52966. See 29 CFR § 1910.1025(k)(ii) (1990). Without negligence, it would be difficult for a court to find liability on the part of the employer. If, under general tort principles, Title VII bans sex-specific fetal-protection policies, the employer fully informs the woman of the risk, and the employer has not acted negligently, the basis for holding an employer liable seems remote at best.

*209 Although the issue is not before us, Justice WHITE observes that "it is far from clear that compliance with Title VII will pre-empt state tort liability." *Post,* at 1211. The cases relied upon by him to support his prediction, however, are inapposite. For example, in *California Federal Savings and Loan Assn. v. Guerra,* 479 U.S. 272, 107 S.Ct. 683, 93 L.Ed.2d 613 (1987), we considered a California statute that expanded upon the requirements of the PDA and concluded that the statute was not pre-empted by Title VII because it was not inconsistent with the purposes of the federal statute and did not require an act that was unlawful under Title **1209 VII. *Id.,* at 291–292, 107 S.Ct., at 694–695. Here, in contrast, the tort liability that Justice WHITE fears will punish employers for *complying* with Title VII's clear command. When it is impossible for an employer to comply with both state and federal requirements, this Court has ruled that federal law pre-empts that of the States. See, e.g., *Florida Lime & Avocado Growers, Inc. v. Paul,* 373 U.S. 132, 142–143, 83 S.Ct. 1210, 1217–1218, 10 L.Ed.2d 248 (1963).

This Court faced a similar situation in *Farmers Union v. WDAY, Inc.,* 360 U.S. 525, 79 S.Ct. 1302, 3 L.Ed.2d 1407 (1959). In *WDAY,* it held that § 315(a) of the Federal Communications Act of 1934 barred a broadcasting station from removing defamatory statements contained in speeches broadcast by candidates for public office. It then considered a libel action that arose as a result of a speech made over the radio and television facilities of WDAY by a candidate for the 1956 senatorial race in North Dakota. It held that the statutory prohibition of censorship carried with it an immunity from liability for defamatory statements made by the speaker. To allow libel actions "would sanction the unconscionable result of permitting civil and perhaps criminal liability to be imposed for the very conduct the statute demands of the licensee." *Id.,* at 531, 79 S.Ct., at 1306. It concluded:

> "We are aware that causes of action for libel are widely recognized throughout the States. But we have not hesitated to abrogate state law where satisfied that *210 its enforcement would stand 'as an obstacle to the accomplishment and execution of the full purposes and objectives of Congress.' " *Id.,* at 535, 79 S.Ct., at 1308, quoting *Bethlehem Steel Co. v. New York State Labor Relations Bd.,* 330 U.S. 767, 773, 67 S.Ct. 1026, 1030, 91 L.Ed. 1234 (1947).

If state tort law furthers discrimination in the workplace and prevents employers from hiring women who are capable of manufacturing the product as efficiently as men, then it will impede the accomplishment of Congress' goals in enacting Title VII. Because Johnson Controls has not argued that it faces any costs from tort liability, not to mention crippling ones, the pre-emption question is not before us. We therefore say no more than that the concurrence's speculation appears unfounded as well as premature.

[9] ☑ The tort liability argument reduces to two equally unpersuasive propositions. First, Johnson Controls attempts to solve the problem of reproductive health hazards by resorting to an exclusionary policy. Title VII plainly forbids illegal sex discrimination as a method of diverting attention from an employer's obligation to police the workplace. Second, the specter of an award of damages reflects a fear that hiring fertile women will cost more. The extra cost of employing members of one sex, however, does not provide an affirmative Title VII defense for a discriminatory refusal to hire members of that gender. See *Manhart,* 435 U.S., at 716–718, and n. 32, 98 S.Ct., at 1379–1380, and n. 32. Indeed, in passing the PDA, Congress considered at length the considerable cost of providing equal treatment of pregnancy and related conditions, but made the "decision to forbid special treatment of pregnancy despite the social costs associated therewith." *Arizona Governing Comm. for Tax Deferred Annuity and Deferred Compensation Plans v. Norris,* 463 U.S. 1073, 1085, n. 14, 103 S.Ct. 3492, 3499, n. 14, 77 L.Ed.2d 1236 (1983) (opinion of MARSHALL, J.). See *Price Waterhouse v. Hopkins,* 490 U.S. 228, 109 S.Ct. 1775, 104 L.Ed.2d 268 (1989).

[10] ☑ We, of course, are not presented with, nor do we decide, a case in which costs would be so prohibitive as to threaten the *211 survival of the employer's business. We merely reiterate our prior holdings that the incremental cost of hiring women cannot justify discriminating against them.

VII

Our holding today that Title VII, as so amended, forbids sex-specific fetal-protection **1210 policies is neither remarkable nor unprecedented. Concern for a woman's existing or potential offspring historically has been the excuse for denying women equal employment opportunities. See, e.g., *Muller v. Oregon,*

208 U.S. 412, 28 S.Ct. 324, 52 L.Ed. 551 (1908). Congress in the PDA prohibited discrimination on the basis of a woman's ability to become pregnant. We do no more than hold that the PDA means what it says.

It is no more appropriate for the courts than it is for individual employers to decide whether a woman's reproductive role is more important to herself and her family than her economic role. Congress has left this choice to the woman as hers to make.

The judgment of the Court of Appeals is reversed, and the case is remanded for further proceedings consistent with this opinion.

It is so ordered.

JUSTICE WHITE, WITH WHOM THE CHIEF JUSTICE AND JUSTICE KENNEDY JOIN, CONCURRING IN PART AND CONCURRING IN THE JUDGMENT.

The Court properly holds that Johnson Controls' fetal-protection policy overtly discriminates against women, and thus is prohibited by Title VII of the Civil Rights Act of 1964 unless it falls within the bona fide occupational qualification (BFOQ) exception, set forth at 42 U.S.C. § 2000e-2(e). The Court erroneously holds, however, that the BFOQ defense is so narrow that it could never justify a sex-specific fetal-protection policy. I nevertheless concur in the judgment of reversal because on the record before us summary judgment in favor of Johnson Controls was improperly entered *212 by the District Court and affirmed by the Court of Appeals.

I

In evaluating the scope of the BFOQ defense, the proper starting point is the language of the statute. Cf. *Demarest v. Manspeaker,* 498 U.S. 184, 190, 111 S.Ct. 599, 603, 112 L.Ed.2d 608 (1991); *Board of Ed. of Westside Community Schools (Dist. 66) v. Mergens,* 496 U.S. 226, 237, 110 S.Ct. 2356, 2365, 110 L.Ed.2d 191 (1990). Title VII forbids discrimination on the basis of sex, except "in those certain instances where ... sex ... is a bona fide occupational qualification reasonably necessary to the normal operation of that particular business or enterprise." 42 U.S.C. § 2000e-2(e)(1). For the fetal-protection policy involved in this case to be a BFOQ, therefore, the policy must be "reasonably necessary" to the "normal operation" of making batteries, which is Johnson Controls' "particular business." Although that is a difficult standard to satisfy, nothing in the statute's language indicates that it could *never* support a sex-specific fetal-protection policy.[FN1]

FN1. The Court's heavy reliance on the word " 'occupational' " in the BFOQ statute, *ante,* at 12, is unpersuasive. *Any* requirement for employment can be said to be an occupational qualification, since "occupational" merely means related to a job. See Webster's Third New International Dictionary 1560 (1976). Thus, Johnson Controls' requirement that employees engaged in battery manufacturing be either male or non-fertile clearly is an "occupational qualification." The issue, of course, is whether that qualification is "reasonably necessary to the normal operation" of Johnson Controls' business. It is telling that the Court offers no case support, either from this Court or the lower federal courts, for its interpretation of the word "occupational."

On the contrary, a fetal-protection policy would be justified under the terms of the statute if, for example, an employer could show that exclusion of women from certain jobs was reasonably necessary to avoid substantial tort liability. Common sense tells us that it is part of the normal operation of business concerns to avoid causing injury to third parties, as well as to employees, if for no other reason than to avoid *213 tort liability and its substantial costs. This possibility of tort liability is not hypothetical; every State currently allows children born alive to recover in tort for prenatal injuries caused by **1211 third parties, see W. Keeton, D. Dobbs, R. Keeton, & D. Owen, Prosser and Keeton on Law of Torts § 55, p. 368 (5th ed. 1984), and an increasing number of courts have recognized a right to recover even for prenatal injuries caused by torts committed prior to conception, see 3 F. Harper, F. James, & O. Gray, Law of Torts § 18.3, pp. 677–678, n. 15 (2d ed. 1986).

The Court dismisses the possibility of tort liability by no more than speculating that if "Title VII bans sex-specific fetal-protection policies, the employer fully informs the woman of the risk, and the employer has not acted negligently, the basis for holding an employer liable seems remote at best." *Ante,* at 1208. Such speculation will be small comfort to employers. First, it is far from clear that compliance with Title VII will pre-empt state tort liability, and the Court offers no support for that proposition.[FN2] Second, although warnings may preclude claims by injured *employees,* they will not preclude claims by injured children because the general rule is that parents cannot waive causes of action on behalf of their children, and the parents' negligence will not be imputed to the children.[FN3] Finally, although state tort liability *214 for prenatal injuries generally requires negligence, it will be difficult for employers to determine in advance what will constitute negligence. Compliance with OSHA standards, for example, has been held not to be a defense to state tort or criminal liability. See *National Solid Wastes Management Assn. v. Killian,* 918 F.2d 671, 680, n. 9 (CA7 1990) (collecting cases); see also 29 U.S.C. § 653(b)(4). Moreover, it is possible that employers will be held strictly liable, if, for example, their manufacturing process is considered "abnormally dangerous." See Restatement (Second) of Torts § 869, Comment *b* (1979).

FN2. Cf. *English v. General Electric Co.,* 496 U.S. 72, 110 S.Ct. 2270, 110 L.Ed.2d 65 (1990) (state law action for intentional infliction of emotional distress not pre-empted by Energy Reorganization Act of 1974); *California Federal Savings and Loan Assn. v. Guerra,* 479 U.S. 272, 290–292, 107 S.Ct. 683, 694–695, 93 L.Ed.2d 613 (1987) (state statute requiring the provision of leave and reinstatement to employees disabled by pregnancy not pre-empted by the Pregnancy Discrimination Act (PDA), 92 Stat. 2076, 42 U.S.C. § 2000e(k)); *Silkwood v. Kerr-McGee Corp.,* 464 U.S. 238, 256, 104 S.Ct. 615, 625–26, 78 L.Ed.2d 443 (1984) (state punitive damages claim not pre-empted by federal laws regulating nuclear powerplants); *Bernstein v. Aetna Life & Casualty,* 843 F.2d 359, 364–365 (CA9 1988) ("It is well-established that Title VII does not preempt state common law remedies"); see also 42 U.S.C. § 2000e-7.

FN3. See, e.g., *In re Estate of Infant Fontaine,* 128 N.H. 695, 700, 519 A.2d 227, 230 (1986); *Collins v. Eli Lilly Co.,* 116 Wis.2d 166, 200, n. 14, 342 N.W.2d 37,

53, n. 14, cert. denied, 469 U.S. 826, 105 S.Ct. 107, 83 L.Ed.2d 51 (1984); *Doyle v. Bowdoin College,* 403 A.2d 1206, 1208, n. 3 (Me.1979); *Littleton v. Jordan,* 428 S.W.2d 472 (Tex.Civ.App.1968); *Fallaw v. Hobbs,* 113 Ga.App. 181, 182–183, 147 S.E.2d 517, 519 (1966); see also Restatement (Second) of Torts § 488(1) (1965).

Relying on *Los Angeles Dept. of Water and Power v. Manhart,* 435 U.S. 702, 98 S.Ct. 1370, 55 L.Ed.2d 657 (1978), the Court contends that tort liability cannot justify a fetal-protection policy because the extra costs of hiring women is not a defense under Title VII. *Ante,* at 1209. This contention misrepresents our decision in *Manhart.* There, we held that a requirement that female employees contribute more than male employees to a pension fund, in order to reflect the greater longevity of women, constituted discrimination against women under Title VII because it treated them as a class rather than as individuals. 435 U.S., at 708, 716–717, 98 S.Ct., at 1379-1380. We did not in that case address in any detail the nature of the BFOQ defense, and we certainly did not hold that cost was irrelevant to the BFOQ analysis. Rather, we merely stated in a footnote that "there has been no showing that sex distinctions are reasonably necessary to the normal operation of the Department's retirement plan." *Id.,* at 716, n. 30, 98 S.Ct., at 1379, n. 30. We further noted that although Title VII does not contain a "cost-justification defense comparable to the affirmative defense available in a price discrimination *215 suit," "no defense based on the *total* cost of employing men and women was **1212 attempted in this case." *Id.,* at 716-717, and n. 32, 98 S.Ct., at 1379-1380, and n. 32.

Prior decisions construing the BFOQ defense confirm that the defense is broad enough to include considerations of cost and safety of the sort that could form the basis for an employer's adoption of a fetal-protection policy. In *Dothard v. Rawlinson,* 433 U.S. 321, 97 S.Ct. 2720, 53 L.Ed.2d 786 (1977), the Court held that being male was a BFOQ for "contact" guard positions in Alabama's maximum-security male penitentiaries. The Court first took note of the actual conditions of the prison environment: "In a prison system where violence is the order of the day, where inmate access to guards is facilitated by dormitory living arrangements, where every institution is understaffed, and where a substantial portion of the inmate population is composed of sex offenders mixed at random with other prisoners, there are few visible deterrents to inmate assaults on women custodians." *Id.,* at 335-336, 97 S.Ct., at 2730. The Court also stressed that "[m]ore [was] at stake" than a risk to individual female employees: "The likelihood that inmates would assault a woman because she was a woman would pose a real threat not only to the victim of the assault but also to the basic control of the penitentiary and protection of its inmates and the other security personnel." *Ibid.* Under those circumstances, the Court observed that "it would be an oversimplification to characterize [the exclusion of women] as an exercise in 'romantic paternalism.' Cf. *Frontiero v. Richardson,* 411 U.S. 677, 684 [93 S.Ct. 1764, 1769, 36 L.Ed.2d 583]." *Id.,* 433 U.S., at 335, 97 S.Ct., at 2729.

We revisited the BFOQ defense in *Western Air Lines, Inc. v. Criswell,* 472 U.S. 400, 105 S.Ct. 2743, 86 L.Ed.2d 321 (1985), this time in the context of the Age Discrimination in Employment Act of 1967 (ADEA). There, we endorsed the two-part inquiry for evaluating a BFOQ defense used by the Court of Appeals for the Fifth Circuit in *Usery v. Tamiami Trail Tours, Inc.,* 531 F.2d 224 (1976). First, the job qualification must not be "so peripheral to the central mission of the employer's business" that no discrimination *216 could be "'reasonably *necessary* to the normal operation of the particular business.'" 472 U.S., at 413, 105 S.Ct., at 2751. Although safety is *not* such a peripheral concern, *id.,* at 413, 419, 105 S.Ct., at 2751, 2754,[FN4] the inquiry "'adjusts to the safety factor'"—"'[t] he greater the safety factor, measured by the likelihood of harm and the probable severity of that harm in case of an accident, the more stringent may be the job qualifications,' " *id.,* at 413, 105 S.Ct., at 2751 (quoting *Tamiami, supra,* at 236). Second, the employer must show either that all or substantially all persons excluded 'would be unable to perform safely and efficiently the duties of the job involved,'" or that it is "'impossible or highly impractical'" to deal with them on an individual basis. 472 U.S., at 414, 105 S.Ct., at 2752 (quoting *Tamiami, supra,* at 235 (quoting *Weeks v. Southern Bell Telephone & Telegraph Co.,* 408 F.2d 228, 235 (CA5 1969))). We further observed that this inquiry properly takes into account an employer's interest in safety—"[w]hen an employer establishes that a job qualification has been carefully formulated to respond to documented concerns for public safety, it will not be overly burdensome to persuade a trier of fact that the qualification is 'reasonably necessary' to safe operation of **1213 the business." 472 U.S., at 419, 105 S.Ct., at 2754.

FN4. An example of a "peripheral" job qualification was in *Diaz v. Pan American World Airways, Inc.,* 442 F.2d 385 (CA5), cert. denied, 404 U.S. 950, 92 S.Ct. 275, 30 L.Ed.2d 267 (1971). There, the Fifth Circuit held that being female was not a BFOQ for the job of flight attendant, despite a determination by the trial court that women were better able than men to perform the "non-mechanical" functions of the job, such as attending to the passengers' psychological needs. The court concluded that such non-mechanical functions were merely "tangential" to the normal operation of the airline's business, noting that "[n]o one has suggested that having male stewards will so seriously affect the operation of an airline as to jeopardize or even minimize its ability to provide safe transportation from one place to another." 442 F.2d, at 388.

Dothard and *Criswell* make clear that avoidance of substantial safety risks to third parties is *inherently* part of both an employee's ability to perform a job and an employer's *217 "normal operation" of its business. Indeed, in both cases, the Court approved the statement in *Weeks v. Southern Bell Telephone & Telegraph Co., supra,* that an employer could establish a BFOQ defense by showing that "all or substantially all women would be unable to perform *safely and efficiently* the duties of the job involved." *Id.,* at 235 (emphasis added). See *Criswell,* 472 U.S., at 414, 105 S.Ct., at 2751-52; *Dothard, supra,* 433 U.S., at 333, 97 S.Ct., at 2728-29. The Court's statement in this case that "the safety exception is limited to instances in which sex or pregnancy actually interferes with the employee's ability to perform the job," *ante,* at 1206, therefore adds no support to its conclusion that a fetal-protection policy could never be justified as a BFOQ. On the facts of this case, for example, protecting fetal safety while carrying out the duties of battery manufacturing is as much a legitimate concern as is safety to third parties in guarding prisons (*Dothard*) or flying airplanes (*Criswell*).[FN5]

FN5. I do not, as the Court asserts, *ante,* at 1205, reject the " 'essence of the business' " test. Rather, I merely reaffirm the obvious—that safety to third parties is part of the "essence" of most if not all businesses. Of course, the BFOQ inquiry " 'adjusts to the safety factor.' " *Criswell,* 472 U.S., at 413, 105 S.Ct., at 2751 (quoting *Usery v. Tamiami Trail Tours, Inc.,* 531 F.2d 224, 236 (CA5 1976). As a result, more stringent occupational qualifications may be justified for jobs involving higher safety risks, such as flying airplanes. But a recognition that the importance of safety varies among businesses does not mean that safety is completely irrelevant to the essence of a job such as battery manufacturing.

Dothard and *Criswell* also confirm that costs are relevant in determining whether a discriminatory policy is reasonably necessary for the normal operation of a business. In *Dothard,* the safety problem that justified exclusion of women from the prison guard positions was largely a result of inadequate staff and facilities. See 433 U.S., at 335, 97 S.Ct., at 2729-30. If the cost of employing women could not be considered, the employer there should have been required to hire more staff and restructure the prison environment rather than exclude women. Similarly, in *Criswell* the airline could have been *218 required to hire more pilots and install expensive monitoring devices rather than discriminate against older employees. The BFOQ statute, however, reflects "Congress' unwillingness to require employers to change the very nature of their operations." *Price Waterhouse v. Hopkins,* 490 U.S. 228, 242, 109 S.Ct. 1775, 1786, 104 L.Ed.2d 268 (1989) (plurality opinion).

The PDA, contrary to the Court's assertion, *ante,* at 1206, did not restrict the scope of the BFOQ defense. The PDA was only an amendment to the "Definitions" section of Title VII, 42 U.S.C. § 2000e, and did not purport to eliminate or alter the BFOQ defense. Rather, it merely clarified Title VII to make it clear that pregnancy and related conditions are included within Title VII's antidiscrimination provisions. As we have already recognized, "the purpose of the PDA was simply to make the treatment of pregnancy consistent with general Title VII principles." *Arizona Governing Comm. for Tax Deferred Annuity and Deferred Compensation Plans v. Norris,* 463 U.S. 1073, 1085, n. 14, 103 S.Ct. 3492, 3499-3500, n. 14, 77 L.Ed.2d 1236 (1983).[FN6]

FN6. Contrary to the Court's assertion, *ante,* at 1206, neither the majority decision nor the dissent in *California Federal Savings and Loan Assn. v. Guerra,* 479 U.S. 272, 107 S.Ct. 683, 93 L.Ed.2d 613 (1987), is relevant to the issue whether the PDA altered the BFOQ standard for pregnancy-related discrimination. In that case, the Court held that the PDA did not pre-empt a state law requiring employers to provide leave and reinstatement to pregnant employees. The Court reasoned that the PDA was not intended to prohibit all employment practices that favor pregnant women. *Id.,* at 284-290, 107 S.Ct., at 691-694. The dissent disagreed with that conclusion, arguing that the state statute was pre-empted because the PDA's language that pregnant employees "shall be treated the same for all employment-related purposes" appeared to forbid preferential treatment of pregnant workers. *Id.,* at 297-298, 107 S.Ct., at 698-699. Obviously, the dispute in that case between the majority and the dissent was purely over what constituted *discrimination* under Title VII, as amended by the PDA, not over the scope of the BFOQ defense.

**1214 This interpretation is confirmed by the PDA's legislative history. As discussed in *Newport News Shipbuilding & Dry Dock Co. v. EEOC,* 462 U.S. 669, 678-679, and n. 17, 103 S.Ct. 2622, 2628-2629, and n. 17, 77 L.Ed.2d 89 (1983), the PDA was designed to overrule the decision in *General Electric Co. v. Gilbert,* 429 U.S. 125, 97 S.Ct. 401, 50 L.Ed.2d 343 (1976), where the Court *219 had held that "an exclusion of pregnancy from a disability-benefits plan providing general coverage is not a gender-based discrimination at all." *Id.,* at 136, 97 S.Ct., at 408. The PDA thus "makes clear that it is discriminatory to treat pregnancy-related conditions less favorably than other medical conditions." *Newport News, supra,* 462 U.S., at 684, 103 S.Ct., at 2631. It does not, however, alter the standards for employer *defenses.* The Senate Report, for example, stated that the PDA "defines sex discrimination, as proscribed in the existing statute, to include these physiological occurrences [pregnancy, childbirth, and related medical conditions] peculiar to women; *it does not change the application of Title VII to sex discrimination in any other way.*" S.Rep. No. 95-331, pp. 3-4 (1977) (emphasis added). Similarly, the House Report stated that "[p]regnancy-based distinctions will be subject to the same scrutiny *on the same terms* as other acts of sex discrimination proscribed in the existing statute." H.R.Rep. No. 95-948, p. 4 (1978), U.S.Code Cong. & Admin.News 1978, p. 4752 (emphasis added).[FN7]

FN7. Even if the PDA *did* establish a separate BFOQ standard for pregnancy-related discrimination, if a female employee could only perform the duties of her job by imposing substantial safety and liability risks, she would not be "similar in [her] ability or inability to work" as a male employee, under the terms of the PDA. See 42 U.S.C. § 2000e(k).

In enacting the BFOQ standard, "Congress did not ignore the public interest in safety." *Criswell,* 472 U.S., at 419, 105 S.Ct., at 2754. The Court's narrow interpretation of the BFOQ defense in this case, however, means that an employer cannot exclude even *pregnant* women from an environment highly toxic to their fetuses. It is foolish to think that Congress intended such a result, and neither the language of the BFOQ exception nor our cases require it.[FN8]

FN8. The Court's cramped reading of the BFOQ defense is also belied by the legislative history of Title VII, in which three examples of permissible sex discrimination were mentioned—a female nurse hired to care for an elderly woman, an all-male professional baseball team, and a masseur. See 110 Cong.Rec. 2718 (1964) (Rep. Goodell); *id.,* at 7212-7213 (interpretive memorandum introduced by Sens. Clark and Case); *id.,* at 2720 (Rep. Multer). In none of those situations would gender "actually interfer[e] with the employee's ability to perform the job," as required today by the Court, *ante,* at 1206.

The Court's interpretation of the BFOQ standard also would seem to preclude considerations of privacy as a basis for sex-based discrimination, since those considerations do not relate directly to an employee's physical ability to perform the duties of the job. The lower federal courts, however, have consistently recognized that privacy interests may justify sex-based requirements for certain jobs. See, e.g., *Fesel v. Masonic Home of Delaware, Inc.,* 447 F.Supp. 1346 (Del. 1978), aff'd, 591 F.2d 1334 (CA3 1979) (nurse's aide in retirement home); *Jones v. Hinds General Hospital,* 666 F.Supp. 933 (SD Miss.1987) (nursing assistant); *Local 567 American Federation of State, County, and Municipal Employees, AFL-CIO v. Michigan Council 25, American Federation of State, County, and Municipal Employees, AFL-CIO,* 635 F.Supp. 1010 (ED Mich.1986) (mental health workers); *Norwood v. Dale Maintenance System, Inc.,* 590 F.Supp. 1410 (ND Ill.1984) (washroom attendant); *Backus v. Baptist Medical Center,* 510 F.Supp. 1191 (ED Ark.1981) (nursing position in obstetrics and gynecology department of hospital), vacated as moot, 671 F.2d 1100 (CA8 1982).

*220 II

Despite my disagreement with the Court concerning the scope of the BFOQ defense, I concur in reversing the Court of Appeals **1215 because that court erred in affirming the District Court's grant of summary judgment in favor of Johnson Controls. First, the Court of Appeals erred in failing to consider the level of risk avoidance that was part of Johnson Controls' "normal operation." Although the court did conclude that there was a "substantial risk" to fetuses from lead exposure in fertile women, 886 F.2d 871, 879-883, 898 (CA7 1989), it merely meant that there was a high risk that *some* fetal injury would occur absent a fetal-protection policy. That analysis, of course, fails to address the *extent* of fetal injury that is likely to occur.[FN9] If the fetal-protection policy insists on a risk-avoidance level substantially higher than other risk levels *221 tolerated by Johnson Controls such as risks to employees and consumers, the policy should not constitute a BFOQ.[FN10]

FN9. Apparently, between 1979 and 1983, only eight employees at Johnson Controls became pregnant while maintaining high blood lead levels, and only one of the babies born to this group later recorded an elevated blood lead level. See *ante,* at 1199; 886 F.2d, at 876-877.

FN10. It is possible, for example, that alternatives to exclusion of women, such as warnings combined with frequent blood testings, would sufficiently minimize the risk such that it would be comparable to other risks tolerated by Johnson Controls.

Second, even without more information about the normal level of risk at Johnson Controls, the fetal-protection policy at issue here reaches too far. This is evident both in its presumption that, absent medical documentation to the contrary, all women are fertile regardless of their age, see *id.,* at 876, n. 8, and in its exclusion of presumptively fertile women from positions that might result in a promotion to a position involving high lead exposure, *id.,* at 877. There has been no showing that either of those aspects of the policy is reasonably necessary to ensure safe and efficient operation of Johnson Controls' battery-manufacturing business. Of course, these infirmities in the company's policy do not warrant invalidating the entire fetal-protection program.

Third, it should be recalled that until 1982 Johnson Controls operated without an exclusionary policy, and it has not identified any grounds for believing that its current policy is reasonably necessary to its normal operations. Although it is now more aware of some of the dangers of lead exposure, *id.,* at 899, it has not shown that the risks of fetal harm or the costs associated with it have substantially increased. Cf. *Manhart,* 435 U.S., at 716, n. 30, 98 S.Ct., at 1379, n. 30, in which we rejected a BFOQ defense because the employer had operated prior to the discrimination with no significant adverse effects.

Finally, the Court of Appeals failed to consider properly petitioners' evidence of harm to offspring caused by lead exposure in males. The court considered that evidence only in its discussion of the business necessity standard, in which it focused on whether *petitioners* had met their burden of proof. 886 F.2d, at 889-890. The burden of proving that a discriminatory qualification is a BFOQ, however, rests with *222 the employer. See, e.g., *Price Waterhouse,* 490 U.S., at 248, 109 S.Ct., at 1789; *Dothard,* 433 U.S., at 333, 97 S.Ct., at 2728-29. Thus, the court should have analyzed whether the evidence was sufficient for petitioners to survive summary judgment in light of *respondent's* burden of proof to establish a BFOQ. Moreover, the court should not have discounted the evidence as "speculative," 886 F.2d, at 889, merely because it was based on animal studies. We have approved the use of animal studies to assess risks, see *Industrial Union Dept. v. American Petroleum Institute,* 448 U.S. 607, 657, n. 64, 100 S.Ct. 2844, 2871-72, n. 64, 65 L.Ed.2d 1010 (1980), and OSHA uses animal studies in establishing its lead control regulations, see *United Steelworkers of America, AFL-CIO-CLC v. Marshall,* 208 U.S.App.D.C. 60, 128, n. 97, 647 F.2d 1189, 1257, n. 97 (1980), cert. denied, 453 U.S. 913, 101 S.Ct. 3148, 69 L.Ed.2d 997 (1981). It seems clear that if the Court of Appeals had properly analyzed that evidence, it would have concluded that summary judgment against petitioners was not appropriate because **1216 there was a dispute over a material issue of fact.

As Judge Posner observed below:

> "The issue of the legality of fetal-protection is as novel and difficult as it is contentious and the most sensible way to approach it at this early stage is on a case-by-case basis, involving careful examination of the facts as developed by the full adversary process of a trial. The record in this case is too sparse. The district judge jumped the gun. By affirming on this scanty basis we may be encouraging incautious employers to adopt fetal-protection policies that could endanger the jobs of millions of women for minor gains in fetal safety and health.
> "But although the defendant did not present enough evidence to warrant the grant of summary judgment in its favor, there is no ground for barring it from presenting additional evidence at trial. Therefore it would be equally precipitate for us to direct the entry of judgment in the plaintiffs' favor...." 886 F.2d, at 908.

*223 JUSTICE SCALIA, CONCURRING IN THE JUDGMENT.

I generally agree with the Court's analysis, but have some reservations, several of which bear mention.

First, I think it irrelevant that there was "evidence in the record about the debilitating effect of lead exposure on the male reproductive system," *ante,* at 1203. Even without such evidence, treating women differently "on the basis of pregnancy" constitutes discrimination "on the basis of sex," because Congress has unequivocally said so. Pregnancy Discrimination Act, 92 Stat. 2076, 42 U.S.C. § 2000e(k).

Second, the Court points out that "Johnson Controls has shown no factual basis for believing that all or substantially all women would be unable to perform safely ... the duties of the job involved," *ante,* at 1208 (internal quotation marks omitted). In my view, this is not only "somewhat academic in light of our conclusion that the company may not exclude fertile women at all," *ibid.;* it is entirely irrelevant. By reason of the Pregnancy Discrimination Act, it would not matter if all pregnant women placed their children at risk in taking these jobs, just as it does not matter if no men do so. As Judge Easterbrook put it in his dissent below: "Title VII gives parents the power to make occupational decisions affecting their families. A legislative forum is available to those who believe that such decisions should be made elsewhere." 886 F. 2d 871, 915 (CA7 1989).

Third, I am willing to assume, as the Court intimates, *ante,* at 1208-1209, that any action required by Title VII cannot give rise to liability under state tort law. That assumption, however, does not answer the question whether an action *is* required by Title VII (including the BFOQ provision) even if it is subject to liability under state tort law. It is perfectly reasonable to believe that Title VII has *accommodated* state tort law through the BFOQ exception. However, all that need be said in the present case is that Johnson has not demonstrated a substantial risk of tort liability—which is *224 alone enough to defeat a tort-based assertion of the BFOQ exception.

Last, the Court goes far afield, it seems to me, in suggesting that increased cost alone—short of "costs ... so prohibitive as to threaten the survival of the employer's business," *ante,* at 1209—cannot support a bfoq defense. See *ante,* at 1207. I agree with JUSTICE WHITE's concurrence, *ante,* at 1211, that nothing in our prior cases suggests this, and in my view it is wrong. I think, for example, that a shipping company may refuse to hire pregnant women as crew members on long voyages because the on-board facilities for foreseeable emergencies, though quite feasible, would be inordinately expensive. **1217 In the present case, however, Johnson has not asserted a cost-based BFOQ.

I concur in the judgment of the Court.

U.S. Wis., 1991.

International Union, United Auto., Aerospace and Agr. Implement Workers of America, UAW v. Johnson Controls, Inc. 499 U.S. 187, 111 S.Ct. 1196, 55 Fair Empl.Prac.Cas. (BNA) 365, 55 Empl. Prac. Dec. P 40,605, 113 L.Ed.2d 158, 59 USLW 4209, 14 O.S.H. Cas. (BNA) 2102, 1991 O.S.H.D. (CCH) P 29,256

12 The Pregnancy Discrimination Act of 1978

One of the ceaseless wonders of the world: The power of a smile.

Malcolm Forbes

Whenever you lend let it be your money, and not your name. Money you may get again, and, if not, you may contrive to do without it; name once lost, you cannot get again, and, if you cannot contrive to do without it, you had better never have been born.

Edward Bulwer-Lytton

LEARNING OBJECTIVES

1. Acquire an understanding of the requirements of the Pregnancy Discrimination Act.
2. Acquire an understanding of the protections afforded under the Pregnancy Discrimination Act.
3. Acquire an understanding of the methods of ensuring compliance with the Pregnancy Discrimination Act.

Safety professionals should be aware that the Pregnancy Discrimination Act of 1978* is an amendment to Title VII of the Civil Rights Act. The Pregnancy Discrimination Act prohibits companies or organizations from discriminating against pregnant employees as it relates to their employment practices. For safety professionals, the Pregnancy Discrimination Act and the protections it provides is often applicable in situations where a female employee becomes pregnant. However, safety professionals should be aware that the Pregnancy Discrimination Act's protections also extend to provide protection in the areas of childbirth and medical conditions related to pregnancy or childbirth.

Additionally, safety professionals should be aware that the protections can be afforded to employees under the Pregnancy Discrimination Act under conditions beyond pregnancy, including such areas as in vitro fertilization. Safety professionals should also be aware that protections under the Pregnancy Discrimination Act may extend to the spouses of the pregnant female employee.†

---

* 42 U.S.C.A. Section 2000c(k)(as amended by Act, Oct. 31, 1978, Pub. Law 95-555, 92 Stat. 2076).
† Newport News Shipbuilding and Dry Dock Co. v. EEOC, 32 FEP 1 (S.Ct. 1983).

Safety professionals should be aware that the basic requirement of the Pregnancy Discrimination Act is to treat a pregnant employee the same as you would treat any other employee with a temporary disability, ascertaining and accommodating their ability or inability to perform the job.

Generally, safety professionals should be aware that it may not constitute discrimination under the Pregnancy Discrimination Act for a pregnant employee to be denied a leave of absence or an extension on a leave of absence so long as related leave of absence or extension of a leave of absence are the same for male employees for related to leave of absence for sickness or health-related reasons.*

Of particular importance for safety professionals is the issue of termination of leave of absence at predetermined times. Safety professionals should be aware that a company's requirement that a maternity leave start or stop on a specific day, without a showing of business or medical necessity, may constitute a violation of the Pregnancy Discrimination Act and/or Title VII if there is no applicable structure for male employee leave of absence.†

A controversial area that safety professionals should be aware of is that of breast feeding when the mother returns to the job. Although not specifically addressed in the Pregnancy Discrimination Act, safety professionals should be aware of the break-time requirement that is now part of the Fair Labor Standards Act, specifically in Section 4207 of the Patient Protection and Affordable Care Act‡ (also referred to as the "Affordable Care Act"). Safety professionals should be aware that this provision requires companies and organizations to provide "reasonable break time for an employee to express breast milk for her nursing child for 1 year after the child's birth each time such employee has need to express the milk."§ Additionally, safety professionals should be aware that this provision requires the company or organization to provide "a place, other than a bathroom, that is shielded from view and free from intrusion from coworkers and the public, which may be used by an employee to express breast milk."¶ This is relatively new, and safety professionals should be aware that this break-time requirement became effective when the Affordable Care Act was signed into law on March 23, 2010.

***

Pregnancy Discrimination

Pregnancy discrimination involves treating a woman (an applicant or employee) unfavorably because of pregnancy, childbirth, or a medical condition related to pregnancy or childbirth.

---

* 29 CFR Section 1604.10(b).
† See, EEOC Decision No 70-68 (1973).
‡ Public Law 111-148.
§ Id.
¶ See 29 U.S.C. 207(r).

PREGNANCY DISCRIMINATION AND WORK SITUATIONS

The law forbids discrimination when it comes to any aspect of employment, including hiring, firing, pay, job assignments, promotions, layoff, training, fringe benefits, such as leave and health insurance, and any other term or condition of employment.

PREGNANCY DISCRIMINATION AND TEMPORARY DISABILITY

If a woman is temporarily unable to perform her job due to a medical condition related to pregnancy or childbirth, the *employer or other covered entity* must treat her the same as any other temporarily disabled employee. For example, the employer may have to provide modified tasks, alternative assignments, disability leave, or unpaid leave.

PREGNANCY DISCRIMINATION AND HARASSMENT

It is unlawful to harass a woman because of pregnancy, childbirth, or a medical condition related to pregnancy or childbirth.

Although the law doesn't prohibit simple teasing, offhand comments, or isolated incidents that are not very serious, harassment is illegal when it is so frequent or severe that it creates a hostile or offensive work environment or when it results in an adverse employment decision (such as the victim being fired or demoted).

The harasser can be the victim's supervisor, a supervisor in another area, a co-worker, or someone who is not an employee of the employer, such as a client or customer.

PREGNANCY AND WORKPLACE LAWS

Pregnant employees may have additional rights under the Family and Medical Leave Act (FMLA), which is enforced by the U.S. Department of Labor. For example, The Wage and Hour Division released a fact sheet, Break Time for Nursing Mothers, under the FLSA. For more information on FMLA, contact the nearest office of the Wage and Hour Division, U.S. Department of Labor. The Wage and Hour Division can be reached at:

202-693-0051 (voice),
202-693-7755 (TTY), or
US Department of Labor—Wage and Hour Division

PREGNANCY, MATERNITY AND PARENTAL LEAVE

Under federal law, if an employee is temporarily unable to perform her job due to pregnancy or childbirth, the employer must treat her the same as any other temporarily disabled employee. For example, if the employer allows temporarily disabled employees to modify tasks, perform alternative assignments, or take disability leave or leave without pay, the employer also must allow an employee who is temporarily disabled due to pregnancy to do the same.

If an employer provides personal leave for other reasons, for example, to take courses or other training, then the employer must grant personal leave for care of a new child.

An employer may not single out pregnancy-related conditions for special procedures to determine an employee's ability to work. However, if an employer requires its employees to submit a doctor's statement concerning their ability to work before granting leave or paying sick benefits, the employer may require employees affected by pregnancy-related conditions to submit such statements.

Further, under the Family and Medical Leave Act (FMLA) of 1993, a new parent (including foster and adoptive parents) may be eligible for 12 weeks of leave (unpaid or paid if the employee has earned or accrued it) that may be used for care of the new child. To be eligible, the employee must have worked for the employer for 12 months prior to taking the leave and the employer must have a specified number of employees.*

Safety professionals should become familiar with your company or organization's policies regarding compliance with the Pregnancy Discrimination Act, as well as Family and Medical Leave Act, Affordable Care Act, and general leave of absence policy and requirements.

Additionally, safety professional should be well aware of the internal complaint reporting mechanism established by your company. And remember, as discussed in detail in Chapter 5, safety professionals should learn to say "How can I help you?" rather than shooting from the hip with decisions that potentially could impact an employee and their family.

***

THE PREGNANCY DISCRIMINATION ACT OF 1978

AN ACT,

To amend Title VII of the Civil Rights Act of 1964 to prohibit sex discrimination on the basis of pregnancy.

Be it enacted by the Senate and House of Representatives of the United States of America in Congress assembled,

That section 701 of the Civil Rights Act of 1964 is amended by adding at the end thereof the following new subsection:

"(k) The terms 'because of sex' or 'on the basis of sex' include, but are not limited to, because of or on the basis of pregnancy, childbirth, or related medical conditions; and women affected by pregnancy, childbirth, or related medical conditions shall be treated the same for all employment-related purposes, including receipt of benefits under fringe benefit programs, as other persons not so affected but similar in their ability or inability to work, and nothing in section 703(h) of this title shall be interpreted to permit otherwise. This subsection shall not require an employer to pay for health insurance benefits for abortion, except where the life of the mother would be endangered if the fetus were carried to term, or except where medical complications

---

* Pregnancy Discrimination, EEOC website (2011).

have arisen from an abortion: Provided, that nothing herein shall preclude an employer from providing abortion benefits or otherwise affect bargaining agreements in regard to abortion."

Sec. 2. (a) Except as provided in subsection (b), the amendment made by this Act shall be effective on the date of enactment.

(b) The provisions of the amendment made by the first section of this Act shall not apply to any fringe benefit program or fund, or insurance program which is in effect on the date of enactment of this Act until 180 days after enactment of this Act.

Sec. 3. Until the expiration of a period of one year from the date of enactment of this Act or, if there is an applicable collective-bargaining agreement in effect on the date of enactment of this Act, until the termination of that agreement, no person who, on the date of enactment of this Act is providing either by direct payment or by making contributions to a fringe benefit fund or insurance program, benefits in violation with this Act shall, in order to come into compliance with this Act, reduce the benefits or the compensation provided any employee on the date of enactment of this Act, either directly or by failing to provide sufficient contributions to a fringe benefit fund or insurance program: Provided, That where the costs of such benefits on the date of enactment of this Act are apportioned between employers and employees, the payments or contributions required to comply with this Act may be made by employers and employees in the same proportion: And provided further, That nothing in this section shall prevent the readjustment of benefits or compensation for reasons unrelated to compliance with this Act.

Approved October 31, 1978.*

---

* The Pregnancy Discrimination Act of 1978, EEOC website (2011).

CHAPTER QUESTIONS

1. The Pregnancy Discrimination Act provides protection for:
   a. Women who are pregnant
   b. Any employee with a temporary disability
   c. Any employee with a permanent disability
   d. None of the above
2. The employer is required under the Pregnancy Discrimination Act to:
   a. Treat the women differently due to her condition
   b. Treat the woman the same as other employees
   c. Provide the pregnant woman with special consideration
   d. None of the above
3. The Pregnancy Discrimination Act was an amendment to:
   a. ADA
   b. ADAAA

   c. Title VII

   d. None of the above

4. The Affordable Care Act addresses:

   a. Pregnancy leave of absence

   b. Extensions on pregnancy leave time

   c. Breastfeeding

   d. None of the above

Answers: 1—a; 2—b; 3—c; 4—c

Supreme Court of the United States

Lilly M. LEDBETTER, Petitioner, v. The GOODYEAR TIRE and RUBBER CO., INC.*

No. 05-1074.

Argued November 27, 2006.
Decided May 29, 2007.

Background: Female retiree sued former employer, alleging that sex dis-crimination-based poor performance evaluations she had received earlier in her tenure with employer had resulted in lower pay than her male colleagues through end of her career, and asserting claims under Title VII and Equal Pay Act. The United States District Court for the Northern District of Alabama, U.W. Clemon, C.J., granted summary judgment for employer on Equal Pay Act claim, but entered judgment on jury verdict for retiree on Title VII claim. The United States Court of Appeals for the Eleventh Circuit reversed, 421 F.3d 1169. Certiorari was granted.

    Holding: The United States Supreme Court, Justice Alito, held that dis-crete discriminatory acts triggering time limit for filing Equal Employment Opportunity Commission (EEOC) charge could only be discriminatory pay decisions, abrogating *Forsyth v. Federation Employment and Guidance Serv.,* 409 F.3d 565.

    Affirmed.

    Justice Ginsburg filed dissenting opinion joined by Justices Stevens, Souter, and Breyer.

**2163 *618 SYLLABUS[FN*]

FN* The syllabus constitutes no part of the opinion of the Court but has been prepared by the Reporter of Decisions for the convenience of the reader. See *United States v. Detroit Timber and Lumber Co.,* 200 U.S. 321, 337, 26 S.Ct. 282, 50 L.Ed. 499.

---

* From Westlaw, modified for the purposes of this text.

During most of the time that petitioner Ledbetter was employed by respondent Goodyear, salaried employees at the plant where she worked were given or denied raises based on performance evaluations. Ledbetter submitted a questionnaire to the Equal Employment Opportunity Commission (EEOC) in March 1998 and a formal EEOC charge in July 1998. After her November 1998 retirement, she filed suit, asserting, among other things, a sex discrimination claim under Title VII of the Civil Rights Act of 1964. The District Court allowed her Title VII pay discrimination claim to proceed to trial. There, Ledbetter alleged that several supervisors had in the past given her poor evaluations because of her sex; that as a result, her pay had not increased as much as it would have if she had been evaluated fairly; that those past pay decisions affected the amount of her pay throughout her employment; and that by the end of her employment, she was earning significantly less than her male colleagues. Goodyear maintained that the evaluations had been nondiscriminatory, but the jury found for Ledbetter, awarding back pay and damages. On appeal, Goodyear contended that the pay discrimination claim was time barred with regard to all pay decisions made before September 26, 1997—180 days before Ledbetter filed her EEOC questionnaire—and that no discriminatory act relating to her pay occurred after that date. The Eleventh Circuit reversed, holding that a Title VII pay discrimination claim cannot be based on allegedly discriminatory events that occurred before the last pay decision that affected the employee's pay during the EEOC charging period, and concluding that there was insufficient evidence to prove that Goodyear had acted with discriminatory intent in making the only two pay decisions during that period, denials of raises in 1997 and 1998.

Held: Because the later effects of past discrimination do not restart the clock for filing an EEOC charge, Ledbetter's claim is untimely. Pp. 2166–2178.

(a) An individual wishing to bring a Title VII lawsuit must first file an EEOC charge within, as relevant here, 180 days "after the alleged unlawful employment practice occurred." 42 U.S.C. § 2000e-2(a)(1). In addressing the issue of an EEOC charge's timeliness, this Court has stressed the need to identify with care the specific employment practice *619 at issue. Ledbetter's arguments—that the paychecks that she received during the charging period and the 1998 raise denial each violated Title VII and triggered a new EEOC charging period—fail because they would require the Court in effect to jettison the defining element of the disparate-treatment claim on which her Title VII **2164 recovery was based, discriminatory intent. *United Air Lines, Inc. v. Evans,* 431 U.S. 553, 97 S.Ct. 1885, 52 L.Ed.2d 571, *Delaware State College v. Ricks,* 449 U.S. 250, 101 S.Ct. 498, 66 L.Ed.2d 431, *Lorance v. AT&T Technologies, Inc.,* 490 U.S. 900, 109 S.Ct. 2261, 104 L.Ed.2d 961, and *National Railroad Passenger Corporation v. Morgan,* 536 U.S. 101, 122 S.Ct. 2061, 153 L.Ed.2d 106, clearly instruct that the EEOC charging period is triggered when a discrete unlawful practice takes place. A new violation does not occur, and a new charging period does not commence, upon the occurrence of subsequent nondiscriminatory acts that entail adverse effects resulting from the past discrimination. But if an employer engages in a series of separately actionable intentionally discriminatory acts, then a fresh violation takes place when each act is committed. Ledbetter makes no claim that intentionally discriminatory conduct occurred during the charging period or that discriminatory decisions occurring before that period were not communicated to her. She argues simply that Goodyear's nondiscriminatory conduct during the charging period gave present effect to discriminatory conduct outside of that period. But current effects alone cannot breathe life into prior, uncharged discrimination. Ledbetter should have filed an EEOC charge within 180 days after each allegedly discriminatory employment decision was made and communicated to her. Her attempt to shift forward the intent associated with prior discriminatory acts to the 1998 pay decision is unsound, for it would shift intent away from the act that consummated the discriminatory employment practice to a later act not performed with bias or discriminatory motive, imposing liability in the absence of the requisite intent. Her argument would also distort Title VII's "integrated, multistep enforcement procedure." *Occidental Life Ins. Co. of Cal. v. EEOC,* 432 U.S. 355, 359, 97 S.Ct. 2447, 53 L.Ed.2d 402. The short EEOC filing deadline reflects Congress' strong preference for the prompt resolution of employment discrimination allegations through voluntary conciliation and cooperation. *Id.,* at 367-368, 97 S.Ct. 2447. Nothing in Title VII supports treating the intent element of Ledbetter's disparate-treatment claim any differently from the employment practice element of the claim. Pp. 2166–2172.

(b) *Bazemore v. Friday,* 478 U.S. 385, 106 S.Ct. 3000, 92 L.Ed.2d 315 *(per curium),* which concerned a disparate-treatment pay claim, is entirely consistent with *Evans, Ricks, Lorance,* and *Morgan. Bazemore's* rule is that an employer violates Title VII and triggers a new EEOC charging period whenever the employer issues paychecks using a discriminatory pay structure. It is not, as Ledbetter contends, a "paycheck accrual rule" under which each paycheck, even if not accompanied by discriminatory intent, triggers a *620 new EEOC charging period during which the complainant may properly challenge any prior discriminatory conduct that impacted that paycheck's amount, no matter how long ago the discrimination occurred. Because Ledbetter has not adduced evidence that Goodyear initially adopted its performance-based pay system in order to discriminate based on sex or that it later applied this system to her within the charging period with discriminatory animus, *Bazemore* is of no help to her. Pp. 2171–2176.

(c) Ledbetter's "paycheck accrual rule" is also not supported by either analogies to the statutory regimes of the Equal Pay Act of 1963, the Fair Labor Standards Act of 1938, or the National Labor Relations Act, or policy arguments for giving special treatment to pay claims. Pp. 2176–2178.

421 F.3d 1169, affirmed.

**2165 ALITO, J., delivered the opinion of the Court, in which ROBERTS, C.J., and SCALIA, KENNEDY, and THOMAS, JJ., joined. GINSBURG, J., filed a dissenting opinion, in which STEVENS, SOUTER, and BREYER, JJ., joined.

JUSTICE **ALITO** DELIVERED THE OPINION OF THE COURT.

[1] *621 This case calls upon us to apply established precedent in a slightly different context. We have previously held that the time for filing a charge of employment discrimination with the Equal Employment Opportunity Commission (EEOC) begins when the discriminatory act occurs. We have explained that this rule applies to any "[d]iscrete ac[t]" of discrimination, including discrimination in "termination, failure to promote, denial of transfer, [and] refusal to hire." *National Railroad Passenger Corporation v. Morgan,* 536 U.S. 101, 114, 122 S.Ct. 2061, 153 L.Ed.2d 106 (2002). Because a pay-setting decision is a "discrete act," it follows that the period for filing an EEOC charge begins when the act occurs. Petitioner, having abandoned her claim under the Equal Pay Act, asks us to deviate from our prior decisions in order to permit her to assert her claim under Title VII. Petitioner also contends that discrimination in pay is different from other types of employment discrimination and thus should be governed by a different rule. But because a pay-setting decision is a discrete act that occurs at a particular point in time, these arguments must be rejected. We therefore affirm the judgment of the Court of Appeals.

I

Petitioner Lilly Ledbetter (Ledbetter) worked for respondent Goodyear Tire and Rubber Company (Goodyear) at its Gadsden, Alabama, plant from 1979 until 1998. During much of this time, salaried employees at the plant were given or denied raises based on their supervisors' evaluation of their performance. In March 1998, Ledbetter submitted a questionnaire to the EEOC alleging certain acts of sex discrimination, and in July of that year she filed a formal EEOC charge. After taking early retirement in November 1998, *622 Ledbetter commenced this action, in which she asserted, among other claims, a Title VII pay discrimination claim and a claim under the Equal Pay Act of 1963 (EPA), 77 Stat. 56, 29 U.S.C. § 206(d).

The District Court granted summary judgment in favor of Goodyear on several of Ledbetter's claims, including her Equal Pay Act claim, but allowed others, including her Title VII pay discrimination claim, to proceed to trial. In support of this latter claim, Ledbetter introduced evidence that during the course of her employment**2166 several supervisors had given her poor evaluations because of her sex, that as a result of these evaluations her pay was not increased as much as it would have been if she had been evaluated fairly, and that these past pay decisions continued to affect the amount of her pay throughout her employment. Toward the end of her time with Goodyear, she was being paid significantly less than any of her male colleagues. Goodyear maintained that the evaluations had been nondiscriminatory, but the jury found for Ledbetter and awarded her back pay and damages.

On appeal, Goodyear contended that Ledbetter's pay discrimination claim was time barred with respect to all pay decisions made prior to September 26, 1997—that is, 180 days before the filing of her EEOC questionnaire.[FN1] And Goodyear argued that no discriminatory act relating to Ledbetter's pay occurred after that date.

FN1. The parties assume that the EEOC charging period runs backwards from the date of the questionnaire, even though Ledbetter's discriminatory pay claim was not added until the July 1998 formal charge. 421 F.3d 1169, 1178 (C.A.11 2005). We likewise assume for the sake of argument that the filing of the questionnaire, rather than the formal charge, is the appropriate date.

The Court of Appeals for the Eleventh Circuit reversed, holding that a Title VII pay discrimination claim cannot be based on any pay decision that occurred prior to the last pay decision that affected the employee's pay during the EEOC *623 charging period. 421 F.3d 1169, 1182-1183 (2005). The Court of Appeals then concluded that there was insufficient evidence to prove that Goodyear had acted with discriminatory intent in making the only two pay decisions that occurred within that time span, namely, a decision made in 1997 to deny Ledbetter a raise and a similar decision made in 1998. *Id.,* at 1186-1187.

Ledbetter filed a petition for a writ of certiorari but did not seek review of the Court of Appeals' holdings regarding the sufficiency of the evidence in relation to the 1997 and 1998 pay decisions. Rather, she sought review of the following question:

> "Whether and under what circumstances a plaintiff may bring an action under Title VII of the Civil Rights Act of 1964 alleging illegal pay discrimination when the disparate pay is received during the statutory limitations period, but is the result of intentionally discriminatory pay decisions that occurred outside the limitations period." Pet. for Cert. i.

In light of disagreement among the Courts of Appeals as to the proper application of the limitations period in Title VII disparate-treatment pay cases, compare 421 F.3d 1169, with *Forsyth v. Federation Employment and Guidance Serv.,* 409 F.3d 565 (C.A.2 2005); *Shea v. Rice,* 409 F.3d 448 (C.A.D.C.2005), we granted certiorari, 548 U.S. 903, 126 S.Ct. 2965, 165 L.Ed.2d 949 (2006).

II

Title VII of the Civil Rights Act of 1964 makes it an "unlawful employment practice" to discriminate "against any individual with respect to his compensation ... because of such individual's ... sex." 42 U.S.C. § 2000e-2(a)(1). An individual wishing to challenge an employment practice under this provision must first file a charge with the EEOC. § 2000e-5(e)(1). Such a charge must be filed within a specified period (either 180 or 300 days, depending on the State) *624 "after the alleged unlawful employment practice occurred," *ibid.,* and if the employee does not submit a timely EEOC charge, the employee may **2167 not challenge that practice in court, § 2000e-5(f)(1).

[2] In addressing the issue whether an EEOC charge was filed on time, we have stressed the need to identify with care the specific employment practice

that is at issue. *Morgan,* 536 U.S., at 110-111, 122 S.Ct. 2061. Ledbetter points to two different employment practices as possible candidates. Primarily, she urges us to focus on the paychecks that were issued to her during the EEOC charging period (the 180-day period preceding the filing of her EEOC questionnaire), each of which, she contends, was a separate act of discrimination. Alternatively, Ledbetter directs us to the 1998 decision denying her a raise, and she argues that this decision was "unlawful because it carried forward intentionally discriminatory disparities from prior years." Reply Brief for Petitioner 20. Both of these arguments fail because they would require us in effect to jettison the defining element of the legal claim on which her Title VII recovery was based.

Ledbetter asserted disparate treatment, the central element of which is discriminatory intent. See *Chardon v. Fernandez,* 454 U.S. 6, 8, 102 S.Ct. 28, 70 L.Ed.2d 6 (1981) *(per curium); Teamsters* v . *United States,* 431 U.S. 324, 335, n. 15, 97 S.Ct. 1843, 52 L.Ed.2d 396 (1977); *Watson v. Fort Worth Bank and Trust,* 487 U.S. 977, 1002, 108 S.Ct. 2777, 101 L.Ed.2d 827 (1988) (Blackmun, J., joined by Brennan, and Marshall, JJ., concurring in part and concurring in judgment) ("[A] disparate-treatment challenge focuses exclusively on the intent of the employer"). However, Ledbetter does not assert that the relevant Goodyear decision makers acted with actual discriminatory intent either when they issued her checks during the EEOC charging period or when they denied her a raise in 1998. Rather, she argues that the paychecks were unlawful because they would have been larger if she had been evaluated in a nondiscriminatory manner *prior to* the EEOC charging period. Brief for Petitioner 22. Similarly, she maintains that the *625 1998 decision was unlawful because it "carried forward" the effects of prior, uncharged discrimination decisions. Reply Brief for Petitioner 20. In essence, she suggests that it is sufficient that discriminatory acts that occurred prior to the charging period had continuing effects during that period. Brief for Petitioner 13 ("[E]ach paycheck that offers a woman less pay than a similarly situated man because of her sex is a separate violation of Title VII with its own limitations period, regardless of whether the paycheck simply implements a prior discriminatory decision made outside the limitations period"); see also Reply Brief for Petitioner 20. This argument is squarely foreclosed by our precedents.

In *United Air Lines, Inc. v. Evans,* 431 U.S. 553, 97 S.Ct. 1885, 52 L.Ed.2d 571 (1977), we rejected an argument that is basically the same as Ledbetter's. Evans was forced to resign because the airline refused to employ married flight attendants, but she did not file an EEOC charge regarding her termination. Some years later, the airline rehired her but treated her as a new employee for seniority purposes. *Id.,* at 554-555, 97 S.Ct. 1885. Evans then sued, arguing that, while any suit based on the original discrimination was time barred, the airline's refusal to give her credit for her prior service gave "present effect to [its] past illegal act and thereby perpetuate[d] the consequences of forbidden discrimination." *Id.,* at 557, 97 S.Ct. 1885.

We agreed with Evans that the airline's "seniority system [did] indeed have a continuing impact on her pay and fringe benefits," *id.,* at 558, 97 S.Ct. 1885, but we noted that "the critical question [was] **2168 whether any present *violation* exist[ed]." *Ibid.* (emphasis in original). We concluded that the continuing effects of the precharging period discrimination did not make out a present violation. As Justice STEVENS wrote for the Court:

> "United was entitled to treat [Evans' termination] as lawful after respondent failed to file a charge of discrimination within the 90 days then allowed by § 706(d). A discriminatory act which is not made the basis for a *626 timely charge ... is merely an unfortunate event in history which has no present legal consequences." *Ibid.*

It would be difficult to speak to the point more directly.

Equally instructive is *Delaware State College v. Ricks,* 449 U.S. 250, 101 S.Ct. 498, 66 L.Ed.2d 431 (1980), which concerned a college professor, Ricks, who alleged that he had been discharged because of national origin. In March 1974, Ricks was denied tenure, but he was given a final, nonrenewable 1-year contract that expired on June 30, 1975. *Id.,* at 252-253, 101 S.Ct. 498. Ricks delayed filing a charge with the EEOC until April 1975, *id.,* at 254, 101 S.Ct. 498, but he argued that the EEOC charging period ran from the date of his actual termination rather than from the date when tenure was denied. In rejecting this argument, we recognized that "one of the *effects* of the denial of tenure," namely, his ultimate termination, "did not occur until later." *Id.,* at 258, 101 S.Ct. 498 (emphasis in original). But because Ricks failed to identify any specific discriminatory act "that continued until, or occurred at the time of, the actual termination of his employment," *id.,* at 257, 101 S.Ct. 498, we held that the EEOC charging period ran from "the time the tenure decision was made and communicated to Ricks," *id.,* at 258, 101 S.Ct. 498.

This same approach dictated the outcome in *Lorance v. AT&T Technologies, Inc.,* 490 U.S. 900, 109 S.Ct. 2261, 104 L.Ed.2d 961 (1989), which grew out of a change in the way in which seniority was calculated under a collective-bargaining agreement. Before 1979, all employees at the plant in question accrued seniority based simply on years of employment at the plant. In 1979, a new agreement made seniority for workers in the more highly paid (and traditionally male) position of "tester" depend on time spent in that position alone and not in other positions in the plant. Several years later, when female testers were laid off due to low seniority as calculated under the new provision, they filed an EEOC charge alleging that the 1979 scheme had been adopted with discriminatory intent, namely, to protect incumbent male testers when women with substantial *627 plant seniority began to move into the traditionally male tester positions. *Id.,* at 902-903, 109 S.Ct. 2261.

We held that the plaintiffs' EEOC charge was not timely because it was not filed within the specified period after the adoption in 1979 of the new seniority rule. We noted that the plaintiffs had not alleged that the new seniority rule treated men and women differently or that the rule had been applied in a discriminatory manner. Rather, their complaint was that the rule was adopted originally with discriminatory intent. *Id.,* at 905, 109 S.Ct. 2261. And as in *Evans* and *Ricks,* we held that the EEOC charging period ran from the time when the discrete act of alleged intentional discrimination occurred, not from the date when the effects of this practice were felt. 490 U.S., at 907-908, 109 S.Ct. 2261. We stated:

> "Because the claimed invalidity of the facially nondiscriminatory and neutrally applied tester seniority system is wholly dependent on the alleged illegality of signing the underlying agreement, it is **2169 the date of that signing which governs the limitations period." *Id.*, at 911[, 109 S.Ct. 2261].[FN2]

FN2. After *Lorance,* Congress amended Title VII to cover the specific situation involved in that case. See 42 U.S.C. § 2000e-5(e)(2) (allowing for Title VII liability arising from an intentionally discriminatory seniority system both at the time of its adoption and at the time of its application). The dissent attaches great significance to this amendment, suggesting that it shows that *Lorance* was wrongly reasoned as an initial matter. *Post,* at 2182–2184 (opinion of GINSBURG, J.). However, the very legislative history cited by the dissent explains that this amendment and the other 1991 Title VII amendments "'*expand[ed]* the scope of relevant civil rights statutes in order to provide adequate protection to victims of discrimination.'" *Post,* at 2183 (emphasis added). For present purposes, what is most important about the amendment in question is that it applied only to the adoption of a discriminatory seniority system, not to other types of employment discrimination. *Evans* and *Ricks,* upon which *Lorance* relied, 490 U.S., at 906-908, 109 S.Ct. 2261, and which employed identical reasoning, were left in place, and these decisions are more than sufficient to support our holding today.

*628 Our most recent decision in this area confirms this understanding. In *Morgan,* we explained that the statutory term "employment practice" generally refers to "a discrete act or single 'occurrence'" that takes place at a particular point in time. 536 U.S., at 110-111, 122 S.Ct. 2061. We pointed to "termination, failure to promote, denial of transfer, [and] refusal to hire" as examples of such "discrete" acts, and we held that a Title VII plaintiff "can only file a charge to cover discrete acts that 'occurred' within the appropriate time period." *Id.*, at 114, 122 S.Ct. 2061.

[3] The instruction provided by *Evans, Ricks, Lorance,* and *Morgan* is clear. The EEOC charging period is triggered when a discrete unlawful practice takes place. A new violation does not occur, and a new charging period does not commence, upon the occurrence of subsequent nondiscriminatory acts that entail adverse effects resulting from the past discrimination. But of course, if an employer engages in a series of acts each of which is intentionally discriminatory, then a fresh violation takes place when each act is committed. See *Morgan, supra,* at 113, 122 S.Ct. 2061.

Ledbetter's arguments here—that the paychecks that she received during the charging period and the 1998 raise denial each violated Title VII and triggered a new EEOC charging period—cannot be reconciled with *Evans, Ricks, Lorance,* and *Morgan.* Ledbetter, as noted, makes no claim that intentionally discriminatory conduct occurred during the charging period or that discriminatory decisions that occurred prior to that period were not communicated to her. Instead, she argues simply that Goodyear's conduct during the charging period gave present effect to discriminatory conduct outside of that period. Brief for Petitioner 13. But current effects alone cannot breathe life into prior, uncharged discrimination; as we held in *Evans,* such effects in themselves have "no present legal consequences." 431 U.S., at 558, 97 S.Ct. 1885. Ledbetter should have filed an EEOC charge within 180 days after each allegedly discriminatory pay decision was made and communicated to her. She did not do so, *629 and the paychecks that were issued to her during the 180 days prior to the filing of her EEOC charge do not provide a basis for overcoming that prior failure.

In an effort to circumvent the need to prove discriminatory intent during the charging period, Ledbetter relies on the intent associated with other decisions made by other persons at other times. Reply Brief for Petitioner 6 ("Intentional discrimination ... occurs when ... differential treatment takes place, even if the **2170 intent to engage in that conduct for a discriminatory purpose was made previously").

Ledbetter's attempt to take the intent associated with the prior pay decisions and shift it to the 1998 pay decision is unsound. It would shift intent from one act (the act that consummates the discriminatory employment practice) to a later act that was not performed with bias or discriminatory motive. The effect of this shift would be to impose liability in the absence of the requisite intent.

Our cases recognize this point. In *Evans,* for example, we did not take the airline's discriminatory intent in 1968, when it discharged the plaintiff because of her sex, and attach that intent to its later act of neutrally applying its seniority rules. Similarly, in *Ricks,* we did not take the discriminatory intent that the college allegedly possessed when it denied Ricks tenure and attach that intent to its subsequent act of terminating his employment when his nonrenewable contract ran out. On the contrary, we held that "the only alleged discrimination occurred—and the filing limitations periods therefore commenced—at the time the tenure decision was made and communicated to Ricks." 449 U.S., at 258, 101 S.Ct. 498.

Not only would Ledbetter's argument effectively eliminate the defining element of her disparate-treatment claim, but it would distort Title VII's "integrated, multistep enforcement procedure." *Occidental Life Ins. Co. of Cal. v. EEOC,* 432 U.S. 355, 359, 97 S.Ct. 2447, 53 L.Ed.2d 402 (1977). We have previously noted the legislative compromises that preceded the enactment of Title VII, *630 *Mohasco Corp. v. Silver,* 447 U.S. 807, 819-821, 100 S.Ct. 2486, 65 L.Ed.2d 532 (1980); *EEOC v. Commercial Office Products Co.,* 486 U.S. 107, 126, 108 S.Ct. 1666, 100 L.Ed.2d 96 (1988) (STEVENS, J., joined by Rehnquist, C.J., and SCALIA, J., dissenting). Respectful of the legislative process that crafted this scheme, we must "give effect to the statute as enacted," *Mohasco, supra,* at 819, 100 S.Ct. 2486, and we have repeatedly rejected suggestions that we extend or truncate Congress' deadlines. See, e.g., *Electrical Workers v. Robbins and Myers, Inc.,* 429 U.S. 229, 236-240, 97 S.Ct. 441, 50 L.Ed.2d 427 (1976) (union grievance procedures do not toll EEOC filing deadline); *Alexander v. Gardner-Denver Co.,* 415 U.S. 36, 47-49, 94 S.Ct. 1011, 39 L.Ed.2d 147 (1974) (arbitral decisions do not foreclose access to court following a timely filed EEOC complaint).

Statutes of limitations serve a policy of repose. *American Pipe & Constr. Co. v. Utah,* 414 U.S. 538, 554-555, 94 S.Ct. 756, 38 L.Ed.2d 713 (1974). They "represent a pervasive legislative judgment that it is unjust to fail to put the adversary on notice to defend within a specified period of time and that 'the right to be free of stale claims in time comes to prevail over the right to prosecute them.' " *United States v. Kubrick,* 444 U.S. 111, 117 [, 100 S.Ct. 352, 62 L.Ed.2d 259] (1979) (quoting *Railroad Telegraphers v. Railway Express Agency, Inc.,* 321 U.S. 342, 349[, 64 S.Ct. 582, 88 L.Ed. 788] (1944)).

The EEOC filing deadline "protect[s] employers from the burden of defending claims arising from employment decisions that are long past." *Ricks, supra,* at 256-257, 101 S.Ct. 498. Certainly, the 180-day EEOC charging deadline, 42 U.S.C. § 2000e-5(e)(1), is short by any measure, but "[b]y choosing what are obviously quite short deadlines, Congress clearly intended to encourage the prompt processing of all charges of employment discrimination." *Mohasco, supra,* at 825, 100 S.Ct. 2486. This short deadline reflects Congress' strong preference for the prompt **2171 resolution of employment discrimination allegations *631 through voluntary conciliation and cooperation. *Occidental Life Ins., supra,* at 367-368, 97 S.Ct. 2447; *Alexander, supra,* at 44, 94 S.Ct. 1011.

A disparate-treatment claim comprises two elements: an employment practice, and discriminatory intent. Nothing in Title VII supports treating the intent element of Ledbetter's claim any differently from the employment practice element.[FN3] If anything, concerns regarding stale claims weigh more heavily with respect to proof of the intent associated with employment practices than with the practices themselves. For example, in a case such as this in which the plaintiff's claim concerns the denial of raises, the employer's challenged acts (the decisions not to increase the employee's pay at the times in question) will almost always be documented and will typically not even be in dispute. By contrast, the employer's intent is almost always disputed, and evidence relating to intent may fade quickly with time. In most disparate-treatment cases, much if not all of the evidence of intent is circumstantial. Thus, the critical issue in a case involving a long-past performance evaluation will often be whether the evaluation was so far off the mark that a sufficient inference of discriminatory intent can be drawn. See *Watson,* 487 U.S., at 1004, 108 S.Ct. 2777 (Blackmun, J., joined by Brennan and Marshall, JJ., concurring in part and concurring in judgment) (noting that in a disparate-treatment claim, the *McDonnell Douglas Corp. v. Green,* 414 U.S. 792 (1973), factors establish discrimination by inference). See also, e.g., *632 *Zhuang v. Datacard Corp.,* 414 F.3d 849 (C.A.8 2005) (rejecting inference of discrimination from performance evaluations); *Cooper v. Southern Co.,* 390 F.3d 695, 732-733 (C.A.11 2004) (same). This can be a subtle determination, and the passage of time may seriously diminish the ability of the parties and the fact finder to reconstruct what actually happened.[FN4]

FN3. Of course, there may be instances where the elements forming a cause of action span more than 180 days. Say, for instance, an employer forms an illegal discriminatory intent towards an employee but does not act on it until 181 days later. The charging period would not begin to run until the employment practice was executed on day 181 because until that point the employee had no cause of action. The act and intent had not yet been joined. Here, by contrast, Ledbetter's cause of action was fully formed and present at the time that the discriminatory employment actions were taken against her, at which point she could have, and should have, sued.

FN4. The dissent dismisses this concern, *post*, at 2185–2186, but this case illustrates the problems created by tardy lawsuits. Ledbetter's claims of sex discrimination turned principally on the misconduct of a single Goodyear supervisor, who, Ledbetter testified, retaliated against her when she rejected his sexual advances during the early 1980s, and did so again in the mid-1990s when he falsified deficiency reports about her work. His misconduct, Ledbetter argues, was "a principal basis for [her] performance evaluation in 1997." Brief for Petitioner 6; see also *id.*, at 5-6, 8, 11 (stressing the same supervisor's misconduct). Yet, by the time of trial, this supervisor had died and therefore could not testify. A timely charge might have permitted his evidence to be weighed contemporaneously.

    Ledbetter contends that employers would be protected by the equitable doctrine of laches, but Congress plainly did not think that laches was sufficient in this context. Indeed, Congress took a diametrically different approach, including in Title VII a provision allowing only a few months in most cases to file a charge with the EEOC. 42 U.S.C. § 2000e-5(e)(1).

    Ultimately, "experience teaches that strict adherence to the procedural requirements specified by the legislature is the best guarantee of evenhanded administration of the law." **2172 *Mohasco,* 447 U.S., at 826, 100 S.Ct. 2486. By operation of §§ 2000e-5(e)(1) and 2000e-5(f)(1), a Title VII "claim is time barred if it is not filed within these time limits." *Morgan,* 536 U.S., at 109, 122 S.Ct. 2061; *Electrical Workers,* 429 U.S., at 236, 97 S.Ct. 441. We therefore reject the suggestion that an employment practice committed with no improper purpose and no discriminatory intent is rendered unlawful nonetheless because it gives some effect to an intentional discriminatory act that occurred outside the charging period. Ledbetter's claim is, for this reason, untimely.

*633 III

A

In advancing her two theories Ledbetter does not seriously contest the logic of *Evans, Ricks, Lorance,* and *Morgan* as set out above, but rather argues that our decision in *Bazemore v. Friday,* 478 U.S. 385, 106 S.Ct. 3000, 92 L.Ed.2d 315 (1986) *(per curium),* requires different treatment of her claim because it relates to pay. Ledbetter focuses specifically on our statement that "[e]ach week's paycheck that delivers less to a black than to a similarly situated white is a wrong actionable under Title VII." *Id.*, at 395, 106 S.Ct. 3000. She argues that in *Bazemore* we adopted a "paycheck accrual rule" under which each paycheck, even if not accompanied by discriminatory intent, triggers a new EEOC charging period during which the complainant may properly challenge any prior discriminatory conduct that impacted the amount of that paycheck, no matter how long ago the discrimination occurred. On this reading, *Bazemore* dispensed with the need to prove actual discriminatory intent in pay cases and, without giving any hint that it was doing so, repudiated the very different approach taken previously in *Evans* and *Ricks*. Ledbetter's interpretation is unsound.

Bazemore concerned a disparate-treatment pay claim brought against the North Carolina Agricultural Extension Service (Service). 478 U.S., at 389-390, 106 S.Ct. 3000. Service employees were originally segregated into "a white branch" and "a 'Negro branch,'" with the latter receiving less pay, but in 1965 the two branches were merged. *Id.,* at 390-391, 106 S.Ct. 3000. After Title VII was extended to public employees in 1972, black employees brought suit claiming that pay disparities attributable to the old dual pay scale persisted. *Id.,* at 391, 106 S.Ct. 3000. The Court of Appeals rejected this claim, which it interpreted to be that the "'discriminatory difference in salaries should have been affirmatively eliminated.'" *Id.,* at 395, 106 S.Ct. 3000.

This Court reversed in a *per curiam* opinion, 478 U.S., at 386-388, 106 S.Ct. 3000, but all of the Members of the Court joined Justice Brennan's *634 separate opinion, see *id.,* at 388, 106 S.Ct. 3000 (opinion concurring in part). Justice Brennan wrote:

> "The error of the Court of Appeals with respect to salary disparities created prior to 1972 and perpetuated thereafter is too obvious to warrant extended discussion: that the Extension Service discriminated with respect to salaries *prior* to the time it was covered by Title VII does not excuse perpetuating that discrimination *after* the Extension Service became covered by Title VII. To hold otherwise would have the effect of exempting from liability those employers who were historically the greatest offenders of the rights of blacks. A pattern or practice that would have constituted a violation of Title VII, but for the fact that the statute had not yet become effective, became a violation upon Title VII's effective date, and to the extent an employer continued to engage in that act **2173 or practice, it is liable under that statute. While recovery may not be permitted for pre-1972 acts of discrimination, to the extent that this discrimination was perpetuated after 1972, liability may be imposed." *Id.,* at 395, 106 S.Ct. 3000 (emphasis in original).

Far from adopting the approach that Ledbetter advances here, this passage made a point that was "too obvious to warrant extended discussion," namely, that when an employer adopts a facially discriminatory pay structure that puts some employees on a lower scale because of race, the employer engages in intentional discrimination whenever it issues a check to one of these disfavored employees. An employer that adopts and intentionally retains such a pay structure can surely be regarded as intending to discriminate on the basis of race as long as the structure is used.

Bazemore thus is entirely consistent with our prior precedents, as Justice Brennan's opinion took care to point out. Noting that *Evans* turned on whether " 'any *present violation* exist[ed],' " Justice Brennan stated that the *Bazemore* *635 plaintiffs were alleging that the defendants "ha[d] not from the date of the

Act forward made all their employment decisions in a wholly nondiscriminatory way," 478 U.S., at 396-397, n. 6, 106 S.Ct. 3000 (emphasis in original; internal quotation marks and brackets omitted)—which is to say that they had engaged in fresh discrimination. Justice Brennan added that the Court's "holding in no sense g[ave] legal effect to the pre-1972 actions, but, consistent with *Evans* ... focuse[d] on the present salary structure, which is illegal if it *is a mere continuation of the pre-1965 discriminatory pay structure.*" *Id.,* at 397, n. 6, 106 S.Ct. 3000 (emphasis added).

The sentence in Justice Brennan's opinion on which Ledbetter chiefly relies comes directly after the passage quoted above, and makes a similarly obvious point:

> "Each week's paycheck that delivers less to a black than to a similarly situated white is a wrong actionable under Title VII, regardless of the fact that this pattern was begun prior to the effective date of Title VII." *Id.,* at 395[, 106 S.Ct. 3000].[FN5]

FN5. That the focus in *Bazemore* was on a current violation, not the carrying forward of a past act of discrimination, was made clearly by the side opinion in the Court of Appeals:

"[T]he majority holds, in effect, that because the pattern of discriminatory salaries here challenged originated before applicable provisions of the Civil Rights Act made their payment illegal, any 'lingering effects' of that earlier pattern cannot (presumably on an indefinitely maintained basis) be considered in assessing a challenge to post-act continuation of that pattern."

"*Hazelwood* and *Evans* indeed made it clear that an employer cannot be found liable, or sanctioned with remedy, for employment decisions made before they were declared illegal or as to which the claimant has lost any right of action by lapse of time. For this reason it is generally true that, as the catch-phrase has it, Title VII imposed 'no obligation to catch-up,' i.e., affirmatively to remedy present effects of pre-Act discrimination, whether in composing a work force or otherwise. But those cases cannot be thought to insulate employment decisions that presently are illegal on the basis that at one time *comparable* decisions were legal when made by the particular employer. It is therefore one thing to say that an employer who upon the effective date of Title VII finds itself with a racially unbalanced work-force need not act affirmatively to redress the balance; and quite another to say that it may also continue to make discriminatory hiring decisions because it was by that means that its present work force was composed. It may not, in short, under the *Hazelwood/Evans* principle continue practices now violative simply because at one time they were not." *Bazemore v. Friday,* 751 F.2d 662, 695-696 (C.A.4 1984) (Phillips, J., concurring in part and dissenting in part) (emphasis in original; footnotes omitted).

**2174 *636 In other words, a freestanding violation may always be charged within its own charging period regardless of its connection to other violations. We repeated this same point more recently in *Morgan:* "The existence of past acts and the employee's prior knowledge of their occurrence ... does not bar employees from filing charges about related discrete acts so long as the acts are independently discriminatory and charges addressing those acts are themselves timely filed." 536 U.S., at 113, 122 S.Ct. 2061.[FN6] Neither of these opinions stands for the proposition that an action not comprising an employment practice and alleged discriminatory intent is separately chargeable, just because it is related to some past act of discrimination.

FN6. The briefs filed with this Court in *Bazemore v. Friday,* 478 U.S. 385, 106 S.Ct. 3000, 92 L.Ed.2d 315 (1986) *(per curium),* further elucidate the point. The petitioners described the Service's conduct as "[t]he continued use of a racially explicit base wage." Brief for Petitioner Bazemore et al. in *Bazemore v. Friday,* O.T.1985, No. 85-93, p. 33. The United States' brief also properly distinguished the commission of a discrete discriminatory act with continuing adverse results from the intentional carrying forward of a discriminatory pay system. Brief for Federal Petitioners in *Bazemore v. Friday,* O.T.1984, Nos. 85-93 and 85-428, p. 17. This case involves the former, not the latter.

Ledbetter attempts to eliminate the obvious inconsistencies between her interpretation of *Bazemore* and the *Evans/Ricks/Lorance/Morgan* line of cases on the ground that none of the latter cases involved pay raises, but the logic of our prior cases is fully applicable to pay cases. To take *Evans* *637 as an example, the employee there was unlawfully terminated; this caused her to lose seniority; and the loss of seniority affected her wages, among other things. 431 U.S., at 555, n. 5, 97 S.Ct. 1885 ("[S]eniority determine[s] a flight attendant's wages; the duration and timing of vacations; rights to retention in the event of layoffs and rights to re-employment thereafter; and rights to preferential selection of flight assignments"). The relationship between past discrimination and adverse present effects was the same in *Evans* as it is here. Thus, the argument that Ledbetter urges us to accept here would necessarily have commanded a different outcome in *Evans.*

[4] *Bazemore* stands for the proposition that an employer violates Title VII and triggers a new EEOC charging period whenever the employer issues paychecks using a discriminatory pay structure. But a new Title VII violation does not occur and a new charging period is not triggered when an employer issues paychecks pursuant to a system that is "facially nondiscriminatory and neutrally applied." *Lorance,* 490 U.S., at 911, 109 S.Ct. 2261. The fact that precharging period discrimination adversely affects the calculation of a neutral factor (like seniority) that is used in determining future pay does not mean that each new paycheck constitutes a new violation and restarts the EEOC charging period.

Because Ledbetter has not adduced evidence that Goodyear initially adopted its performance-based pay system in order to discriminate on the basis of sex or that it later applied this system to her within the charging period with any discriminatory animus, *Bazemore* is of no help to her. Rather, all Ledbetter has alleged is that Goodyear's agents discriminated against her individually in the past and that this discrimination reduced the amount of later paychecks. Because Ledbetter did not file timely EEOC charges relating to her employer's discriminatory pay decisions in the past, she cannot maintain a suit based on that past discrimination at this time.

*638 **2175 B

The dissent also argues that pay claims are different. Its principal argument is that a pay discrimination claim is like a hostile work environment claim because both types of claims are "'based on the cumulative effect of individual acts,'" *post,* at 2180-2181, but this analogy overlooks the critical conceptual distinction between these two types of claims. And although the dissent relies heavily on *Morgan,* the dissent's argument is fundamentally inconsistent with *Morgan's* reasoning.

Morgan distinguished between "discrete" acts of discrimination and a hostile work environment. A discrete act of discrimination is an act that in itself "constitutes a separate actionable 'unlawful employment practice'" and that is temporally distinct. *Morgan,* 536 U.S., at 114, 117, 122 S.Ct. 2061. As examples we identified "termination, failure to promote, denial of transfer, or refusal to hire." *Id.,* at 114, 122 S.Ct. 2061. A hostile work environment, on the other hand, typically comprises a succession of harassing acts, each of which "may not be actionable on its own." In addition, a hostile work environment claim "cannot be said to occur on any particular day." *Id.,* at 115-116, 122 S.Ct. 2061. In other words, the actionable wrong is the environment, not the individual acts that, taken together, create the environment.[FN7]

FN7. Moreover, the proposed hostile salary environment claim would go far beyond *Morgan's* limits. *Morgan* still required at least some of the discriminatorily-motivated acts predicate to a hostile work environment claim to occur within the charging period. 536 U.S., at 117, 122 S.Ct. 2061 ("Provided that *an act contributing to the claim occurs within the filing period,* the entire time period of the hostile environment may be considered by a court" (emphasis added)). But the dissent would permit claims where no one acted in any way with an improper motive during the charging period. *Post,* at 2181, 2186.

[5] Contrary to the dissent's assertion, *post,* at 2180–2181, what Ledbetter alleged was not a single wrong consisting of a succession of acts. Instead, she alleged a series of discrete discriminatory *639 acts, see Brief for Petitioner 13, 15 (arguing that payment of each paycheck constituted a separate violation of Title VII), each of which *was* independently identifiable and actionable, and *Morgan* is perfectly clear that when an employee alleges "serial violations," i.e., a series of actionable wrongs, a timely EEOC charge must be filed with respect to each discrete alleged violation. 536 U.S., at 113, 122 S.Ct. 2061.

While this fundamental misinterpretation of *Morgan* is alone sufficient to show that the dissent's approach must be rejected, it should also be noted that the dissent is coy as to whether it would apply the same rule to all pay discrimination claims or whether it would limit the rule to cases like Ledbetter's, in which multiple discriminatory pay decisions are alleged. The dissent relies on the fact that Ledbetter was allegedly subjected to a series of discriminatory pay decisions over a period of time, and the dissent suggests that she did not realize for some time that she had been victimized. But not all pay cases share these characteristics.

If, as seems likely, the dissent would apply the same rule in all pay cases, then, if a single discriminatory pay decision made 20 years ago continued to affect an employee's pay today, the dissent would presumably hold that the employee could file a timely EEOC charge today. And the dissent would presumably allow this even if the employee had full knowledge of all the circumstances relating to the 20-year-old decision at the time it was made.[FN8] **2176 The dissent, it appears, proposes that we adopt a special rule for pay cases based on the particular characteristics of one case that is *640 certainly not representative of all pay cases and may not even be typical. We refuse to take that approach.

FN8. The dissent admits as much, responding only that an employer could resort to equitable doctrines such as laches. *Post*, at 2186. But first, as we have noted, Congress has already determined that defense to be insufficient. *Supra*, at 2184–2185. Second, it is far from clear that a suit filed under the dissent's theory, alleging that a paycheck paid recently within the charging period was itself a freestanding violation of Title VII because it reflected the effects of 20-year-old discrimination, would even be barred by laches.

IV

In addition to the arguments previously discussed, Ledbetter relies largely on analogies to other statutory regimes and on extra statutory policy arguments to support her "paycheck accrual rule."

A

Ledbetter places significant weight on the EPA, which was enacted contemporaneously with Title VII and prohibits paying unequal wages for equal work because of sex. 29 U.S.C. § 206(d). Stating that "the lower courts routinely hear [EPA] claims challenging pay disparities that first arose outside the limitations period," Ledbetter suggests that we should hold that Title VII is violated each time an employee receives a paycheck that reflects past discrimination. Brief for Petitioner 34-35.

The simple answer to this argument is that the EPA and Title VII are not the same. In particular, the EPA does not require the filing of a charge with the EEOC or proof of intentional discrimination. See § 206(d)(1) (asking only whether the alleged inequality resulted from "any other factor other than sex"). Ledbetter originally asserted an EPA claim, but that claim was dismissed by the District Court and is not before us. If Ledbetter had pursued her EPA claim, she would not face the Title VII obstacles that she now confronts.[FN9]

FN9. The Magistrate Judge recommended dismissal of Ledbetter's EPA claim on the ground that Goodyear had demonstrated that the pay disparity resulted from Ledbetter's consistently weak performance, not her sex. App. to Pet. for Cert. 71a-77a. The Magistrate Judge also recommended dismissing the Title VII disparate-pay claim on the same basis. *Id.,* at 65a-69a. Ledbetter objected to the Magistrate Judge's disposition of the Title VII and EPA claims, arguing that the Magistrate Judge had improperly resolved a disputed factual issue. See Plaintiff's Objections to Magistrate Judge's Report and Recommendation, 1 Record in No. 03-15246-G (CA11), Doc. 32. The District Court sustained this objection as to the "disparate pay" claim, but without specifically mentioning the EPA claim, which had been dismissed by the Magistrate Judge on the same basis. See App.

to Pet. for Cert. 43a-44a. While the record is not entirely clear, it appears that at this point Ledbetter elected to abandon her EPA claim, proceeding to trial with only the Title VII disparate-pay claim, thus giving rise to the dispute the Court must now resolve.

*641 Ledbetter's appeal to the Fair Labor Standards Act of 1938 (FLSA) is equally unavailing. Stating that it is "well established that the statute of limitations for violations of the minimum wage and overtime provisions of the [FLSA] runs anew with each paycheck," Brief for Petitioner 35, Ledbetter urges that the same should be true in a Title VII pay case. Again, however, Ledbetter's argument overlooks the fact that an FLSA minimum wage or overtime claim does not require proof of a specific intent to discriminate. See 29 U.S.C. § 207 (establishing overtime rules); cf. § 255(a) (establishing 2-year statute of limitations for FLSA claims, except for claims of a "willful violation," which may be commenced within 3 years).

**2177 Ledbetter is on firmer ground in suggesting that we look to cases arising under the National Labor Relations Act (NLRA) since the NLRA provided a model for Title VII's remedial provisions and, like Title VII, requires the filing of a timely administrative charge (with the National Labor Relations Board) before suit may be maintained. *Lorance,* 490 U.S., at 909, 109 S.Ct. 2261; *Ford Motor Co. v. EEOC,* 458 U.S. 219, 226, n. 8, 102 S.Ct. 3057, 73 L.Ed.2d 721 (1982). Cf. 29 U.S.C. § 160(b) ("[N]o complaint shall issue based upon any unfair labor practice occurring more than six months prior to the filing of the charge with the Board").

Ledbetter argues that the NLRA's 6-month statute of limitations begins anew for each paycheck reflecting a prior violation of the statute, but our precedents suggest otherwise. In *Machinists v. NLRB,* 362 U.S. 411, 416-417, 80 S.Ct. 822, 4 L.Ed.2d 832 (1960), we *642 held that "where conduct occurring within the limitations period can be charged to be an unfair labor practice only through reliance on an earlier unfair labor practice[,] the use of the earlier unfair labor practice [merely] serves to cloak with illegality that which was otherwise lawful." This interpretation corresponds closely to our analysis in *Evans* and *Ricks* and supports our holding in the present case.

B

Ledbetter, finally, makes a variety of policy arguments in favor of giving the alleged victims of pay discrimination more time before they are required to file a charge with the EEOC. Among other things, she claims that pay discrimination is harder to detect than other forms of employment discrimination.[FN10]

FN10. We have previously declined to address whether Title VII suits are amenable to a discovery rule. *National Railroad Passenger Corporation v. Morgan,* 536 U.S. 101, 114, n. 7, 122 S.Ct. 2061, 153 L.Ed.2d 106 (2002). Because Ledbetter does not argue that such a rule would change the outcome in her case, we have no occasion to address this issue.

We are not in a position to evaluate Ledbetter's policy arguments, and it is not our prerogative to change the way in which Title VII balances the interests of aggrieved employees against the interest in encouraging the "prompt processing of all charges of employment discrimination," *Mohasco,* 447 U.S., at 825, 100 S.Ct. 2486, and the interest in repose.

Ledbetter's policy arguments for giving special treatment to pay claims find no support in the statute and are inconsistent with our precedents.[FN11] We apply the statute as written, *643 and this means that any unlawful employment practice, including those involving compensation, must be presented to the EEOC within the period prescribed by statute.

FN11. Ledbetter argues that the EEOC's endorsement of her approach in its Compliance Manual and in administrative adjudications merits deference. But we have previously declined to extend *Chevron U.S.A. Inc. v. Natural Resources Defense Council, Inc.,* 407 U.S. 837 (1984), deference to the Compliance Manual, *Morgan, supra,* at 111, n. 6, 122 S.Ct. 2061, and similarly decline to defer to the EEOC's adjudicatory positions. The EEOC's views in question are based on its misreading of *Bazemore.* See, e.g., *Amft v. Mineta,* No. 07A40116, 2006 WL 985183, *5 (EEOC Office of Fed. Operations, Apr. 6, 2006); *Albritton v. Postmaster General,* No. 01A44063, 2004 WL 2983682, *2 (EEOC Office of Fed. Operations, Dec. 17, 2004). Agencies have no special claim to deference in their interpretation of our decisions. *Reno v. Bossier Parish School Bd.,* 528 U.S. 320, 336, n. 5, 120 S.Ct. 866, 145 L.Ed.2d 845 (2000). Nor do we see reasonable ambiguity in the statute itself, which makes no distinction between compensation and other sorts of claims and which clearly requires that discrete employment actions alleged to be unlawful be motivated "because of such individual's ... sex." 42 U.S.C. § 2000e-2(a)(1).

**2178 * * *

For these reasons, the judgment of the Court of Appeals for the Eleventh Circuit is affirmed.

It is so ordered.

JUSTICE **GINSBURG,** WITH WHOM JUSTICE **STEVENS,** JUSTICE **SOUTER,** AND JUSTICE **BREYER** JOIN, DISSENTING.

Lilly Ledbetter was a supervisor at Goodyear Tire and Rubber's plant in Gadsden, Alabama, from 1979 until her retirement in 1998. For most of those years, she worked as an area manager, a position largely occupied by men. Initially, Ledbetter's salary was in line with the salaries of men performing substantially similar work. Over time, however, her pay slipped in comparison to the pay of male area managers with equal or less seniority. By the end of 1997, Ledbetter was the only woman working as an area manager, and the pay discrepancy between Ledbetter and her 15 male counterparts was stark: Ledbetter was paid $3,727 per month; the lowest paid male area manager received $4,286 per month, the highest paid, $5,236. See 421 F.3d 1169, 1174 (C.A.11 2005); Brief for Petitioner 4.

Ledbetter launched charges of discrimination before the Equal Employment Opportunity Commission (EEOC) in March 1998. Her formal administrative complaint specified that, in violation of Title VII, Goodyear paid her a

discriminatorily *644 low salary because of her sex. See 42 U.S.C. § 2000e-2(a)
(1) (rendering it unlawful for an employer "to discriminate against any individual with respect to [her] compensation ... because of such individual's ... sex").
That charge was eventually tried to a jury, which found it "more likely than not
that [Goodyear] paid [Ledbetter] a[n] unequal salary because of her sex." App.
102. In accord with the jury's liability determination, the District Court entered
judgment for Ledbetter for back pay and damages, plus counsel fees and costs.

The Court of Appeals for the Eleventh Circuit reversed. Relying on Goodyear's
system of annual merit-based raises, the court held that Ledbetter's claim, in
relevant part, was time barred. 421 F.3d, at 1171, 1182-1183. Title VII provides
that a charge of discrimination "shall be filed within [180] days after the alleged
unlawful employment practice occurred." 42 U.S.C. § 2000e-5(e)(1).[FN1] Ledbetter
charged, and proved at trial, that within the 180-day period, her pay was substantially less than the pay of men doing the same work. Further, she introduced
evidence sufficient to establish that discrimination against female managers at
the Gadsden plant, not performance inadequacies on her part, accounted for the
pay differential. See, e.g., App. 36-47, 51-68, 82-87, 90-98, 112-113. That evidence was unavailing, the Eleventh Circuit held, and the Court today agrees,
because it was incumbent on Ledbetter to file charges year-by-year, each time
Goodyear failed to increase her salary commensurate with the salaries of male
peers. Any annual pay decision not contested immediately (within 180 days), the
Court affirms, becomes grandfathered, a *fait accompli* beyond the province of
Title VII ever to repair.

FN1. If the complainant has first instituted proceedings with a state or local
agency, the filing period is extended to 300 days or 30 days after the denial of
relief by the agency. 42 U.S.C. § 2000e-5(e)(1). Because the 180-day period applies
to Ledbetter's case, that figure will be used throughout. See *ante,* at 2166, 2167.

*645 The Court's insistence on immediate contest overlooks common characteristics of pay discrimination. Pay disparities often occur, as they did in
Ledbetter's case, in small increments; cause to suspect that discrimination is at
work develops only **2179 over time. Comparative pay information, moreover,
is often hidden from the employee's view. Employers may keep under wraps the
pay differentials maintained among supervisors, no less the reasons for those
differentials. Small initial discrepancies may not be seen as meet for a federal
case, particularly when the employee, trying to succeed in a nontraditional environment, is averse to making waves.

Pay disparities are thus significantly different from adverse actions "such as
termination, failure to promote ... or refusal to hire," all involving fully communicated discrete acts, "easy to identify" as discriminatory. See *National
Railroad Passenger Corporation v. Morgan,* 536 U.S. 101, 114, 122 S.Ct. 2061,
153 L.Ed.2d 106 (2002). It is only when the disparity becomes apparent and sizable, for example, through future raises calculated as a percentage of current salaries, that an employee in Ledbetter's situation is likely to comprehend her plight
and, therefore, to complain. Her initial readiness to give her employer the benefit
of the doubt should not preclude her from later challenging the then-current and
continuing payment of a wage depressed on account of her sex.

On questions of time under Title VII, we have identified as the critical inquiries: "What constitutes an 'unlawful employment practice' and when has that practice 'occurred'?" *Id.,* at 110, 122 S.Ct. 2061. Our precedent suggests, and lower courts have overwhelmingly held, that the unlawful practice is the *current payment* of salaries infected by gender-based (or race-based) discrimination— a practice that occurs whenever a paycheck delivers less to a woman than to a similarly situated man. See *Bazemore v. Friday,* 478 U.S. 385, 395, 106 S.Ct. 3000, 92 L.Ed.2d 315 (1986) (Brennan, J., joined by all other Members of the Court, concurring in part).

*646 I

Title VII proscribes as an "unlawful employment practice" discrimination "against any individual with respect to his compensation ... because of such individual's race, color, religion, sex, or national origin." 42 U.S.C. § 2000e-2(a)(1). An individual seeking to challenge an employment practice under this proscription must file a charge with the EEOC within 180 days "after the alleged unlawful employment practice occurred." § 2000e-5(e)(1). See *ante,* at 2166; *supra,* at 2178, n. 1.

Ledbetter's petition presents a question important to the sound application of Title VII: What activity qualifies as an unlawful employment practice in cases of discrimination with respect to compensation. One answer identifies the pay-setting decision, and that decision alone, as the unlawful practice. Under this view, each particular salary-setting decision is discrete from prior and subsequent decisions, and must be challenged within 180 days on pain of forfeiture. Another response counts both the pay-setting decision and the actual payment of a discriminatory wage as unlawful practices. Under this approach, each payment of a wage or salary infected by sex-based discrimination constitutes an unlawful employment practice; prior decisions, outside the 180-day charge-filing period, are not themselves actionable, but they are relevant in determining the lawfulness of conduct within the period. The Court adopts the first view, see *ante,* at 2165, 2166-2167, 2169-2170, but the second is more faithful to precedent, more in tune with the realities of the workplace, and more respectful of Title VII's remedial purpose.

A

In *Bazemore,* we unanimously held that an employer, the North Carolina Agricultural Extension Service, committed an unlawful employment practice each time it **2180 paid black employees less than similarly situated white employees. 478 U.S., at 395, 106 S.Ct. 3000 (opinion of Brennan, J.). Before 1965, the Extension *647 Service was divided into two branches: a white branch and a "Negro branch." *Id.,* at 390, 106 S.Ct. 3000. Employees in the "Negro branch" were paid less than their white counterparts. In response to the Civil Rights Act of 1964, which included Title VII, the State merged the two branches into a single organization, made adjustments to reduce the salary disparity, and began giving annual raises based on nondiscriminatory factors. *Id.,* at 390-391, 394-395, 106 S.Ct. 3000. Nonetheless, "some pre-existing salary disparities continued to linger on." *Id.*, at 394, 106 S.Ct. 3000 (internal quotation marks omitted). We rejected the Court of Appeals' conclusion that the plaintiffs could not prevail because the lingering disparities were simply a continuing effect of a decision lawfully made prior to the effective date of Title VII. See *id.*, at 395-396, 106 S.Ct. 3000. Rather, we reasoned, "[e]ach week's paycheck that delivers less to a black than to a similarly situated white is a wrong actionable under Title VII." *Id.*, at 395, 106 S.Ct. 3000. Paychecks perpetuating past discrimination, we thus recognized, are actionable not simply because they are "related" to a decision made outside the charge-filing period, cf. *ante*, at 2174, but because they discriminate anew each time they issue, see *Bazemore*, 478 U.S., at 395-396, and n. 6, 106 S.Ct. 3000; *Morgan*, 536 U.S., at 111-112, 122 S.Ct. 2061.

Subsequently, in *Morgan*, we set apart, for purposes of Title VII's timely filing requirement, unlawful employment actions of two kinds: "discrete acts" that are "easy to identify" as discriminatory, and acts that recur and are cumulative in impact. See *id.*, at 110, 113-115, 122 S.Ct. 2061. "[A] [d]iscrete ac[t] such as termination, failure to promote, denial of transfer, or refusal to hire," *id.*, at 114, 122 S.Ct. 2061, we explained, "'occur[s]' on the day that it 'happen[s].' A party, therefore, must file a charge within ... 180 ... days of the date of the act or lose the ability to recover for it." *Id.*, at 110, 122 S.Ct. 2061; see *id.*, at 113, 122 S.Ct. 2061 ("[D]iscrete discriminatory acts are not actionable if time barred, even when they are related to acts alleged in timely filed charges. Each discrete discriminatory act starts a new clock for filing charges alleging that act.").

*648 "[D]ifferent in kind from discrete acts," we made clear, are "claims ... based on the cumulative effect of individual acts." *Id.*, at 115, 122 S.Ct. 2061. The *Morgan* decision placed hostile work environment claims in that category. "Their very nature involves repeated conduct." *Ibid.* "The unlawful employment practice" in hostile work environment claims, "cannot be said to occur on any particular day. It occurs over a series of days or perhaps years and, in direct contrast to discrete acts, a single act of harassment may not be actionable on its own." *Ibid.* (internal quotation marks omitted). The persistence of the discriminatory conduct both indicates that management should have known of its existence and produces a cognizable harm. *Ibid.* Because the very nature of the hostile work environment claim involves repeated conduct, "[I]t does not matter, for purposes of the statute, that some of the component acts of the hostile work environment fall outside the statutory time period. Provided that an act contributing to the claim occurs within the filing period, the entire time period of the hostile environment may be considered by a court for the purposes of determining liability." *Id.*, at 117[, 122 S.Ct. 2061].

Consequently, although the unlawful conduct began in the past, "a charge may be **2181 filed at a later date and still encompass the whole." *Ibid.*

Pay disparities, of the kind Ledbetter experienced, have a closer kinship to hostile work environment claims than to charges of a single episode of discrimination. Ledbetter's claim, resembling Morgan's, rested not on one particular paycheck, but on "the cumulative effect of individual acts." See *id.*, at 115, 122 S.Ct. 2061. See also Brief for Petitioner 13, 15-17, and n. 9 (analogizing Ledbetter's claim to the recurring and cumulative harm at issue in *Morgan*); Reply Brief for

Petitioner 13 (distinguishing pay discrimination from "easy to identify" discrete acts (internal quotation marks omitted)). *649 She charged insidious discrimination building up slowly but steadily. See Brief for Petitioner 5-8. Initially in line with the salaries of men performing substantially the same work, Ledbetter's salary fell 15% to 40% behind her male counterparts only after successive evaluations and percentage-based pay adjustments. See *supra,* at 2178. Over time, she alleged and proved, the repetition of pay decisions undervaluing her work gave rise to the current discrimination of which she complained. Though component acts fell outside the charge-filing period, with each new paycheck, Goodyear contributed incrementally to the accumulating harm. See *Morgan,* 536 U.S., at 117, 122 S.Ct. 2061; *Bazemore,* 478 U.S., at 395-396, 106 S.Ct. 3000; cf. *Hanover Shoe, Inc. v. United Shoe Machinery Corp.,* 392 U.S. 481, 502, n. 15, 88 S.Ct. 2224, 20 L.Ed.2d 1231 (1968).[FN2]

FN2. *National Railroad Passenger Corporation v. Morgan,* 536 U.S. 101, 117, 122 S.Ct. 2061, 153 L.Ed.2d 106 (2002), the Court emphasizes, required that "an act contributing to the claim occu[r] within the [charge-]filing period." *Ante,* at 2175, and n. 7 (emphasis deleted; internal quotation marks omitted). Here, each paycheck within the filing period compounded the discrimination Ledbetter encountered, and thus contributed to the "actionable wrong," i.e., the succession of acts composing the pattern of discriminatory pay, of which she complained.

B

The realities of the workplace reveal why the discrimination with respect to compensation that Ledbetter suffered does not fit within the category of singular discrete acts "easy to identify." A worker knows immediately if she is denied a promotion or transfer, if she is fired or refused employment. And promotions, transfers, hirings, and firings are generally public events, known to co-workers. When an employer makes a decision of such open and definitive character, an employee can immediately seek out an explanation and evaluate it for pretext. Compensation disparities, in contrast, are often hidden from sight. It is not unusual, decisions in point illustrate, for management to decline to publish *650 employee pay levels, or for employees to keep private their own salaries. See, e.g., *Goodwin v. General Motors Corp.,* 275 F.3d 1005, 1008-1009 (C.A.10 2002) (plaintiff did not know what her colleagues earned until a printout listing of salaries appeared on her desk, seven years after her starting salary was set lower than her co-workers' salaries); *McMillan v. Massachusetts Soc. for the Prevention of Cruelty to Animals,* 140 F.3d 288, 296 (C.A.1 1998) (plaintiff worked for employer for years before learning of salary disparity published in a newspaper). [FN3] Tellingly, as the record in **2182 this case bears out, Goodyear kept salaries confidential; employees had only limited access to information regarding their colleagues' earnings. App. 56-57, 89.

FN3. See also Bierman and Gely, "Love, Sex and Politics? Sure. Salary? No Way": Workplace Social Norms and the Law, 25 Berkeley J. Emp. and Lab. L. 167, 168, 171 (2004) (one-third of private sector employers have adopted specific rules prohibiting employees from discussing their wages with co-workers; only one in ten employers has adopted a pay openness policy).

The problem of concealed pay discrimination is particularly acute where the disparity arises not because the female employee is flatly denied a raise but because male counterparts are given larger raises. Having received a pay increase, the female employee is unlikely to discern at once that she has experienced an adverse employment decision. She may have little reason even to suspect discrimination until a pattern develops incrementally and she ultimately becomes aware of the disparity. Even if an employee suspects that the reason for a comparatively low raise is not performance but sex (or another protected ground), the amount involved may seem too small, or the employer's intent too ambiguous, to make the issue immediately actionable or winnable.

Further separating pay claims from the discrete employment actions identified in *Morgan*, an employer gains from sex-based pay disparities in a way it does not from a discriminatory denial of promotion, hiring, or transfer. When a *651 male employee is selected over a female for a higher level position, someone still gets the promotion and is paid a higher salary; the employer is not enriched. But when a woman is paid less than a similarly situated man, the employer reduces its costs each time the pay differential is implemented. Furthermore, decisions on promotions, like decisions installing seniority systems, often implicate the interests of third-party employees in a way that pay differentials do not. Cf. *Teamsters v. United States,* 431 U.S. 324, 352-353, 97 S.Ct. 1843, 52 L.Ed.2d 396 (1977) (recognizing that seniority systems involve "vested ... rights of employees" and concluding that Title VII was not intended to "destroy or water down" those rights). Disparate pay, by contrast, can be remedied at any time solely at the expense of the employer who acts in a discriminatory fashion.

C

In light of the significant differences between pay disparities and discrete employment decisions of the type identified in *Morgan,* the cases on which the Court relies hold no sway. See *ante,* at 2167-2170 (discussing *United Air Lines, Inc. v. Evans,* 431 U.S. 553, 97 S.Ct. 1885, 52 L.Ed.2d 571 (1977), *Delaware State College v. Ricks,* 449 U.S. 250, 101 S.Ct. 498, 66 L.Ed.2d 431 (1980), and *Lorance v. AT & T Technologies, Inc.,* 490 U.S. 900, 109 S.Ct. 2261, 104 L.Ed.2d 961 (1989)). *Evans* and *Ricks* both involved a single, immediately identifiable act of discrimination: in *Evans,* a constructive discharge, 431 U.S., at 554, 97 S.Ct. 1885; in *Ricks,* a denial of tenure, 449 U.S., at 252, 101 S.Ct. 498. In each case, the employee filed charges well after the discrete discriminatory act occurred: When United Airlines forced Evans to resign because of its policy barring married female flight attendants, she filed no charge; only four years later, when Evans was rehired, did she allege that the airline's former no-marriage rule was unlawful and therefore should not operate to deny her seniority credit for her prior service. See *Evans,* 431 U.S., at 554-557, 97 S.Ct. 1885. Similarly, when Delaware State College denied Ricks tenure, he did not object until his terminal contract came to an end, one year later. *Ricks,* 449 U.S., at 253-254, 257-258, 101 S.Ct. 498. *652 No repetitive, cumulative discriminatory employment practice was at issue in either case. See **2183 *Evans,* 431 U.S., at 557-558, 97 S.Ct. 1885; *Ricks,* 449 U.S., at 258, 101 S.Ct. 498.[FN4]

FN4. The Court also relies on *Machinists v. NLRB,* 362 U.S. 411, 80 S.Ct. 822, 4 L.Ed.2d 832 (1960), which like *Evans* and *Ricks,* concerned a discrete act: the execution of a collective bargaining agreement containing a union security clause. 362 U.S., at 412, 417, 80 S.Ct. 822. In *Machinists,* it was undisputed that under the National Labor Relations Act (NLRA), a union and an employer may not agree to a union security clause "if at the time of original execution the union does not represent a majority of the employees in the [bargaining] unit." *Id.,* at 412-414, 417, 80 S.Ct. 822. The complainants, however, failed to file a charge within the NLRA's six-month charge filing period; instead, they filed charges 10 and 12 months after the execution of the agreement, objecting to its subsequent enforcement. See *id.,* at 412, 414, 80 S.Ct. 822. Thus, as in *Evans* and *Ricks,* but in contrast to Ledbetter's case, the employment decision at issue was easily identifiable and occurred on a single day.

Lorance is also inapposite, for, in this Court's view, it too involved a one-time discrete act: the adoption of a new seniority system that "had its genesis in sex discrimination." See 490 U.S., at 902, 905, 109 S.Ct. 2261 (internal quotation marks omitted). The Court's extensive reliance on *Lorance, ante,* at 2168–2170, 2172, 2174, moreover, is perplexing for that decision is no longer effective: In the 1991 Civil Rights Act, Congress superseded *Lorance's* holding. § 112, 105 Stat. 1079 (codified as amended at 42 U.S.C. § 2000e-5(e)(2)). Repudiating our judgment that a facially neutral seniority system adopted with discriminatory intent must be challenged immediately, Congress provided:

"For purposes of this section, an unlawful employment practice occurs ... when the seniority system is adopted, when an individual becomes subject to the seniority system, or when a person aggrieved is injured by the application of the seniority system or provision of the system." *Ibid.*

Congress thus agreed with the dissenters in *Lorance* that "the harsh reality of [that] decision," was "glaringly at odds with the purposes of Title VII." *653 490 U.S., at 914, 109 S.Ct. 2261 (opinion of Marshall, J.). See also § 3, 105 Stat. 1071 (1991 Civil Rights Act was designed "to respond to recent decisions of the Supreme Court by expanding the scope of relevant civil rights statutes in order to provide adequate protection to victims of discrimination").

True, § 112 of the 1991 Civil Rights Act directly addressed only seniority systems. See *ante,* at 2169, and n. 2. But Congress made clear (1) its view that this Court had unduly *contracted* the scope of protection afforded by Title VII and other civil rights statutes, and (2) its aim to generalize the ruling in *Bazemore.* As the Senate Report accompanying the proposed Civil Rights Act of 1990, the precursor to the 1991 Act, explained:

"Where, as was alleged in *Lorance,* an employer adopts a rule or decision with an unlawful discriminatory motive, each application of that rule or decision is a new violation of the law. In *Bazemore* ..., for example, ... the Supreme Court properly held that each application of th[e] racially motivated salary structure, i.e., each new paycheck, constituted a distinct violation of Title VII. Section 7(a)(2) generalizes the result correctly reached in *Bazemore*." Civil Rights Act of 1990, S.Rep. No. 101-315, p. 54 (1990).[FN5]

FN5. No Senate Report was submitted with the Civil Rights Act of 1991, which was in all material respects identical to the proposed 1990 Act.

See also 137 Cong. Rec. 29046, 29047 (1991) (Sponsors' Interpretative Memorandum) ("This legislation should be interpreted **2184 as disapproving the extension of *[Lorance]* to contexts outside of seniority systems."). But cf. *ante,* at 2174 (relying on *Lorance* to conclude that "when an employer issues paychecks pursuant to a system that is facially nondiscriminatory and neutrally applied" a new Title VII violation does not occur (internal quotation marks omitted)).

Until today, in the more than 15 years since Congress amended Title VII, the Court had not once relied upon *654 *Lorance*. It is mistaken to do so now. Just as Congress' "goals in enacting Title VII ... never included conferring absolute immunity on discriminatorily adopted seniority systems that survive their first [180] days," 490 U.S., at 914, 109 S.Ct. 2261 (Marshall, J., dissenting), Congress never intended to immunize forever discriminatory pay differentials unchallenged within 180 days of their adoption. This assessment gains weight when one comprehends that even a relatively minor pay disparity will expand exponentially over an employee's working life if raises are set as a percentage of prior pay.

A clue to congressional intent can be found in Title VII's back pay provision. The statute expressly provides that back pay may be awarded for a period of up to two years before the discrimination charge is filed. 42 U.S.C. § 2000e-5(g)(1) ("Back pay liability shall not accrue from a date more than two years prior to the filing of a charge with the Commission."). This prescription indicates that Congress contemplated challenges to pay discrimination commencing before, but continuing into, the 180-day filing period. See *Morgan,* 536 U.S., at 119, 122 S.Ct. 2061 ("If Congress intended to limit liability to conduct occurring in the period within which the party must file the charge, it seems unlikely that Congress would have allowed recovery for two years of back pay."). As we recognized in *Morgan,* "the fact that Congress expressly limited the amount of recoverable damages elsewhere to a particular time period [i.e., two years] indicates that the [180-day] timely filing provision was not meant to serve as a specific limitation ... [on] the conduct that may be considered." *Ibid.*

D

In tune with the realities of wage discrimination, the Courts of Appeals have overwhelmingly judged as a present violation the payment of wages infected by discrimination: Each paycheck less than the amount payable had the employer adhered to a nondiscriminatory compensation regime, courts have held, constitutes a cognizable harm. See, e.g., *655 *Forsyth v. Federation Employment and Guidance Serv.,* 409 F.3d 565, 573 (C.A.2 2005) ("Any paycheck given within the [charge-filing] period ... would be actionable, even if based on a discriminatory pay scale set up outside of the statutory period."); *Shea v. Rice,* 409 F.3d 448, 452-453 (C.A.D.C.2005) ("[An] employer commit[s] a separate unlawful employment practice each time he pa[ys] one employee less than another for a discriminatory reason" (citing *Bazemore,* 478 U.S., at 396, 106 S.Ct. 3000)); *Goodwin v. General Motors Corp.,* 275 F.3d 1005, 1009-1010 (C.A.10 2002) ("*[Bazemore]* has taught a crucial distinction with respect to discriminatory disparities in pay, establishing that a discriminatory salary is not merely a lingering effect of past discrimination—instead it is itself a continually recurring violation .... [E]ach race-based discriminatory salary payment constitutes a fresh violation of Title VII." (footnote omitted)); *Anderson v. Zubieta,* 180 F.3d 329, 335 (C.A.D.C. 1999) ("The Courts of Appeals have repeatedly reached the ... conclusion" that pay discrimination is "actionable upon receipt of each paycheck."); accord **2185 *Hildebrandt v. Illinois Dept. of Natural Resources,* 347 F.3d 1014, 1025-1029 (C.A.7 2003); *Cardenas v. Massey,* 269 F.3d 251, 257 (C.A.3 2001); *Ashley v. Boyle's Famous Corned Beef Co.,* 66 F.3d 164, 167-168 (C.A.8 1995) (en banc); *Brinkley-Obu v. Hughes Training, Inc.,* 36 F.3d 336, 347-349 (C.A.4 1994); *Gibbs v. Pierce County Law Enforcement Support Agency,* 785 F.2d 1396, 1399-1400 (C.A.9 1986).

Similarly in line with the real-world characteristics of pay discrimination, the EEOC—the federal agency responsible for enforcing Title VII, see, e.g., 42 U.S.C. §§ 2000e-5(f), 2000e-12(a)—has interpreted the Act to permit employees to challenge disparate pay each time it is received. The EEOC's Compliance Manual provides that "repeated occurrences of the same discriminatory employment action, such as discriminatory paychecks, can be challenged as long as one discriminatory act occurred within the charge filing period." 2 EEOC Compliance Manual § 2-IV-C(1)(a), p. 605:0024, and n. 183 (2006); cf. *id.,* § 10-III, p. 633:0002 *656 (Title VII requires an employer to eliminate pay disparities attributable to a discriminatory system, even if that system has been discontinued).

The EEOC has given effect to its interpretation in a series of administrative decisions. See *Albritton v. Potter,* No. 01A44063, 2004 WL 2983682, *2 (EEOC Office of Fed. Operations, Dec. 17, 2004) (although disparity arose and employee became aware of the disparity outside the charge-filing period, claim was not time barred because "[e]ach paycheck that complainant receives which is less than that of similarly situated employees outside of her protected classes could support a claim under Title VII if discrimination is found to be the reason for the pay discrepancy." (citing *Bazemore,* 478 U.S., at 396, 106 S.Ct. 3000)). See also *Bynum-Doles v. Winter,* No. 01A53973, 2006 WL 2096290 (EEOC Office of Fed. Operations, July 18, 2006); *Ward v. Potter,* No. 01A60047, 2006 WL 721992 (EEOC Office of Fed. Operations, Mar. 10, 2006). And in this very case, the EEOC urged the Eleventh Circuit to recognize that Ledbetter's failure to challenge any particular pay-setting decision when that decision was made "does not deprive her of the right to seek relief for discriminatory paychecks she received in 1997 and 1998." Brief of EEOC in Support of Petition for Rehearing and Suggestion for Rehearing En Banc, in No. 03-15264-GG (CA11), p. 14 (hereinafter EEOC Brief) (citing *Morgan,* 536 U.S., at 113, 122 S.Ct. 2061).[FN6]

FN6. The Court dismisses the EEOC's considerable "experience and informed judgment," *Firefighters v. Cleveland,* 478 U.S. 501, 518, 106 S.Ct. 3063, 92 L.Ed.2d 405 (1986) (internal quotation marks omitted), as unworthy of any deference in this case, see *ante,* at 2177, n. 11. But the EEOC's interpretations mirror workplace realities and merit at least respectful attention. In any event, the level of deference due the EEOC here is an academic question, for the agency's conclusion that Ledbetter's claim is not time barred is the best reading of the statute even if the Court "were interpreting [Title VII] from scratch." See *Edelman v. Lynchburg College,* 535 U.S. 106, 114, 122 S.Ct. 1145, 152 L.Ed.2d 188 (2002); see *supra,* at 2166–2172.

*657 II

The Court asserts that treating pay discrimination as a discrete act, limited to each particular pay-setting decision, is necessary to "protec[t] employers from the burden of defending claims arising from employment decisions that are long past." *Ante,* at 2170 (quoting *Ricks,* 449 U.S., at 256-257, 101 S.Ct. 498). But the discrimination of which Ledbetter complained is *not* long past. As she alleged, and as the jury found, Goodyear continued to treat **2186 Ledbetter differently because of sex each pay period, with mounting harm. Allowing employees to challenge discrimination "that extend[s] over long periods of time," into the charge-filing period, we have previously explained, "does not leave employers defenseless" against unreasonable or prejudicial delay. *Morgan,* 536 U.S., at 121, 122 S.Ct. 2061. Employers disadvantaged by such delay may raise various defenses. *Id.,* at 122, 122 S.Ct. 2061. Doctrines such as "waiver, estoppel, and equitable tolling" "allow us to honor Title VII's remedial purpose without negating the particular purpose of the filing requirement, to give prompt notice to the employer." *Id.,* at 121, 122 S.Ct. 2061 (quoting *Zipes v. Trans World Airlines, Inc.,* 455 U.S. 385, 398, 102 S.Ct. 1127, 71 L.Ed.2d 234 (1982)); see 536 U.S., at 121, 122 S.Ct. 2061 (defense of laches may be invoked to block an employee's suit "if he unreasonably delays in filing [charges] and as a result harms the defendant"); EEOC Brief 15 ("[I]f Ledbetter unreasonably delayed challenging an earlier decision, and that delay significantly impaired Goodyear's ability to defend itself ... Goodyear can raise a defense of laches ... ").[FN7]

FN7. Further, as the EEOC appropriately recognized in its brief to the Eleventh Circuit, Ledbetter's failure to challenge particular pay raises within the charge-filing period "significantly limit[s] the relief she can seek. By waiting to file a charge, Ledbetter lost her opportunity to seek relief for any discriminatory paychecks she received between 1979 and late 1997." EEOC Brief 14. See also *supra,* at 2184–2185.

In a last-ditch argument, the Court asserts that this dissent would allow a plaintiff to sue on a single decision made *658 20 years ago "even if the employee had full knowledge of all the circumstances relating to the ... decision at the time it was made." *Ante,* at 2175. It suffices to point out that the defenses just noted would make such a suit foolhardy. No sensible judge would tolerate such inexcusable neglect. See *Morgan,* 536 U.S., at 121, 122 S.Ct. 2061 ("In such cases, the federal courts have the discretionary power ... to locate a just result in light of the circumstances peculiar to the case." (internal quotation marks omitted)).

Ledbetter, the Court observes, *ante,* at 2176, n. 9, dropped an alternative remedy she could have pursued: Had she persisted in pressing her claim under the Equal Pay Act of 1963 (EPA), 29 U.S.C. § 206(d), she would not have encountered a time bar.[FN8] See *ante,* at 2176 ("If Ledbetter had pursued her EPA claim, she would not face the Title VII obstacles that she now confronts."); cf. *Corning Glass Works v. Brennan,* 417 U.S. 188, 208-210, 94 S.Ct. 2223, 41 L.Ed.2d 1 (1974). Notably, the EPA provides no relief when the pay discrimination charged is based on race, religion, national origin, age, or disability. Thus, in truncating the Title VII rule this Court announced in *Bazemore,* the Court does not disarm female workers from achieving redress for unequal pay, but it does impede racial and other minorities from gaining similar relief.[FN9]

FN8. Under the EPA 29 U.S.C. § 206(d), which is subject to the Fair Labor Standards Act's time prescriptions, a claim charging denial of equal pay accrues anew with each paycheck. 1 B. Lindemann and P. Grossman, Employment Discrimination Law 529 (3d ed.1996); cf. 29 U.S.C. § 255(a) (prescribing a two-year statute of limitations for violations generally, but a three-year limitation period for willful violations).

FN9. For example, under today's decision, if a black supervisor initially received the same salary as his white colleagues, but annually received smaller raises, there would be no right to sue under Title VII outside the 180-day window following each annual salary change, however strong the cumulative evidence of discrimination might be. The Court would thus force plaintiffs, in many cases, to sue too soon to prevail, while cutting them off as time barred once the pay differential is large enough to enable them to mount a winnable case.

**2187 *659 Furthermore, the difference between the EPA's prohibition against paying unequal wages and Title VII's ban on discrimination with regard to compensation is not as large as the Court's opinion might suggest. See *ante,* at 2176. The key distinction is that Title VII requires a showing of intent. In practical effect, "if the trier of fact is in equipoise about whether the wage differential is motivated by gender discrimination," Title VII compels a verdict for the employer, while the EPA compels a verdict for the plaintiff. 2 C. Sullivan, M. Zimmer, and R. White, Employment Discrimination: Law and Practice § 7.08[F] [3], p. 532 (3d ed.2002). In this case, Ledbetter carried the burden of persuading the jury that the pay disparity she suffered was attributable to intentional sex discrimination. See *supra,* at 2178; *infra,* at 2187.

III

To show how far the Court has strayed from interpretation of Title VII with fidelity to the Act's core purpose, I return to the evidence Ledbetter presented at trial. Ledbetter proved to the jury the following: She was a member of a protected class; she performed work substantially equal to work of the dominant class (men); she was compensated less for that work; and the disparity was attributable to gender-based discrimination. See *supra,* at 2178.

Specifically, Ledbetter's evidence demonstrated that her current pay was discriminatorily low due to a long series of decisions reflecting Goodyear's pervasive discrimination against women managers in general and Ledbetter in particular. Ledbetter's former supervisor, for example, admitted to the jury that Ledbetter's pay, during a particular one-year period, fell below Goodyear's minimum threshold for her position. App. 93-97. Although Goodyear claimed the pay disparity was due to poor performance, the supervisor acknowledged that Ledbetter received a "Top Performance Award" in 1996. *Id.,* at 90-93. The jury also heard testimony that another supervisor—who evaluated Ledbetter in *660 1997 and whose evaluation led to her most recent raise denial—was openly biased against women. *Id.,* at 46, 77-82. And two women who had previously worked as managers at the plant told the jury they had been subject to pervasive discrimination and were paid less than their male counterparts. One was paid less than the men she supervised. *Id.,* at 51-68. Ledbetter herself testified about the discriminatory animus conveyed to her by plant officials. Toward the end of her career, for instance, the plant manager told Ledbetter that the "plant did not need women, that [women] didn't help it, [and] caused problems." *Id.,* at 36.[FN10] After weighing all the evidence, the jury found for Ledbetter, concluding that the pay disparity was due to intentional discrimination.

FN10. Given this abundant evidence, the Court cannot tenably maintain that Ledbetter's case "turned principally on the misconduct of a single Goodyear supervisor." See *ante,* at 2171, n. 4.

Yet, under the Court's decision, the discrimination Ledbetter proved is not redressable under Title VII. Each and every pay decision she did not immediately challenge wiped the slate clean. Consideration may not be given to the cumulative effect of a series of decisions that, together, set her pay well below that of every male area manager. Knowingly carrying past pay **2188 discrimination forward must be treated as lawful conduct. Ledbetter may not be compensated for the lower pay she was in fact receiving when she complained to the EEOC. Nor, were she still employed by Goodyear, could she gain, on the proof she presented at trial, injunctive relief requiring, prospectively, her receipt of the same compensation men receive for substantially similar work. The Court's approbation of these consequences is totally at odds with the robust protection against workplace discrimination Congress intended Title VII to secure. See, e.g., *Teamsters v. United States,* 431 U.S., at 348, 97 S.Ct. 1843 ("The primary purpose of Title VII was to assure equality of employment opportunities and to eliminate ... discriminatory practices*661 and devices ...." (internal quotation marks omitted)); *Albemarle Paper Co. v. Moody,* 422 U.S. 405, 418, 95 S.Ct. 2362, 45 L.Ed.2d 280 (1975) ("It is ... the purpose of Title VII to make persons whole for injuries suffered on account of unlawful employment discrimination.").

This is not the first time the Court has ordered a cramped interpretation of Title VII, incompatible with the statute's broad remedial purpose. See *supra,* at 2183–2184. See also *Wards Cove Packing Co. v. Atonio,* 490 U.S. 642, 109 S.Ct. 2115, 104 L.Ed.2d 733 (1989) (superseded in part by the Civil Rights Act of 1991); *Price Waterhouse v. Hopkins,* 490 U.S. 228, 109 S.Ct. 1775, 104 L.Ed.2d 268 (1989) (plurality opinion) (same); 1 B. Lindemann and P. Grossman,

Employment Discrimination Law 2 (3d ed. 1996) ("A spate of Court decisions in the late 1980s drew congressional fire and resulted in demands for legislative change[,]" culminating in the 1991 Civil Rights Act (footnote omitted)). Once again, the ball is in Congress' court. As in 1991, the Legislature may act to correct this Court's parsimonious reading of Title VII.

* * *

For the reasons stated, I would hold that Ledbetter's claim is not time barred and would reverse the Eleventh Circuit's judgment.

13 Equal Pay Act

Put all the good eggs in one basket and then watch that basket.

Andrew Carnegie

When you come right down to it, almost any problem eventually becomes a financial problem.

Frederic G. Donner

LEARNING OBJECTIVES

1. Acquire an understanding of the requirements of the Equal Pay Act.
2. Acquire an understanding of the claims process, defenses, and remedies under the Equal Pay Act.
3. Acquire an understanding of the Lilly Ledbetter Fair Pay Act.

HYPOTHETICAL SITUATION

As you are conducting your weekly safety inspection, a female employee asks you, "Hey, is this fair? I'm doing the same work as these guys and my hourly rate of pay is a dollar less per hour." How would you respond to this employee and what would you do?

OVERVIEW

Safety professionals should be aware that the Equal Pay Act was an amendment to the Fair Labor Standards Act and became law in 1963.* In general, the Equal Pay Act makes it unlawful for any company or organization to discriminate in the area of wage rates between male and female employees on the basis of their sex while the employees are doing jobs that require equal skill, effort, and responsibility, and which are performed under the same or similar working conditions, unless the wage differential is justified by one of the exceptions identified in the Equal Pay Act.[†] More specifically, the EEOC identifies that "the Equal Pay Act requires that men and women in the same workplace be given equal pay for equal work. The jobs need not be identical, but they must be substantially equal. Job content (not job titles) determines whether jobs are substantially equal. All forms of pay are covered by this law, including salary, overtime pay, bonuses, stock options, profit sharing and bonus plans, life insurance, vacation and holiday pay, cleaning or gasoline allowances, hotel accommodations, reimbursement for travel expenses, and benefits. If

---

* 29 U.S.C.A. Section 206(d).
[†] Id.

there is an inequality in wages between men and women, employers may not reduce the wages of either sex to equalize their pay."*

Safety professionals should be aware that since the Reorganization Plan No 1 took place in 1978, the EEOC became responsible for the administration and enforcement of the Equal Pay Act (formally with the U.S. Department of Labor). In general, safety professionals should be aware that most of the companies and organization that are sufficiently large enough to employ safety professionals have the requisite number of employees to be covered under the Equal Pay Act.

What does the word "equal" mean? Safety professionals should be aware that the term "equal" as applied to the Equal Pay Act has been interpreted by the courts to mean "substantially equal" and not identical.† Generally, when a position is being evaluated for Equal Pay purposes, the four work standards of skill, effort, responsibility, and working conditions are utilized as the basis of the evaluation of the position. In the area of skill, safety professionals should look at the factors such as experience, training education, and ability when assessing an equal pay situation.‡ Additionally, safety professionals should be aware that the evaluation should focus on the performance requirements of the job rather than simply the possession of the skill when measuring the equality of the skill factor.§ For the work standard of effort, the EEOC and most courts usually evaluate and measure both the mental and physical exertion required to meet the job requirements.¶ For the work standard of responsibility, the degree of accountability required in the performance of the job is the primary measurement to determine equality in the job function.** For the working conditions standard, the EEOC and most courts focus on the difference in conditions that are customarily taken into consideration when establishing a wage rate, and whether any wage differential between jobs are based upon a bona fide job evaluation performed by the company or organization.††

Safety professionals should be aware that the Equal Pay Act only prohibits wage differentials for equal job functions based only on sexual differences.‡‡ Wage differences based on other factors such as merit pay, incentive pay, in accordance with a seniority system, shift differentials, or training programs are usually outside of the requirements of the Equal Pay Act. Safety professionals should also be aware that the Equal Pay Act requires all documents related to the payment of wages, wage rates, job evaluations, merit and incentive pay systems, seniority systems to be maintained for a period not less than two years.§§

A good method for safety professionals to consider when ensuring compliance with the Equal Pay Act is the "Splendid" Approach identified next, which was assembled for private sector employers:

---

* Equal Pay Act, EEOC website (2011).
† See, for example, Shultz v. Wheaton Glass Co., 421 F.2d 259 (1970).
‡ 29 CFR Section 800.125.
§ Id.
¶ 29 CFR Section 800.127.
**29 CFR Section 800.129.
†† See, for example, Corning Glass Works v. Brennan, 417 U.S. 188 (1974).
‡‡ 29.CFR Section 900.142.
§§ 20 CFR Section 1620.21©

THE "SPLENDID" APPROACH

STUDY—Since one cannot solve problems that one doesn't know exists, know the law, the standards that define one's obligations, and the various barriers to EEO and diversity. Assistance can be obtained from EEOC, professional consultants, associations or groups, etc.

PLAN—Know one's own circumstances (workforce and demographic—locally, nationally, and globally). Define one's problem(s); propose solutions; and develop strategies for achieving them.

LEAD—Senior, middle, and lower management must champion the cause of diversity as a business imperative, and provide leadership for successful attainment of the vision of a diverse workforce at all levels of management.

ENCOURAGE—Companies should encourage the attainment of diversity by all managers, supervisors, and employees, and structure their business practices and reward systems to reinforce those corporate objectives. Link pay and performance not only for technical competencies, but also for how employees interact, support, and respect each other.

NOTICE—Take notice of the impact of your practices, after monitoring and assessing company progress. Self-analysis is a key part of this process. Ensure that a corrective strategy does not cause or result in unfairness.

DISCUSSION—Communicate and reinforce the message that diversity is a business asset and a key element of business success in a national and global market.

INCLUSION—Bring everyone into this process, including white males. Help them understand that EEO initiatives are good for the company and, thus, good for everyone in the company. Include them in the analysis, planning, and implementation.

DEDICATION—Stay persistent in your quest. Long-term gains from these practices may cost in the short term. Invest the needed human and capital resources.*

Different than other antidiscrimination laws, safety professionals should be aware that an employee may file a complaint with the U.S. Department of Labor, Wage and Hour Division, or with the EEOC.† However, safety professionals should be aware that the EEOC will conduct the investigation and gather the documents from the employer and other sources regarding the wages, hours, employment practices, conditions in order to determine whether or not a violation of the Equal Pay Act has occurred.‡ Safety professionals should also be aware that the EEOC may request to inspect the company's operations, conduct interviews of employees and acquire copies or take notes from various company documents which may be on file at the company or organization.§

Safety professionals should also be aware that the Equal Pay Act is different from other antidiscrimination laws in that an employee may file an action in both

---

* Best Practices of Private Sector Employers, EEOC website (2011).
† 29 CFR Section 1620.21(c).
‡ 29 CFR Section 1620.19.
§ Id.

federal and state court and the EEOC may also bring an action on behalf of the employee against the company or organization.* Additionally, safety professionals should be aware that the employee and the EEOC may pursue a class action type suit against the company or organization with the employee's approval and consent.† Jury trials are permitted in an action under the Equal Pay Act brought by the employee to recover back pay or other damages, however, actions brought by the EEOC are not permitted to have jury trials.‡ Safety professionals should be aware that employees' damages under the Equal Pay Act can include back pay, liquidated damages, attorney's fees, and cost of the action.§ The EEOC may also seek injunctive relief and back pay for the employee; companies or organizations can be exposed to criminal prosecution for willful violations with fines of up to $10,000 and possible imprisonment.¶

In defending an Equal Pay Act action, safety professionals should start with the fact that the initial burden of proof is on the employee or the EEOC. If the EEOC or employee can show that the female employee was paid less than a correlating male employee performing substantially equal work, the burden shifts to the company or organization to prove that the reason for the equal pay differential was justified. Safety professionals should be aware that justification can be in the form of a seniority system, merit pay procedure, training programs, or other systems or factors beyond simply the sex of the employee. Other possible defenses can include, but not be limited to, procedural defects in the action or filings by the employee, good faith reliance on an administrative rule or regulation, or discharge in bankruptcy.

Safety professionals should also be aware of the Lilly Ledbetter Fair Pay Act, which became law in 2009. This law amended Title VII, the ADEA, the ADA, and the Rehabilitation Act to clarify discriminatory practices or decisions affecting compensation are unlawful each time the decision or practice takes place. This law was enacted in response to the U.S. Supreme Court decision in *Ledbetter v. Goodyear Tire & Rubber Company*** case where the Supreme Court found that an employee must file an EEOC charge within 180 days after the alleged unlawful employment practice occurs and later effects of past discrimination do not restart the clock for filing an EEOC charge. Under this new law, employees who may not realize the discriminatory practice was taking place will have 180 days (or 300 days, depending on the state, from the last event to file with the EEOC. Safety professionals should be aware that employees can recover back pay so long as they file within the 180 or 300 day requirements and can acquire back pay for up to two years prior to the filing with the EEOC or state agency.

---

* 29 U.S.C.A. Section 216(b).

† 29 USCA Section 215(b).

‡ See, Altman v. Stevens Fashion Fabrics, 441 F.Supp 1318 (1977) (permitting jury trials); and Brennan v. J.C. Penney Co., 61 FRD 66 (1974) (prohibiting jury trials).

§ 29 U.S.C.A. Section 216(b) and (c).

¶ 29 CFR Section 1620.22.

**550 U.S. 618 (2007).

NOTICE CONCERNING THE LILLY
LEDBETTER FAIR PAY ACT OF 2009

On January 29, 2009, President Obama signed the Lilly Ledbetter Fair Pay Act of 2009 ("Act"), which supersedes the Supreme Court's decision in *Ledbetter v. Goodyear Tire & Rubber Co., Inc.*, 550 U.S. 618 (2007). *Ledbetter* had required a compensation discrimination charge to be filed within 180 days of a discriminatory pay-setting decision (or 300 days in jurisdictions that have a local or state law prohibiting the same form of compensation discrimination).

The Act restores the pre-*Ledbetter* position of the EEOC that each paycheck that delivers discriminatory compensation is a wrong actionable under the federal EEO statutes, regardless of when the discrimination began. As noted in the Act, it recognizes the "reality of wage discrimination" and restores "bedrock principles of American law."

Under the Act, an individual subjected to compensation discrimination under Title VII of the Civil Rights Act of 1964, the Age Discrimination in Employment Act of 1967, or the Americans with Disabilities Act of 1990 may file a charge within 180 (or 300) days of any of the following:

- when a discriminatory compensation decision or other discriminatory practice affecting compensation is adopted;
- when the individual becomes subject to a discriminatory compensation decision or other discriminatory practice affecting compensation; or
- when the individual's compensation is affected by the application of a discriminatory compensation decision or other discriminatory practice, including each time the individual receives compensation that is based in whole or part on such compensation decision or other practice.

The Act has a retroactive effective date of May 28, 2007, and applies to all claims of discriminatory compensation pending on or after that date.*

---

* Notice Concerning Lilly Ledbetter Fair Pay Act of 2009, EEOC website (2011).

Although safety professionals are not often involved in compensation issues with most companies or organizations, it is important for safety professionals to know the requirements of these laws and to identify situations where the laws may be in play in order to proactively notify their companies or organization to take corrective actions prior to any legal actions by the employee.

Safety professionals should be familiar with their company or organizations' reporting policy and procedures and take the necessary and appropriate action any time an employee contacts the safety professional or the safety professional identifies situations which may be in violation of the Equal Pay Act or Lilly Ledbetter Fair Pay Act.

*****

CHAPTER QUESTIONS

1. The Equal Pay Act addresses:
    a. Compensation differentials between male and female employees
    b. Compensation differentials between job functions
    c. The rate of pay for all employees
    d. None of the above
2. Equal job assessments includes evaluation of:
    a. Skills
    b. Responsibilities
    c. Effort
    d. All of the above
3. Lilly Ledbetter Fair Pay Act addresses:
    a. Restores pre-Ledbetter position addressing filing with the EEOC
    b. Payment of future compensation
    c. Balancing of job responsibilities
    d. None of the above
4. Under the Equal Pay Act, jobs need to be:
    a. Identical
    b. Closely aligned
    c. Evaluated by skill, responsibilities, and other factors
    d. None of the above
5. How would you respond to the hypothetical situation above?

Answers: 1—a; 2—d; 3—a; 4—c; 5—various responses possible.

United States Court of Appeals,

Sixth Circuit.

Heather SPEES, Plaintiff-Appellant, v. JAMES MARINE, INC. and JamesBuilt, LLC, Defendants-Appellees.*

No. 09-5839.

Argued: April 28, 2010.
Decided and Filed: August 10, 2010.

Background: Employee brought action against former employer, seeking relief for pregnancy and disability discrimination. The United States District Court

---

* Case from Westlaw and modified for the purposes of this text.

for the Western District of Kentucky, Thomas B. Russell, Chief Judge, 2009 WL 1097559, granted partial summary judgment in favor of employer. Employee appealed.

Holdings: The Court of Appeals, Ronald Lee Gilman, Circuit Judge, held that:

(1) fact issue precluded summary judgment on employee's pregnancy-discrimination claim based on employee's transfer to tool room;

(2) employee failed to establish pregnancy-discrimination claim based on her termination;

(3) fact issue precluded summary judgment on employee's disability claim based on her transfer to tool room; and

(4) employee failed to establish disability claim based on her termination.

Affirmed in part, reversed in part, and remanded.

Zatkoff, District Judge, sitting by designation, filed an opinion concurring in part and dissenting in part.

BEFORE: CLAY AND GILMAN, CIRCUIT JUDGES; ZATKOFF, DISTRICT JUDGE. [FN*]

FN* The Honorable Lawrence P. Zatkoff, United States District Judge for the Eastern District of Michigan, sitting by designation.

*384 GILMAN, J., delivered the opinion of the court, in which CLAY, J., joined. ZATKOFF, D.J. (pp. 399-401), delivered a separate opinion concurring in part and dissenting in part.

OPINION

RONALD LEE GILMAN, CIRCUIT JUDGE.

Shortly after being employed as a welder for James Marine, Inc. (JMI), Heather Spees discovered that she was pregnant. At the direction of her foreman, Spees obtained a note from her physician restricting her to light-duty work, which resulted in JMI reassigning her to a position in the company's tool room. JMI terminated Spees two months later when a second doctor placed her on bed rest for the duration of her pregnancy. Spees then sued JMI and its subsidiary, JamesBuilt, LLC, seeking relief for (among other things) pregnancy and disability discrimination.

The district court granted summary judgment in favor of JMI and James Built on these claims, which Spees now challenges on appeal. For the following reasons, we **AFFIRM** the judgment of the district court with regard to Spees's pregnancy-discrimination claim and her disability-discrimination claim as they pertain to the termination of her employment, **REVERSE** the district court's grant of summary judgment on Spees's pregnancy-discrimination claim and disability-discrimination claim to the extent that they are based on her reassignment to the tool room, and **REMAND** the case for further proceedings on these latter two claims.

I. BACKGROUND

A. FACTUAL BACKGROUND

JMI owns and operates a construction and repair facility for inland waterway vessels on the banks of the Tennessee River near Calvert City, Kentucky. On May 11, 2007, Spees was hired to work at JMI's JamesBuilt facility, which focuses largely on constructing deck and tank barges, towboats, and dry-docks for the river-shipping industry. JamesBuilt, LLC is a subsidiary of JMI, and the two share the same Human Resources Department. (For convenience, JMI and JamesBuilt are hereinafter collectively referred to as JMI.)

Despite having no prior experience working in a manual-labor position, Spees was hired by JMI as a welder. Spees, like other newly hired welders without welding experience, was required by JMI to complete a 30-day in-house training course. She successfully completed the training program and was promoted to a welder-trainee position in early June 2007.

At this point in time, JMI's 935 nonoffice positions were overwhelmingly male, with only four of these positions filled by female employees. Spees was the only female assigned to the JamesBuilt facility.

Welding work at the JamesBuilt facility is physically demanding. It requires heavy lifting, climbing up ladders and stairs, maneuvering into barge tanks, and, occasionally, the overhead handling of equipment. The summer of 2007 was also particularly hot, with temperatures reaching 100 degrees Fahrenheit or more on multiple occasions.

In addition, welders are exposed to fumes, dust, and organic vapors in the course of their work. To limit the inhalation of these substances, JMI provides welders with respirators to wear while on the job. Spees was fitted with a respirator during orientation, although she often opted not to wear it once she became a welder because she "didn't feel like [she] needed one."

*385 Tony Milam, Spees's foreman, described her as "a good employee" and "a good welder," and he "ribbed" other male employees about "her coming in there and welding as good as what she done." Spees enjoyed her work and believed that her supervisors saw her as "a good employee" and "a hardworking employee."

Shortly after she started working at JMI, Spees became pregnant. This was Spees's third pregnancy; she had given birth to a daughter in 1999 and had suffered a miscarriage in 2005. Fearing that the pregnancy would cause her to lose her job, Spees was "hysterical" when she became aware of her condition. By this point, Spees was roughly five to six weeks' pregnant.

Spees's first course of action was to telephone her brother, Christopher Gunder, who was a JMI foreman. Gunder, in turn, recommended that she call Milam. While talking to Milam, Spees expressed her concern that she would be terminated from her position due to the pregnancy. Milam responded by noting that he "had concerns about her being around the chemicals, the welding smoke, [and] climbing around on some of the jobs" while pregnant, and he told her to see a doctor to "find out exactly what she did or didn't need to be doing or be around."

On June 19, 2007, the day following her telephone conversation with Milam, Spees saw Dr. Jorge Cardenas, an obstetrician in Paducah, Kentucky. Dr. Cardenas had been Spees's physician for a number of years, including when she had suffered her miscarriage two years earlier. During her appointment with Dr. Cardenas, Spees discussed her past miscarriage and described her job duties as a welder. Dr. Cardenas replied that "there was no problem" with Spees resuming her work as a welder while pregnant. Although Dr. Cardenas did not know of any health problems that welding fumes could pose for a fetus, he recommended that Spees wear a respirator while working. At the end of her appointment, Spees received a "Certificate to Return to Work" from Dr. Cardenas that did not list any restrictions on her ability to weld.

Spees left her appointment with the intent of returning to work that same day. On her way back to JMI, she called Milam to inform him that she had received clearance from Dr. Cardenas to resume welding. Milam, in the meantime, had discussed with his supervisor Kenneth Colbert the possibility of moving Spees to a nonwelding "light-duty" position. He therefore asked Spees to read him Dr. Cardenas's Certificate to Return to Work. Despite Dr. Cardenas's having cleared Spees to work, Milam believed, based on "common sense," that "there was some questions about her being pregnant and being able to safely perform the job that she was required to do." Milam's concerns were also in part driven by the fact that Spees "had complications with other pregnancies before."

Spees testified that, upon hearing her read Dr. Cardenas's note, Milam told Spees that she "needed something more descriptive or else they were going to get rid" of her. According to Spees, Milam requested that she obtain a second note from Dr. Cardenas mentioning "toxic fumes" and limiting her to "light duty." Milam told Spees that such a note would help her get a transfer to a position in the tool room, thereby allowing her to retain her employment with JMI during her pregnancy.

At Milam's direction, Spees returned to Dr. Cardenas that same day to ask for a second note that limited her to light duty. Dr. Cardenas, complying with her request, wrote her a work order that read "patient requires light duty & avoid [sic] toxic *386 fumes." He testified that although there was no medical reason to limit Spees's job duties, he wrote the note "to allay some of [Spees]'s concerns" and "for the purpose of reducing her anxiety." According to Dr. Cardenas, Spees did not inform him that her superiors at JMI had requested that the note be written.

Spees returned to JMI and showed Milam the second note from Dr. Cardenas that limited her to light duty. Milam informed Spees that he had consulted further with Colbert and that they had already decided that Spees could no longer weld. He did, however, tell Spees that she could work in the tool room, noting that "for right now, we don't know what to do with you." Milam believed that the transfer would be temporary and that Spees could resume welding after the pregnancy. Despite voicing her desire to continue welding, Spees accepted the change.

Gunder also participated in the decision to reassign Spees to the tool room. He testified that, while JMI was deliberating where Spees should be working, he and Milam "went to [Colbert], and we just decided that it wouldn't be a good idea for her to [weld]." Gunder added that he was motivated by concern for the health of his sister's unborn child, and he believed that the job duties of a welder—the "constant[ ] dragging [and] pulling" as well as inhaling the welding fumes— should not be performed by Spees while pregnant.

Spees began working in the tool room on June 20, 2007, the day after she visited Dr. Cardenas. Her primary duties in that capacity were to dispense tools to other employees and to ensure that none of the equipment was lost or stolen. Spees found the job to be as physically demanding as welding, with the working conditions being "just as hot," and her having to lift the same tools and materials as she did when welding. The main difference was that she did not have to per- form any overhead work while in the tool room. Spees received the same salary for her work in the tool room as she did when welding.

Shortly after she began working in the tool room, Spees encountered Tom Freeman, the head of JMI's Safety Department, to whom she had never before spoken. According to Spees, Freeman told her that working at JMI "was not women's work" and that she needed to go back to her doctor to ask for a "descrip- tive note." He added that the descriptive note should specify "everything you can and cannot do," including whether Spees could climb ladders, lift heavy objects, and work in the heat. Freeman then told Spees that "I am your boss," and "I am requesting this." When Spees noted that she had already submitted a work order from Dr. Cardenas, Freeman told her the note "wasn't good enough" and that she "needed to understand that this [is] a man's world."

Freeman's comments upset Spees, causing her to seek out Milam in his office. Milam, upon hearing about Freeman's remarks, told her "I'm your boss. They are trying to fire you. Do not waste your time. I have already been told by [Colbert] that you are not going to weld no matter what your doctor is going to do." He later reprimanded Freeman for his comments to Spees.

Spees worked the daytime shift in the tool room for approximately one week. This assignment was temporary in nature and was made to give JMI time to "get everything straightened out" regarding Spees's pregnancy. Milam then informed Spees that there was a night-shift position available in the tool room, indicating that such a transfer would allow her to maintain her employment with JMI dur- ing the pregnancy.

*387 Acting on this information, Spees went to see Pam DeWeese, an employee in JMI's Human Resources Department. Spees thought that the sched- ule change would be difficult given her status as a single mother, but she also believed that the change was necessary in order to keep her job. During their con- versation, Chad Walker, the Human Resources Director, stopped by DeWeese's office and told Spees that he could fire her because she was not injured at work, but that he was going to work with her and allow her to work in the evening. Following Speess conversation with DeWeese and Walker, the change in shifts was approved and Spees began working nights. Gunder, Spees's brother, was the night foreman at JMI.

Spees worked the night shift in the tool room without incident for approxi- mately one month. But in early August 2007, temperatures in Calvert City reached 106 degrees Fahrenheit, and Spees was experiencing significant swelling due to the heat. She also had vomited on multiple occasions while commuting to work. Spees again visited Dr. Cardenas, who wrote her a note stating that she should take off one week from work due to the heat.

Milam and another manager with JMI knew of the problems that Spees was experiencing and encouraged her to "go on some medical leave." At this point, Spees transferred her medical care from Dr. Cardenas to Dr. Susan Mueller, another obstetrician in Paducah. Spees had an appointment with Dr. Mueller on August 16, 2007 and, during this visit, Dr. Mueller discovered that Spees had an "incompetent cervix." (An incompetent cervix, according to www.medterms. com, is a cervix that is "abnormally liable to dilate and so is not competent to keep the fetus up in the uterus and keep it from being spontaneously aborted.") In light of this complication, Dr. Mueller placed Spees on bed rest for the remainder of her pregnancy and wrote her a note to that effect. Spees did not request to be placed on bedrest nor did she inform Dr. Mueller that her supervisor had encouraged her to go on medical leave. Although Dr. Mueller believed that Spees was not physically capable of doing her job, Spees disagreed and thought that she could have continued working. Spees nevertheless submitted Dr. Mueller's note to Gunder.

At this point, Spees had been absent from work for more than 14 days due to her various doctors' appointments and the one-week rest recommended by Dr. Cardenas. Because Spees had not yet worked 90 days for JMI, she was entitled to only two weeks of approved leave and was not eligible for leave at all under the Family and Medical Leave Act (FMLA), 29 U.S.C. § 2611 *et seq.* Gunder, who was aware that Spees had exhausted her leave, called DeWeese to discuss what to do. DeWeese told Gunder to terminate Spees. She further instructed Gunder to note in Spees's termination paperwork that the company would rehire her after her baby was born.

Gunder called Spees to tell her about the termination. Spees was surprised by her firing, having believed that she would be placed on long-term medical leave. According to Spees, Gunder told her that she "was being fired for being pregnant." Spees further contends that Gunder failed to tell her that she would be rehired following the conclusion of her pregnancy.

Gunder filled out Spees's termination form. It stated that "Dr. put Heather off work until after her delivery date. Not enough time in for [medical leave of absence]. Rehire after delivery." Despite this statement on the form, Spees did not know until after taking legal action against JMI that the company intended to rehire her following the pregnancy.

*388 B. Procedural Background

Spees filed suit against JMI in April 2008, alleging unlawful discrimination under Title VII of the Civil Rights Act of 1964 (Title VII), 42 U.S.C. § 2000e *et seq.*, under the Americans with Disabilities Act (ADA), 42 U.S.C. § 12101 *et seq.*, and under the Kentucky Civil Rights Act (KCRA), Ky.Rev.Stat. Ann. § 344.010 *et seq.* Her claims included (1) pregnancy discrimination on the basis of her transfer to the tool-room position, (2) pregnancy discrimination on the basis of her termination, and (3) disability discrimination. Spees also brought disparate-treatment and disparate-impact gender-discrimination claims on the basis of allegedly inferior women's restroom and locker facilities at JamesBuilt, but neither of these claims is raised by Spees on appeal.

Following discovery, Spees filed a motion for summary judgment on all of her claims as to JMI's liability. JMI then responded to Spees's motion and filed its own counter-motion for summary judgment. The district court granted summary judgment in favor of JMI on Spees's pregnancy-discrimination and disability-discrimination claims. Regarding Spees's tool-room transfer claim, the court concluded that Spees's reassignment did not constitute an adverse employment action. It similarly determined that Spees could not succeed on her job-termination claim because she could not show that JMI's justification for firing her—Dr. Mueller's note placing her on bed rest and the fact that she had exhausted her medical leave—was a pretext designed to mask discrimination. Finally, the court was not persuaded that pregnancy constituted a disability for ADA purposes and, accordingly, granted summary judgment in favor of JMI on Spees's disability-discrimination claim.

The remaining two claims of disparate-treatment and disparate-impact gender discrimination—based on the allegedly inferior restroom and locker facilities for women—were tried to a jury, which found in favor of JMI on both claims. Spees now appeals the district court's grant of summary judgment in favor of JMI on her claims of pregnancy discrimination and disability discrimination.

II. ANALYSIS

A. STANDARD OF REVIEW

[1] We review de novo a district court's grant of summary judgment. *ACLU of Ky. v. Grayson County,* 591 F.3d 837, 843 (6th Cir.2010). Summary judgment is proper where no genuine issue of material fact exists and the moving party is entitled to judgment as a matter of law. Fed.R.Civ.P. 56(c)(2). In considering a motion for summary judgment, we must draw all reasonable inferences in favor of the nonmoving party. *Matsushita Elec. Indus. Co. v. Zenith Radio Corp.,* 475 U.S. 574, 587, 106 S.Ct. 1348, 89 L.Ed.2d 538 (1986). The central issue is "whether the evidence presents a sufficient disagreement to require submission to a jury or whether it is so one-sided that one party must prevail as a matter of law." *Anderson v. Liberty Lobby, Inc.,* 477 U.S. 242, 251-52, 106 S.Ct. 2505, 91 L.Ed.2d 202 (1986).

B. STANDARD FOR PREGNANCY-DISCRIMINATION CLAIMS

Spees contends that JMI discriminated against her on the basis of her pregnancy on two separate occasions: (1) when she was transferred to work in the tool room, and (2) when she was terminated. The district court concluded that Spees had failed to make out a prima facie case of discrimination on the first claim because the tool-room transfer was not deemed an *389 adverse employment action. It further determined that, although Spees established a prima facie case of discrimination regarding her termination, she was unable to show that JMI's proffered justification for the firing—Dr. Mueller's note restricting Spees to bedrest and Spees's exhaustion of her medical leave—was pretextual. The court thus granted summary judgment to JMI on both claims.

Spees brought her pregnancy-discrimination claims pursuant to Title VII and the KCRA. Title VII makes it an "unlawful employment practice for an employer ... to discriminate against any individual with respect to his compensation, terms, conditions, or privileges of employment, because of such individual's ... sex." 42 U.S.C. § 2000e-2(a). "Because of sex" as used in Title VII includes "because of or on the basis of pregnancy, childbirth, or related medical conditions." 42 U.S.C. § 2000e(k). "[W]omen affected by pregnancy, childbirth, or related medical conditions shall be treated the same for all employment-related purposes ... as other persons not so affected but similar in their ability or inability to work." *Id.*

[2] The KCRA likewise prohibits discrimination against pregnant women. *See* Ky.Rev.Stat. §§ 344.030(8), 344.040(1). And the KCRA is "similar to Title VII of the 1964 federal Civil Rights Act and should be interpreted consistently with federal law." *Ammerman v. Bd. of Educ. of Nicholas County,* 30 S.W.3d 793, 797-98 (Ky.2000); *see also Jefferson County v. Zaring,* 91 S.W.3d 583, 586 (Ky.2002) (observing that because "the provisions of the KCRA are virtually identical to those of the Federal act[,] ... in this particular area we must consider the way the Federal act has been interpreted" [citation and internal quotation marks omitted]).

As an initial matter, we must determine the proper analytical framework to apply to Spees's pregnancy-discrimination claims at the summary judgment stage of the case. The district court analyzed both claims pursuant to the burden-shifting framework first announced in *McDonnell Douglas Corp. v. Green,* 411 U.S. 792, 802-03, 93 S.Ct. 1817, 36 L.Ed.2d 668 (1973), as amended by *Texas Department of Community Affairs v. Burdine,* 450 U.S. 248, 252-54, 101 S.Ct. 1089, 67 L.Ed.2d 207 (1981). Under that familiar tripartite analysis, a plaintiff seeking to survive summary judgment on a Title VII claim must overcome the following hurdles:

First, the plaintiff has the burden of proving by the preponderance of the evidence a prima facie case of discrimination. Second, if the plaintiff succeeds in proving the prima facie case, the burden shifts to the defendant "to articulate some legitimate, nondiscriminatory reason for the employee's rejection." Third, should the defendant carry this burden, the plaintiff must then have an opportunity to prove by a preponderance of the evidence that the legitimate reasons offered by the defendant were not its true reasons, but were a pretext for discrimination.

Burdine, 450 U.S. at 252-53, 101 S.Ct. 1089 (citations omitted). "The ultimate burden of persuading the trier of fact that the defendant intentionally discriminated against the plaintiff remains at all times with the plaintiff." *Id.* at 253, 101 S.Ct. 1089.

[3] Subsequent cases, however, have held that a different standard applies to so-called "mixed-motive" claims. Such claims are based on the plaintiff's allegation that "race, color, religion, sex, or national origin was *a* motivating factor for any employment practice, even though other factors also motivated the practice." Title VII, 42 U.S.C. § 2000e-2(m) (emphasis added). Allegations of discriminatory *390 conduct thus fall into one of two categories: single-motive claims, "where an illegitimate reason motivated an employment decision," or mixed-motive claims, "where both legitimate and illegitimate reasons motivated the employer's decision." *White v. Baxter Healthcare Corp.,* 533 F.3d 381, 396 (6th Cir.2008).

[4] This court in *White* held that the *McDonnell Douglas/Burdine* framework does not apply to mixed-motive claims. *Id.* at 400. Instead, "a Title VII plaintiff asserting a mixed-motive claim need only produce evidence sufficient to convince a jury that: (1) the defendant took an adverse employment action against the plaintiff; and (2) race, color, religion, sex, or national origin was *a* motivating factor for the defendant's adverse employment action." *Id.* (emphasis in original) (citation and internal quotation marks omitted). The plaintiff's burden of producing evidence to support a mixed-motive claim "is not onerous and should preclude sending the case to the jury only where the record is devoid of evidence that could reasonably be construed to support the plaintiff's claim." *Id.*

[5] Although portions of the *McDonnell Douglas/Burdine* framework might be "useful" in presenting a mixed-motive claim, the *White* court made clear that "compliance with the ... shifting burdens of production is *not* required in order to demonstrate that the defendant's adverse employment action was motivated in part by a consideration of the plaintiff's race, color, religion, sex, or national origin." *Id.* at 401 (emphasis in original) (citation and internal quotation marks omitted). The "ultimate question" in a mixed-motive analysis is simply "whether there are any genuine issues of material fact concerning the defendant's motivation for its adverse employment decision, and, if none are present, whether the law ... supports a judgment in favor of the moving party on the basis of the undisputed facts." *Id.* at 402. Inquiries into what motivated an employer's decision "are very fact intensive" and "will generally be difficult to determine at the summary judgment stage." *Id.* (citation omitted).

This relatively lenient summary judgment standard is counterbalanced by potential restrictions on a plaintiff's recovery for a mixed-motive claim. Under Title VII, a plaintiff asserting a mixed-motive claim is entitled only to declaratory relief, limited injunctive relief, and attorney fees and costs where the employer demonstrates that it would have taken the same employment action in the absence of an impermissible motivating factor. 42 U.S.C. § 2000e-5(g)(2)(B).

[6] [7] Plaintiffs must give proper notice when bringing mixed-motive claims. *Hashem-Younes v. Danou Enters. Inc.,* 311 Fed.Appx. 777, 779 (6th Cir.2009) (affirming the district court's application of the *McDonnell Douglas/Burdine* framework where the plaintiff failed to raise a mixed-motive claim in her complaint or in her response to the defendants' summary judgment motion, and the record was "utterly silent as to mixed motives"). Spees provided such notice of

her mixed-motive claims in the district court. As stated in her complaint, both discrimination claims alleged that Spees's pregnancy "was *a* motivating factor in [JMI]'s treatment of her." (Emphasis added.) She also specified in a footnote to her motion for summary judgment that she was bringing mixed-motive claims and was using the *McDonnell Douglas/Burdine* framework in her motion only because of uncertainty regarding the proper analysis of mixed-motive claims on a *plaintiff's* motion for summary judgment. Finally, Spees reiterated that she was pursuing mixed-motive claims under Title VII in her reply in *391 support of her motion for summary judgment/response to JMI's motion for summary judgment. We therefore conclude that Spees provided adequate notice of her mixed-motive claims.

In light of this notice, the district court's failure to apply the *White* analytical framework was in error. To properly analyze Spees's claims, we need determine only whether JMI took an adverse employment action against Spees and whether her pregnancy was a motivating factor for the adverse action. *See White,* 533 F.3d at 400. Each of Spees's claims is addressed below with the *White* framework in mind.

C. Pregnancy-Discrimination Claim based on Spees's Transfer to the Tool Room

[8] Spees first claims that JMI discriminated against her by transferring her to a tool-room position once it learned of her pregnancy. The district court granted JMI's motion for summary judgment on this claim, finding that the transfer did not constitute an adverse employment action.

1. The Transfer as an Adverse Employment Action

[9] [10] An adverse employment action has been defined as "a materially adverse change in the terms and conditions of [a plaintiff's] employment." *White v. Burlington N. & Santa Fe Ry. Co.,* 364 F.3d 789, 795 (6th Cir.2004) (en banc) (citation omitted). A "bruised ego" or a "mere inconvenience or an alteration of job responsibilities" is not sufficient to constitute an adverse employment action. *Id.* at 797. Adverse employment actions are typically marked by a "significant change in employment status," including "hiring, firing, failing to promote, reassignment with significantly different responsibilities, or a decision causing a significant change in benefits." *Id.* at 798 (quoting *Burlington Indus. v. Ellerth,* 524 U.S. 742, 761, 118 S.Ct. 2257, 141 L.Ed.2d 633 (1998)).

[11] Reassignments and position transfers can qualify as adverse employment actions, particularly where they are accompanied by "salary or work hour changes." *See Kocsis v. Multi-Care Mgmt., Inc.,* 97 F.3d 876, 885-86 (6th Cir.1996) (holding that a job transfer was not an adverse employment action because the plaintiff "enjoyed the same ... rate of pay and benefits, and her duties were not materially modified"). And even if a reassignment is not paired with a salary or work-hour change, it can nonetheless be considered an adverse employment action where there is evidence that the employee received "a less distinguished title, a material loss of benefits, significantly diminished material

responsibilities, or other indices that might be unique to a particular situation."
Id. at 886 (citation omitted).

Upon learning that Spees was pregnant, JMI transferred her from a daytime welding position to a daytime position in the tool room, where she worked for approximately one week before being transferred to a nighttime shift in order to keep her job. Some evidence indicates that the transfer was not a materially adverse change in her employment. For instance, Spees received the same salary while working in the tool room and did not lose any of her benefits. And, as JMI points out in its brief, the working conditions in the tool room were in some ways better than those while welding. JMI contends, for example, that the summer heat was more tolerable in the tool room because Spees could wear two fewer pieces of gear than when welding, and JMI provided a small fan for Spees's personal use. Spees was also not subject to the toxic fumes from welding while working in the tool room.

*392 But the record contains other evidence to suggest that Spees's transfer was a materially adverse change. In many ways, the tool-room transfer can be seen as a demotion. Spees was required to complete a 30-day training course to become a welder, but there is no evidence that a tool-room position required any specific training or skill. In addition, Spees appears to have felt unchallenged by her tool-room position, testifying that she found it to be "more boring" than welding. This contrast weighs in favor of finding the change in job assignments to be materially adverse. *See White v. Burlington N.,* 364 F.3d at 803 (concluding that an employee's transfer from a forklift operator to a standard railroad track laborer job was an adverse employment action because, in part, "the forklift operator position required more qualifications, which is an indication of prestige").

[12] Moreover, Spees was soon assigned to the night shift, which adversely affected her ability to raise her daughter as a single mother. An "inconvenience resulting from a less favorable schedule can render an employment action 'adverse' even if the employee's responsibilities and wages are left unchanged." *Ginger v. District of Columbia,* 527 F.3d 1340, 1344 (D.C.Cir.2008) (holding that switching police officers to a rotating morning/afternoon/night shift from a permanent night shift was an adverse employment action because it "severely affected their sleep schedules and made it more difficult for them to work overtime and part-time day jobs"). Although Spees did not describe in detail how the schedule change affected her, she did state that she "wasn't happy" about transferring to nights because she was a single mother. And the fact that Spees "requested" the night-shift position does not diminish JMI's responsibility for the schedule change because Spees was constructively forced to work nights. Both Milam and Spees testified that Milam told her to pursue the night-shift because it was the only option available that would allow her to retain her employment with JMI. This evidence supports the conclusion that she suffered an adverse employment action.

Nor does the evidence conclusively indicate that the tool-room position was a more pleasant working environment. Spees testified that working in the tool room was "just as hot" and as "physically demanding" as welding, the only difference being that she did not need to do any overhead handling of the welding equipment. And although Spees was not exposed to toxic fumes while working in the tool room, she could have avoided such fumes by wearing a respirator while welding, as first recommended by Dr. Cardenas.

On balance, Spees's transfer to the tool room resulted in her working a more inconvenient shift in a position that was less challenging and that required fewer qualifications. Viewing this evidence collectively in Spees's favor, a reasonable jury could find that her transfer to the tool room constituted an adverse employment action.

2. Spees's Pregnancy as a Motivating Factor for the Transfer

We must next determine whether Spees presented sufficient evidence from which a reasonable jury could find that her pregnancy was a motivating factor in transferring her to the tool room. In *International Union, UAW v. Johnson Controls, Inc.*, 499 U.S. 187, 111 S.Ct. 1196, 113 L.Ed.2d 158 (1991), the Supreme Court set forth the parameters regarding the acceptable treatment of female employees with childbearing capacity. Johnson Controls had barred all fertile females from its lead-battery plant out of concern for the health *393 of the fetuses that the women might conceive. *Id.* at 191-92, 111 S.Ct. 1196. The Court struck down the company's policy as violating Title VII because the policy discriminated against female employees based on their capacity to become pregnant, even though the employees' "reproductive potential" did not prevent them from being able to perform their jobs. *Id.* at 206, 111 S.Ct. 1196. It added that an employer's safety concerns were a permissible ground for restricting a female employee's job opportunities only where a pregnancy "actually interfere[d] with the employee's ability to perform the job." *Id.* at 204, 111 S.Ct. 1196. The Court concluded that "Congress made clear that the decision to become pregnant or to work while being ... pregnant ... was reserved for each individual woman to make for herself." *Id.* at 206, 111 S.Ct. 1196.

In the present case, Spees presented considerable evidence demonstrating that her pregnancy was at least *a* motivating factor, if not *the* motivating factor, in JMI's decision to transfer her to the tool room. Milam testified that when he first learned of Spees's pregnancy, he had "concerns" that she would not be able to weld. When Spees read him Dr. Cardenas's first note clearing her to return to welding, Milam said that "there was some question about her being pregnant and being able to safely perform the job that she was required to do." He based these concerns on his perception of "common sense." And according to Spees, Milam told her to obtain a second note from Dr. Cardenas limiting her to light duty and instructing her to avoid toxic fumes. JMI then relied on this note in transferring Spees to the tool room.

Other JMI employees superior to Spees exhibited a similar attitude. Tom Freeman, the head of JMI's Safety Department, told Spees that "this is a man's world" and that the notes from Dr. Cardenas were "not acceptable." Freeman's statement that he "didn't know what he was going to do" with Spees could be construed as further questioning her ability to weld while pregnant. Gunder, the night foreman and Spees's brother, also partook in the decision to transfer Spees.

He stated in his deposition that he did not want Spees welding "because she was carrying my niece." Gunder and Milam discussed where Spees should be working, and they "just decided that it wouldn't be a good idea for her to [weld]." In contrast, Spees never told her supervisors at JMI that she was unable to weld. She instead believed that she could weld, with only minimal restrictions, up until the full term of her pregnancy.

Furthermore, there is no evidence to suggest that Spees requested a transfer to the tool room. Milam, on the other hand, knew that there was a night-shift opening in the tool room and recommended that Spees seek a transfer. He and other managers made the decision to move Spees to the tool room, allegedly telling her "we don't know what to do with you." In addition, Milam later indicated to Spees that JMI management had taken unilateral action. According to Spees, Milam informed her that "I have already been told by [Colbert] that you are not going to weld no matter what your doctor is going to do."

JMI defends its decision to transfer Spees by relying on Dr. Cardenas's second note that restricts Spees to light duty and indicates that she should avoid toxic fumes. The company also points to Dr. Cardenas's testimony in which he states that his recommendations were independent of any motivations that JMI may have had. But this evidence does not shield JMI's transfer decision in light of Milam's apparent opinion that Spees should be transferred *394 even before Dr. Cardenas had written the light-duty note. Furthermore, Spees testified that Milam instructed her to obtain the note so limiting her. Dr. Cardenas's statement that he was not influenced by JMI when writing the second note is similarly inconclusive because Spees might have chosen not to inform him (or forgotten to inform him) that she was seeking that note at JMI's request. In sum, evidence exists from which a reasonable jury could find that JMI had decided that Spees was unable to weld due to her pregnancy and had instructed her to get a doctor's note to that effect.

JMI also argues that it would have been subject to a tort claim for negligence if it had permitted Spees to continue welding contrary to the orders contained in Dr. Cardenas's second note. But this argument again overlooks the evidence that Milam told Spees to obtain the restrictive note in the first place. Moreover, as the Supreme Court noted in *Johnson Controls,* JMI's risk of tort liability in this situation would be remote if it "fully inform[ed]" Spees of the risk inherent to welding while pregnant and did not otherwise act negligently. *See id.* at 208, 111 S.Ct. 1196. Summary judgment is accordingly inappropriate on this ground.

As a whole, the evidence is sufficient to raise a genuine issue of material fact as to whether JMI management, rather than undertaking an objective evaluation to determine whether Spees could perform her welding job while pregnant, instead subjectively viewed Spees's pregnancy as rendering her unable to weld. This would allow a reasonable jury to find that JMI's decision to transfer Spees was made out of concern for her pregnancy and the well-being of her unborn child rather than because Spees was unable to perform her job as a welder. Such concerns, though laudatory, do not justify an adverse employment action. *See id.*

at 206, 111 S.Ct. 1196. The district court therefore erred in granting summary judgment in favor of JMI on Spees's transfer claim.

D. Pregnancy-discrimination claim based on Spees's termination

[13] In addition to alleging that JMI violated the antidiscrimination laws in transferring her to work in the tool room, Spees contends that JMI discriminated against her by terminating her employment after she was placed on bedrest. This claim is also a mixed-motive claim, so Spees must show that she was subject to an adverse employment action for which her pregnancy was a motivating factor. *See White v. Baxter Healthcare Corp.,* 533 F.3d 381, 401 (6th Cir.2008). JMI concedes the obvious—that firing an employee constitutes an adverse employment action. (See, e.g., id. at 402.) And, to be sure, Spees's pregnancy played a role in her termination, with complications stemming from the pregnancy causing Spees to be placed on bedrest, which in turn led to her firing due to the fact that she had exhausted all of her available medical leave. But the *White* analysis does not hinge on whether Spees's pregnancy was a link in the chain of events that resulted in her firing. Rather, *White* directs us to examine whether there is evidence that JMI was *motivated* by Spees's pregnancy in making its decision to terminate her.

JMI's justifies its termination decision by pointing to the fact that Spees presented it with Dr. Mueller's note placing Spees on bedrest. Unlike the note from Dr. Cardenas, which Spees claims was obtained at the direction of Milam in order to restrict her to light-duty work, there is no evidence that JMI influenced Dr. Mueller's writing Spees the bedrest note. This restriction instead stemmed from Dr. Mueller's diagnosis of Spees's incompetent-cervix *395 medical condition. Dr. Mueller's assessment was also arrived at independently of any request by Spees, who testified that she neither asked to be placed on bedrest nor told Dr. Mueller that JMI had told her to seek medical leave.

Pursuant to the bed rest note from Dr. Mueller, Spees was unable to work in any capacity at JMI, a point that Spees herself recognizes. Spees also acknowledges that although she would have been placed on medical leave under normal circumstances, she was not eligible for FMLA leave as a recently hired employee who had already exhausted all of the regular leave to which she was entitled. JMI's decision to terminate Spees was thus based on a combination of her being unable to work and her lack of any available medical leave, not upon her pregnancy per se.

But Spees maintains that she would not have submitted the bed rest note from Dr. Mueller to JMI if she had known that she was ineligible for any additional medical leave. She adds that she would have preferred to continue working in defiance of Dr. Mueller's advice. But there is no evidence that Spees resisted being placed on bed rest. To the contrary, it was Spees who submitted the bed rest note to JMI. Spees argues that she did so only because she had been encouraged to "go on some medical leave," but she also conceded in her deposition that no one at JMI guaranteed her that she was eligible for such leave. Her reliance on any expectation of medical leave was therefore unjustified.

In short, absent any evidence that JMI played a role in having Spees placed on bed rest—the event that directly led to her termination—there is no support for Spees's contention that her pregnancy in and of itself was a motivation behind JMI's decision to fire her. Without the bed rest note, the record supports the conclusion that JMI would have allowed Spees to continue working in the tool room despite being pregnant. Spees was terminated, in other words, not because she was pregnant, but because she voluntarily submitted to JMI the bed rest note advising her not to work for the duration of her pregnancy. We therefore conclude that summary judgment was proper on this claim.

E. DISABILITY CLAIM BASED ON SPEES'S TRANSFER TO THE TOOL ROOM

[14] Spees next appeals the entry of summary judgment for JMI on her claim that the company prohibited her from welding and transferred her to a tool-room position because it wrongfully perceived her pregnancy to be a disability. The ADA prohibits discrimination by a covered entity "against a qualified individual on the basis of disability in regard to job application procedures, the hiring, advancement, or discharge of employees, employee compensation, job training, and other terms, conditions, and privileges of employment." 42 U.S.C. § 12112(a). To make out a prima facie case of discrimination under the ADA, a plaintiff must show "(1) that she or he is an individual with a disability, (2) who was otherwise qualified to perform a job's requirements, with or without reasonable accommodation; and (3) who was discriminated against solely because of the disability." *Talley v. Family Dollar Stores of Ohio, Inc.,* 542 F.3d 1099, 1105 (6th Cir.2008) (citation omitted). "The third element requires that the plaintiff suffer an adverse employment action." *Id.*

In this case, the central dispute over Spees's ADA claim revolves around whether she meets the definition of a "disabled" person. A "disability" is defined as "(A) a physical or mental impairment that substantially limits one or more of the major life activities of such individual; (B) a record of such an impairment; or (C) being *396 regarded as having such an impairment." 42 U.S.C. § 12102(2) (2006). This section of the Act was amended in 2009, subsequent to the events giving rise to Spees's lawsuit. But we must analyze Spees's claims pursuant to the earlier version (provided above) because the amendments to the ADA do not apply retroactively. *See Milholland v. Sumner County Bd. of Educ.,* 569 F.3d 562, 567 (6th Cir.2009) (holding that "the ADA Amendments Act does not apply to preamendment conduct").

[15] Spees does not argue that her pregnancy qualified as a disability under subsections (A) or (B); rather, she brings her claim pursuant to the "regarded-as" provision in subsection (C). Moreover, Spees acknowledges that pregnancy, by itself, does not constitute a disability under the ADA and thus cannot form the basis of a regarded-as claim. This concession comports with the unanimous holdings of the federal courts that have addressed the issue. (See, e.g., *Richards v. City of Topeka,* 173 F.3d 1247, 1250 n. 2 [10th Cir.1999].) ("[W]e do note that numerous district courts have concluded that a normal pregnancy without complications is not a disability under 42 U.S.C. § 12102[2][A].") (listing cases); *Navarro-Pomares v. Pfizer Corp.,* 97 F.Supp.2d 208, 212 n. 5 (D.P.R.2000)

(observing that the only district judge to have held that pregnancy, by itself, was a disability under the ADA reversed himself in a subsequent case), *rev'd on other grounds, Navarro v. Pfizer Corp.*, 261 F.3d 90 (1st Cir.2001). Likewise, the interpretive guideline for the term "disability" issued by the Equal Employment Opportunity Commission in its Compliance Manual excludes pregnancy from its definition of disability. EEOCCM § 902.2(c)(3), 2009 WL 4782107 (Nov. 21, 2009) ("Because pregnancy is not the result of a physiological disorder, it is not an impairment.").

[16] Spees's ADA claim instead hinges on her contention that JMI erroneously perceived her to be disabled "based on her history of conditions with a previous pregnancy." This type of claim exists where "(1) an employer mistakenly believes that an employee has a physical impairment that substantially limits one or more major life activities, or (2) an employer mistakenly believes that an actual, nonlimiting impairment substantially limits one or more of an employee's major life activities." *Gruener v. Ohio Cas. Ins. Co.*, 510 F.3d 661, 664 (6th Cir.2008) (brackets and citation omitted). "Either application requires that the employer entertain misperceptions about the employee." *Id.* (brackets, citation, and internal quotation marks omitted). Spees has not alleged that she suffered an actual impairment, so she therefore must show that JMI mistakenly regarded her as having "a physical or mental impairment that substantially limit[ed] one or more of [her] major life activities." *See* 42 U.S.C. § 12102(2) (2006).

Our first step in evaluating Spees's ADA claim is to determine whether her prior miscarriage, or a potentially higher risk of having a future miscarriage, could constitute an impairment. Whereas no court has held that pregnancy by itself is an impairment under the ADA, many district courts have held that pregnancy-related conditions can qualify as such. (See, e.g., *Navarro*, 261 F.3d at 97.) ("While pregnancy itself may not be an impairment, the decided ADA cases tend to classify complications resulting from pregnancy as impairments."). The EEOC interpretive guidelines also recognize that pregnancy-related conditions can constitute impairments under the ADA. EEOCCM § *397 902.2(c)(3), 2009 WL 4782107 (Nov. 21, 2009) ("Complications resulting from pregnancy ... are impairments.").

Pregnancy-related conditions have typically been found to be impairments where they are not part of a "normal" pregnancy. *See Serednyj v. Beverly Healthcare LLC*, No. 2:08-CV-4 RM, 2010 WL 1568606, at * 14 (N.D.Ind. Apr.16, 2010) (surveying cases and noting that "only abnormal complications might qualify as impairments" under the ADA). Susceptibility to a miscarriage, moreover, has been deemed by some courts to be such a condition. *See Cerrato v. Durham*, 941 F.Supp. 388, 393 (S.D.N.Y.1996) (adopting the American Medical Association's Council on Scientific Affairs' conclusion that a "threatened ... miscarriage" is a "substantial complication" not part of an "entirely normal, healthy pregnancy"); *Soodman v. Wildman, Harrold, Allen & Dixon*, No. 95 C 3834, 1997 WL 106257, at *6 (N.D.Ill. Feb.10, 1997) (holding that "the inability or significantly impaired ability to carry a viable fetus to term is ... a 'substantial impairment' " under the ADA).

Although other courts have held that pregnancy complications related to miscarriages are not disabilities, the analysis in those cases did not hinge on the question of whether there was an impairment, but rather on whether the condition was sufficiently severe to substantially limit a major life activity. (See, e.g., *LaCoparra v. Pergament Home Ctrs.*, 982 F.Supp. 213, 228 [S.D.N.Y.1997]) (concluding that the plaintiff's "history of infertility" and prior miscarriage were not disabilities where the "evidence suggests that, if anything, the existence and impact of the complications were temporary"), *overruled on other grounds by Kosakow v. New Rochelle Radiology Assocs., P.C.*, 274 F.3d 706, 724 (2d Cir.2001). There thus appears to be a general consensus that an increased risk of having a miscarriage at a minimum constitutes an impairment falling outside the range of a normal pregnancy.

In the present case, there is evidence that JMI regarded Spees as having an impairment. Milam testified that because Spees had experienced "complications with other pregnancies before," he thought that she should not be working, and he had "concerns about her being around the chemicals, the welding smoke, [and] climbing around on some of the jobs." This statement suggests that Milam believed Spees to be especially sensitive to miscarriages in light of the fact that she had experienced one in the past. Milam's testimony therefore constitutes evidence that JMI perceived Spees as having an impairment.

[17] [18] Spees must next show that JMI viewed her impairment as substantially limiting a major life activity. The only major life activity Spees points to is that of working. A claim that an employer perceived an employee as being unable to work requires "proof that the employer regarded the employee as significantly restricted in the ability to perform either a class of jobs or a broad range of jobs in various classes." *Daugherty v. Sajar Plastics, Inc.*, 544 F.3d 696, 704 (6th Cir.2008) (citation and internal quotation marks omitted). "The inability to perform a single, particular job does not constitute a substantial limitation in the major life activity of working." *Id.* (citation omitted).

In evaluating this issue, the regulations accompanying the ADA direct us to consider certain factors. These factors include those used to determine whether an impairment substantially limits any major life activity, namely:

   (i)   The nature and severity of the impairment;
  (ii) *398 The duration or expected duration of the impairment; and
 (iii) The permanent or long-term impact, or the expected permanent or long-term impact of or resulting from the impairment.

29 C.F.R. § 1630.2(J)(2). The regulations further provide three additional factors where, as here, the major life activity is working:

(A) The geographical area to which the individual has reasonable access;
(B) The job from which the individual has been disqualified because of an impairment, and the number and types of jobs utilizing similar training, knowledge, skills or abilities, within that geographical area, from

which the individual is also disqualified because of the impairment (class of jobs); and/or

(C) The job from which the individual has been disqualified because of an impairment, and the number and types of other jobs not utilizing similar training, knowledge, skills or abilities, within that geographical area, from which the individual is also disqualified because of the impairment (broad range of jobs in various classes).

29 C.F.R. § 1630.2(j)(3)(ii).

In the present case, the evidence supports the conclusion that the tool-room transfer precluded Spees from working in "a class of jobs" for two reasons. First, JMI viewed Spees as being unable to weld in any capacity, thereby precluding her from employing the skills that she had acquired during the one-month training program for welding. The tool-room position, unlike a welding position, did not require any special training, meaning that Spees was effectively removed to an unskilled position and precluded from utilizing any of the welding training that she had received. JMI's belief that Spees could not perform any type of welding work thus weighs in favor of concluding that she was precluded from working in a class of jobs. *Cf. Dutcher v. Ingalls Shipbuilding,* 53 F.3d 723, 727 (5th Cir.1995) (holding that a plaintiff was not precluded from working in a class of jobs where an arm injury restricted her from performing any climbing while welding, but did not prevent her from welding in general).

The second reason supporting the conclusion that JMI prevented Spees from working in a class of jobs is the fact that it restricted her to light-duty work. In this regard, we are persuaded by the EEOC Compliance Manual, which states, as an example of an employee being unable to work in a class of jobs, that "a charging party is substantially limited in working if (s)he has a back impairment that precludes him/her from heavy lifting and, therefore, from the class of heavy labor jobs." EEOCCM § 902.4(c)(3)(ii), 2009 WL 4782109 (Nov. 21, 2009). This dividing line between light-duty and medium- or heavy-duty work for purposes of determining what constitutes a class of jobs has also been previously recognized by this court. *See Henderson v. Ardco, Inc.,* 247 F.3d 645, 652 (6th Cir.2001) (denying summary judgment on an ADA claim where the employer perceived an employee "as unable to perform anything but 'light duty' work, and [ ] perceived that medium to heavy manual labor constituted a majority of the jobs available to her").

Here, JMI put Spees on light-duty work immediately upon learning that she was pregnant. Milam was instrumental in transferring Spees to the tool room, going so far as to instruct Spees to obtain a note from Dr. Cardenas restricting her to light-duty work even though Dr. Cardenas had already cleared her to return to welding. And although Spees testified that she found the tool-room work to be as physically demanding as welding, both parties *399 clearly considered it to be light-duty work. The light-duty nature of the tool-room work, viewed in conjunction with the fact that the tool-room position did not utilize any of the skills that Spees had acquired as a result of her welder training, supports the determination that JMI precluded Spees from working in a class of jobs. We therefore conclude that the first element of Spees's prima facie ADA claim based on the tool-room transfer has been met.

[19] Moreover, Spees has satisfied the remaining two elements of this claim. One of these elements hinges on whether she was qualified to weld, with or without reasonable accommodation. Spees fulfilled this element by presenting considerable evidence, none of which is disputed by JMI, that she successfully completed the training course for welding and performed competently as a welder prior to being transferred to the tool room. And the final element has likewise been met because, as discussed in Part II. C.1. above, the tool-room transfer constituted an adverse employment action. The district court therefore erred in granting summary judgment in favor of JMI on Spees's ADA claim to the extent it is based on the tool-room transfer.

F. Disability Claim Based on Spees's Termination

[20] Spees's final challenge is to the grant of summary judgment in favor of JMI on her ADA claim with regard to the termination of her employment. But Spees has presented no evidence that JMI regarded her as having an impairment that precluded her from working in the tool room. Rather, as discussed above in Part II.D. above, JMI's decision was based on Dr. Mueller's note—obtained independently of any influence by JMI—that restricted Spees to bed rest for the duration of her pregnancy. Spees thus cannot show that JMI's decision to terminate her stemmed from a mistaken belief that she suffered an impairment precluding her ability to work in general.

III. CONCLUSION

For all of the reasons set forth above, we **AFFIRM** the judgment of the district court with regard to Spees's pregnancy-discrimination claim and her disability-discrimination claim as they pertain to the termination of her employment, **REVERSE** the district court's grant of summary judgment on Spees's pregnancy-discrimination claim and disability-discrimination claim to the extent that they are based on her reassignment to the tool room, and **REMAND** the case for further proceedings on these latter two claims.

ZATKOFF, DISTRICT JUDGE, CONCURRING IN PART AND DISSENTING IN PART.

I concur with the factual description, analysis and conclusions set forth in Sections I., II.A., II.B., II.D. and II.F of Judge Gilman's opinion. As such, I join with my colleagues in affirming the district court's judgment with regard to Spees's pregnancy-discrimination claim and her disability-discrimination claim as they pertain to the termination of her employment. I disagree, however, that Spees suffered an adverse employment action when she was transferred to the tool room. Accordingly, I write separately and respectfully dissent from the majority's decision to reverse the district court's judgment on Spees's pregnancy-discrimination and her disability-discrimination claims as they pertain to her transfer to the tool room.

As stated by the majority in Section II.C.1.:

An adverse employment action has been defined as "a materially adverse *400 change in the terms and conditions of [a plaintiff's] employment." *White v. Burlington N. & Santa Fe Ry. Co.*, 364 F.3d 789, 795 (6th Cir.2004) (en banc) (citation omitted). A "bruised ego" or a "mere inconvenience or an alteration of job responsibilities" is not sufficient to constitute an adverse employment action. *Id.* at 797. Adverse employment actions are typically marked by a "significant change in employment status," including "hiring, firing, failing to promote, reassignment with significantly different responsibilities, or a decision causing a significant change in benefits." *Id.* at 798 (quoting *Burlington Indus. v. Ellerth*, 524 U.S. 742, 761, 118 S.Ct. 2257, 141 L.Ed.2d 633 (1998)).

Reassignments and position transfers can qualify as adverse employment actions, particularly where they are accompanied by "salary or work hour changes." *See Kocsis v. Multi-Care Mgmt., Inc.*, 97 F.3d 876, 885-86 (6th Cir. 1996) (holding that a job transfer was not an adverse employment action because the plaintiff "enjoyed the same ... rate of pay and benefits, and her duties were not materially modified"). And even if a reassignment is not paired with a salary or work-hour change, it can nonetheless be considered an adverse employment action where there is evidence that the employee received "a less distinguished title, a material loss of benefits, significantly diminished material responsibilities, or other indices that might be unique to a particular situation." *Id.* at 886 (citation omitted).

The majority also recognizes that "[t]he third element [of a disability-discrimination claim] requires that a plaintiff suffer an adverse employment action." *Talley v. Family Dollar Stores of Ohio, Inc.*, 542 F.3d 1099, 1105 (6th Cir.2008).

I find the foregoing recitation of the law accurate and applicable in this case, just as I agree with the majority's conclusion that:

Some evidence indicates that the transfer was not a materially adverse change in her employment. For instance, Spees received the same salary while working in the tool room and did not lose any of her benefits. And, as JMI points out in its brief, the working conditions in the tool room were in some ways better than those while welding. JMI contends, for example, that the summer heat was more tolerable in the tool room because Spees could wear two fewer pieces of gear than when welding, and JMI provided a small fan for Spees's personal use. Spees was also not subject to the toxic fumes from welding while working in the tool room.

Nonetheless, the majority ultimately concludes that "Spees's transfer to the tool room resulted in her working in a more inconvenient shift in a position that was less challenging and that required fewer qualifications [,] ... [such that] a reasonable jury could find that her transfer to the tool room constituted an adverse employment action."

I am not persuaded that any of those reasons (i.e., inconvenience to Spees, a less challenging position and fewer required qualifications), individually or in the aggregate, could support a finding that Spees suffered an adverse employment action when she was transferred to the tool room. Most significantly, Spees was not subject to "salary or work hour changes[,]" she "enjoyed the same ... rate of pay and benefits, and her duties were not materially modified." *Kocsis,* 97 F.3d at 885-86. Although the tool-room position did not require any specific training, as the welding position did, I do not believe that difference suffices to enable a reasonable jury to find Spees suffered an adverse *401 employment action. This is particularly true because Spees was a "welder-trainee" and that title does not carry any greater prestige than the tool room position Spees assumed upon her transfer.

The majority notes that Spees felt unchallenged in the tool room and found the tool room work "more boring" than welding. Likewise, the majority notes, "Spees did not describe in detail how the schedule change affected her, [but] she did state that she 'wasn't happy' about transferring to nights because she was a single mother." Spees's unhappiness with the change to the night shift and the fact that she felt unchallenged and "more bor[ed]" by the tasks she performed in the tool room, while not irrelevant to her, reflect a "mere inconvenience" and not a "significant change in employment status." *White,* 364 F.3d at 797-98. Rather, those feelings are more indicative of a "bruised ego" than a "significant change in employment status." *Id.* In addition, it is undisputed that Spees approached Pam DeWeese, an employee in JMI's Human Resource Department, about changing to the night-shift position (although this apparently was done after Tony Milam, her foreman, indicated that such a transfer would allow Spees to maintain her employment during the pregnancy).

Finally, I disagree with the majority's conclusion that the evidence does not "conclusively indicate that the tool-room position was a more pleasant working environment" as it pertained to heat and the physical demand on Spees. The evidence, in fact, did conclusively establish that the tool-room position offered a more pleasant working environment. As the majority recognizes, "welders are exposed to fumes, dust, and organic vapors in the course of their work[,]" and, in order to limit the inhalation of those substances, JMI provides welders with respirators to wear while on the job (though Spees often opted not to wear it while working as a welder because she "didn't feel like [she] needed one"). Such conditions are not present in the tool room, and there is no need for a respirator to work in the tool room. Welding work requires climbing up ladders and stairs, maneuvering into barge tanks and occasionally overhead handling of equipment, none of which was required to work in the tool room. Welders also wear a welding helmet and welding gloves, neither of which are necessary to work in the tool room. In addition, working in the tool room did not require Spees to work in a confined space, as the welding position often necessitated, and, unlike the welding jobs, the tool room had a fan. All of these differences are significant, particularly when temperatures reached 100 degrees Fahrenheit or more on multiple occasions during the summer Spees worked for JMI.

For the foregoing reasons, I conclude that Spees did not suffer an adverse employment action when she was transferred to the tool room. Therefore, as a matter of law, Spees cannot prevail on her pregnancy-discrimination claim or her disability-discrimination claim with respect to her transfer to the tool room.

Accordingly, I would affirm the judgment of the district court as it pertains to Spees's pregnancy-discrimination and disability-discrimination claims with regard to her transfer to the tool room.

C.A.6 (Ky.),2010.

Spees v. James Marine, Inc.

617 F.3d 380, 109 Fair Empl.Prac.Cas. (BNA) 1748, 23 A.D. Cases 972, 41 NDLR P 173.

14 Other Antidiscrimination Laws

Revolt and terror pay a price. Order and law have a cost.

Carl Sandburg

We always exempt ourselves from the common laws. When I was a boy and the dentist pulled out a second tooth, I thought to myself that I would grow a third if I needed it. Experience discouraged this prophecy.

Oliver Wendell Holmes, Jr.

LEARNING OBJECTIVES

1. Acquire an understanding of the numerous federal antidiscrimination laws.
2. Acquire a general understanding of the correlating state antidiscrimination laws.

HYPOTHETICAL SITUATION

An employee returns from an extended leave of absence. When the employee left for the leave of absence, he was a male employee. When the employee returned from the leave of absence, she now identifies herself as female. How would you address this situation?

OVERVIEW

Most companies and organizations go the extra mile to ensure that their workplace is free of any type of discrimination or harassment. Safety professionals often place a key role in ensuring that the workplace is free of discrimination or harassment due to the fact that the safety professional is involved in every aspect of the operation and thus serves as the "eyes and ears" of the company or organization.

Whenever a safety professional identifies any situation or issue within the operation that possesses any aspect of potential discrimination or harassment, it is important for the safety professional to be able to report this situation or issue to the appropriate managerial team member to investigate and address the situation or issue. Taking a proactive approach to create a diverse workplace which is free of discrimination and harassment is far less costly than to overlook any potential issue or situation of discrimination or harassment.

Safety professionals should be aware that virtually every law usually possesses a prohibition against harassment and retaliation. From the OSH Act to individual

259

state workers' compensation laws, safety professionals should be aware that when an employee files a complaint with OSHA or files a workers' compensation claim, the employee should always be treated the same as all other employees.

Any type of harassment or retaliation usually triggers a response by the employee and the correlating governmental agency.

Although we covered a broad spectrum of the major federal antidiscrimination laws earlier in this text, safety professionals should be aware that there are other federal laws and directives as well as state and local laws that prohibit discrimination and harassment in the workplace. These laws include, but are not limited to, the following:

FEDERAL ANTIDISCRIMINATION LAWS

- Civil Rights Act of 1866 (Also see Section 1981 actions)*—This law is primarily utilized to address racial discrimination.
- Civil Rights Act of 1871 (Also see Section 1983 actions and Section 1985 actions)†—Section 1983 actions provides "every person who, under color of statute, ordinance, regulation, custom, or usage, subjects or causes to be subjected, any person, within the jurisdiction of the United States, to deprive any rights, privileges, or immunities secured by the Constitution and laws shall be held civilly liable for such actions."‡ Section 1985 actions "declares liable two or more persons who conspire for the purposes of depriving, either directly or indirectly, any person or class of persons of the equal protection of the law, or of equal privileges and immunities under the law."§
- Federal Election Campaign Act¶—This law prohibits corporations and labor organizations from securing political contributions from employee or members by reason of threats or job discrimination.
- Sherman Anti-Trust Act—Employee can utilize the federal anti-trust laws** to redress certain types of work-related discrimination, such as agreements not to hire a former employee or blacklisting a former employee.

STATE ANTIDISCRIMINATION LAWS

Safety professionals should be aware that many of the state laws may vary in the protections afforded to applicants and employees in the areas of discrimination and harassment. However, it should be noted that virtually every state possesses some type of antidiscrimination laws providing protections which parallel or exceed the protections afforded under the federal laws.

For example, the Commonwealth of Kentucky possesses seven statutes which require employers with 8 or more employees to comply and protections are afforded

---

* 42 U.S.C.A. Section 1981.
† 42 U.S.C.A. Section 1983.
‡ 42 U.S.C.A. Section 1983.
§ 42 U.S.C.A. Section 1985(3).
¶ 2 U.S.C.A. Section 437.
** 15 U.S.C.A> Sections 1 and 12.

to individuals in the areas of age, race, national origin, gender, pregnancy, religion, disability, and HIV/AIDS. Conversely, the District of Columbia possesses one law (DC Code Ann. Sections 2-1401-01 through 1-14022-13) which provide protections in the areas of age, race, national origin, gender, sexual orientation, gender identity, disability, religion, HIV/AIDS, genetic testing, marital status, family duties, personal appearance, political affiliation, and victims of intrafamily offices.

It is important for safety professionals to become familiar with the specific laws of the states in which your company or organization possesses operations or performs work to ensure that compliance is achieved and maintained with all applicable laws.

LOCAL LAWS

Safety professionals with company or organizational operations should identify if your local city, county, or community government possesses any laws or ordinances providing protections to employees or applicants in areas that may not be encompassed within the federal or state laws and regulations.

Two areas of potential discrimination where local governments have taken the lead in enacting laws and ordinances providing protection where federal and state laws do not provide protection are in the areas of sexual orientation and gender identity discrimination.

Safety professionals should be aware that the federal law, primarily Title VII, does not specifically prohibit discrimination in the private sector based on sexual orientation. However, by executive order, employees of the federal government are provided protection against discrimination based on sexual orientation. Twenty states and the District of Columbia have laws prohibiting discrimination based upon sexual orientation in the public and private sectors.* However, a substantial number of cities and counties nationwide have adopted laws and ordinances which prohibit sexual orientation discrimination in the public sector and a smaller number in the private sector.†

Often paralleling sexual orientation laws at the local and state levels are laws prohibiting discrimination based upon an individual's gender identity. Federal law, specifically title VII, does not provide explicit prohibition against discrimination based on gender identity. However, safety professionals should be aware that companies and organizations cannot discriminate based upon stereotypes. Gender identity often refers to an individual's self-identified gender which may be different from the individual's anatomical gender provided at birth. Under many of the state and local laws, the individual does not have to undergo sex reassignment surgery to be protected under the specific state or local law.

It is important that safety professionals, as the eyes and ears of the company on the shop floor, be aware of the protection against discrimination and harassment that may be provided under federal, state, and local laws. Although safety professionals

---

* States providing state laws prohibiting discrimination based on sexual orientation include: California, Colorado, Connecticut, Hawaii, Illinois, Iowa, Maine, Massachusetts, Minnesota, Nevada, New Hampshire, New Jersey, New Mexico, New York, Oregon, Rhode Island, Vermont, Washington, and Wisconsin.
† Preventing Sexual Orientation Discrimination in the Workplace, www.nolo.com/legal (2011).

are not expected to be an expert in all of these laws, a good barometer for safety professionals to consider is "How would I feel if this happened to me?" If a safety professional observes a situation that could constitute discrimination or harassment, it is most safety professional's duty, as a managerial team member and employee of the company or organization, to act and notify the appropriate managerial team member of this situation in accordance with company policies and procedures. If there is any doubt, it is often in the best interest of the safety professional, the management team, and the company or organization to notify the appropriate management team member and react to eliminate the harassment or discrimination immediately in-house before the laws discussed are being utilized to enforce the proscribed protection before a federal or state agency or in a court of law.

Prohibited Employment Policies/Practices

Under the laws enforced by EEOC, it is illegal to discriminate against someone (applicant or employee) because of that person's race, color, religion, sex (including pregnancy), national origin, age (40 or older), disability or genetic information. It is also illegal to retaliate against a person because he or she complained about discrimination, filed a charge of discrimination, or participated in an employment discrimination investigation or lawsuit.

The law forbids discrimination in every aspect of employment.

The laws enforced by EEOC prohibit an *employer or other covered entity* from using neutral employment policies and practices that have a disproportionately negative effect on applicants or employees of a particular race, color, religion, sex (including pregnancy), or national origin, or on an individual with a disability or class of individuals with disabilities, if the polices or practices at issue are not job related and necessary to the operation of the business. The laws enforced by EEOC also prohibit an employer from using neutral employment policies and practices that have a disproportionately negative impact on applicants or employees age 40 or older, if the policies or practices at issue are not based on a reasonable factor other than age.

JOB ADVERTISEMENTS

It is illegal for an employer to publish a job advertisement that shows a preference for or discourages someone from applying for a job because of his or her race, color, religion, sex (including pregnancy), national origin, age (40 or older), disability, or genetic information. For example, a help-wanted ad that seeks "females" or "recent college graduates" may discourage men and people over 40 from applying and may violate the law.

RECRUITMENT

It is also illegal for an employer to recruit new employees in a way that discriminates against them because of their race, color, religion, sex (including pregnancy), national origin, age (40 or older), disability, or genetic information.

For example, an employer's reliance on word-of-mouth recruitment by its mostly Hispanic work force may violate the law if the result is that almost all new hires are Hispanic.

APPLICATION AND HIRING

It is illegal for an employer to discriminate against a job applicant because of his or her race, color, religion, sex (including pregnancy), national origin, age (40 or older), disability, or genetic information. For example, an employer may not refuse to give employment applications to people of a certain race.

An employer may not base hiring decisions on stereotypes and assumptions about a person's race, color, religion, sex (including pregnancy), national origin, age (40 or older), disability, or genetic information.

If an employer requires job applicants to take a test, the test must be necessary and related to the job and the employer may not exclude people of a particular race, color, religion, sex (including pregnancy), national origin, or individuals with disabilities. In addition, the employer may not use a test that excludes applicants age 40 or older if the test is not based on a reasonable factor other than age.

If a job applicant with a disability needs an accommodation (such as a sign language interpreter) to apply for a job, the employer is required to provide the accommodation, so long as the accommodation does not cause the employer significant difficulty or expense.

JOB REFERRALS

It is illegal for an employer, employment agency, or union to take into account a person's race, color, religion, sex (including pregnancy), national origin, age (40 or older), disability, or genetic information when making decisions about job referrals.

JOB ASSIGNMENTS AND PROMOTIONS

It is illegal for an employer to make decisions about job assignments and promotions based on an employee's race, color, religion, sex (including pregnancy), national origin, age (40 or older), disability, or genetic information. For example, an employer may not give preference to employees of a certain race when making shift assignments and may not segregate employees of a particular national origin from other employees or from customers.

An employer may not base assignment and promotion decisions on stereotypes and assumptions about a person's race, color, religion, sex (including pregnancy), national origin, age (40 or older), disability, or genetic information.

If an employer requires employees to take a test before making decisions about assignments or promotions, the test may not exclude people of a particular race, color, religion, sex (including pregnancy), or national origin, or individuals with disabilities, unless the employer can show that the test is necessary and related to the job. In addition, the employer may not use a test that excludes employees age 40 or older if the test is not based on a reasonable factor other than age.

PAY AND BENEFITS

It is illegal for an employer to discriminate against an employee in the payment of wages or employee benefits on the bases of race, color, religion, sex (including pregnancy), national origin, age (40 or older), disability, or genetic information. Employee benefits include sick and vacation leave, insurance, access to overtime as well as overtime pay, and retirement programs. For example, an employer many not pay Hispanic workers less than African-American workers because of their national origin, and men and women in the same workplace must be given equal pay for equal work.

In some situations, an employer may be allowed to reduce some employee benefits for older workers, but only if the cost of providing the reduced benefits is the same as the cost of providing benefits to younger workers.

DISCIPLINE AND DISCHARGE

An employer may not take into account a person's race, color, religion, sex (including pregnancy), national origin, age (40 or older), disability, or genetic information when making decisions about discipline or discharge. For example, if two employees commit a similar offense, an employer many not discipline them differently because of their race, color, religion, sex (including pregnancy), national origin, age (40 or older), disability, or genetic information.

When deciding which employees will be laid off, an employer may not choose the oldest workers because of their age.

Employers also may not discriminate when deciding which workers to recall after a layoff.

EMPLOYMENT REFERENCES

It is illegal for an employer to give a negative or false employment reference (or refuse to give a reference) because of a person's race, color, religion, sex (including pregnancy), national origin, age (40 or older), disability, or genetic information.

REASONABLE ACCOMMODATION AND DISABILITY

The law requires that an employer provide reasonable accommodation to an employee or job applicant with a disability, unless doing so would cause significant difficulty or expense for the employer.

A reasonable accommodation is any change in the workplace (or in the ways things are usually done) to help a person with a disability apply for a job, perform the duties of a job, or enjoy the benefits and privileges of employment.

Reasonable accommodation might include, for example, providing a ramp for a wheelchair user or providing a reader or interpreter for a blind or deaf employee or applicant.

REASONABLE ACCOMMODATION AND RELIGION

The law requires an employer to reasonably accommodate an employee's religious beliefs or practices, unless doing so would cause difficulty or expense for the employer. This means an employer may have to make reasonable adjustments at work that will allow the employee to practice his or her religion, such as allowing an employee to voluntarily swap shifts with a co-worker so that he or she can attend religious services.

TRAINING AND APPRENTICESHIP PROGRAMS

It is illegal for a training or apprenticeship program to discriminate on the bases of race, color, religion, sex (including pregnancy), national origin, age (40 or older), disability, or genetic information. For example, an employer may not deny training opportunities to African-American employees because of their race.

In some situations, an employer may be allowed to set age limits for participation in an apprenticeship program.

HARASSMENT

It is illegal to harass an employee because of race, color, religion, sex (including pregnancy), national origin, age (40 or older), disability, or genetic information.

It is also illegal to harass someone because they have complained about discrimination, filed a charge of discrimination, or participated in an employment discrimination investigation or lawsuit.

Harassment can take the form of slurs, graffiti, offensive or derogatory comments, or other verbal or physical conduct. Sexual harassment (including unwelcome sexual advances, requests for sexual favors, and other conduct of a sexual nature) is also unlawful. Although the law does not prohibit simple teasing, offhand comments, or isolated incidents that are not very serious, harassment is illegal if it is so frequent or severe that it creates a hostile or offensive work environment or if it results in an adverse employment decision (such as the victim being fired or demoted).

The harasser can be the victim's supervisor, a supervisor in another area, a coworker, or someone who is not an employee of the employer, such as a client or customer.

Harassment outside of the workplace may also be illegal if there is a link with the workplace. For example, if a supervisor harasses an employee while driving the employee to a meeting. (See further information on harassment at http://www.eeoc.gov/laws/practices/harassment.cfm)

TERMS AND CONDITIONS OF EMPLOYMENT

The law makes it illegal for an employer to make any employment decision because of a person's race, color, religion, sex (including pregnancy), national origin, age (40 or older), disability, or genetic information. That means an employer may not discriminate when it comes to such things as hiring, firing, promotions, and pay. It also means an employer may not discriminate, for example, when granting breaks, approving leave, assigning work stations, or setting any other term or condition of employment—however small.

PREEMPLOYMENT INQUIRIES (GENERAL)

As a general rule, the information obtained and requested through the preemployment process should be limited to those essential for determining if a person is qualified for the job; whereas, information regarding race, sex, national origin, age, and religion are irrelevant in such determinations.

Employers are explicitly prohibited from making preemployment inquiries about disability.

Although state and federal equal opportunity laws do not clearly forbid employers from making preemployment inquiries that relate to, or disproportionately screen out members based on race, color, sex, national origin, religion, or age, such inquiries may be used as evidence of an employer's intent to discriminate unless the questions asked can be justified by some business purpose.

Therefore, inquiries about organizations, clubs, societies, and lodges of which an applicant may be a member. Any other questions that may indicate the applicant's race, sex, national origin, disability status, age, religion, color or ancestry if answered, should generally be avoided.

Similarly, employers should not ask for a photograph of an applicant. If needed for identification purposes, a photograph may be obtained after an offer of employment is made and accepted.

Further information may be found on the EEOC website on preemployment and

- Race
- Height and weight
- Credit rating or economic status
- Religious affiliation or beliefs
- Citizenship
- Marital status, number of children
- Gender
- Arrest and conviction
- Security/background checks for certain religious or ethnic groups
- Disability
- Medical questions and examinations

DRESS CODE

In general, an employer may establish a dress code that applies to all employees or employees within certain job categories. However, there are a few possible exceptions.

While an employer may require all workers to follow a uniform dress code even if the dress code conflicts with some workers' ethnic beliefs or practices, a dress code must not treat some employees less favorably because of their national origin. For example, a dress code that prohibits certain kinds of ethnic dress, such as traditional African or East Indian attire, but otherwise permits casual dress would treat some employees less favorably because of their national origin.

Moreover, if the dress code conflicts with an employee's religious practices and the employee requests an accommodation, the employer must modify the dress code or permit an exception to the dress code unless doing so would result in undue hardship.

Similarly, if an employee requests an accommodation to the dress code because of his disability, the employer must modify the dress code or permit an exception to the dress code, unless doing so would result in undue hardship.

CONSTRUCTIVE DISCHARGE/FORCED TO RESIGN

Discriminatory practices under the laws EEOC enforces also include constructive discharge or forcing an employee to resign by making the work environment so intolerable a reasonable person would not be able to stay.*

CHAPTER QUESTIONS

1. In addition to Title VII, other federal laws prohibiting discrimination include:
    a. ADA
    b. ADEA

---

* Prohibited Employment Policies/Practices, EEOC website (2011).

    c. Civil Rights Act of 1871

    d. All of the above

2. The federal law specifically prohibiting discrimination based on sexual orientation is:

    a. ADA

    b. ADEA

    c. GINA

    d. None of the above

3. Federal law supercedes _____ law and state law supercedes _____ laws.

    a. Federal and state

    b. State and local

    c. Local and county

    d. None of the above

4. The Civil Rights Act of 1866 is used primarily for

    a. Disability discrimination

    b. Age discrimination

    c. Race discrimination

    d. All of the above

5. How would you respond to the hypothetical situation at the start of the chapter?

Answers: 1—a; 2—d; 3—b; 4—c; 5—various responses possible.

Title VII of the Civil Rights Act of 1964

AN ACT,

To enforce the constitutional right to vote, to confer jurisdiction upon the district courts of the United States to provide injunctive relief against discrimination in public accommodations, to authorize the Attorney General to institute suits to protect constitutional rights in public facilities and public education, to extend the Commission on Civil Rights, to prevent discrimination in federally assisted programs, to establish a Commission on Equal Employment Opportunity, and for other purposes.

Be it enacted by the Senate and House of Representatives of the United States of America in Congress assembled, That this Act may be cited as the "Civil Rights Act of 1964".

DEFINITIONS

SEC. 2000e. [Section 701]
   For the purposes of this subchapter-

(a) The term "person" includes one or more individuals, governments, governmental agencies, political subdivisions, labor unions, partnerships, associations, corporations, legal representatives, mutual companies, joint-stock companies, trusts, unincorporated organizations, trustees, trustees in cases under Title 11 *[originally, bankruptcy]*, or receivers.

(b) The term "employer" means a person engaged in an industry affecting commerce who has fifteen or more employees for each working day in each of twenty or more calendar weeks in the current or preceding calendar year, and any agent of such a person, but such term does not include (1) the United States, a corporation wholly owned by the Government of the United States, an Indian tribe, or any department or agency of the District of Columbia subject by statute to procedures of the competitive service (as defined in section 2102 of Title 5 *[United States Code]*), or

(2) a bona fide private membership club (other than a labor organization) which is exempt from taxation under section 501(c) of Title 26 *[the Internal Revenue Code of 1986]*, except that during the first year after March 24, 1972 *[the date of enactment of the Equal Employment Opportunity Act of 1972]*, persons having fewer than twenty-five employees (and their agents) shall not be considered employers.

(c) The term "employment agency" means any person regularly undertaking with or without compensation to procure employees for an employer or to procure for employees opportunities to work for an employer and includes an agent of such a person.

(d) The term "labor organization" means a labor organization engaged in an industry affecting commerce, and any agent of such an organization, and includes any organization of any kind, any agency, or employee representation committee, group, association, or plan so engaged in which employees participate and which exists for the purpose, in whole or in part, of dealing with employers concerning grievances, labor disputes, wages, rates of pay, hours, or other terms or conditions of employment, and any conference, general committee, joint or system board, or joint council so engaged which is subordinate to a national or international labor organization.

(e) A labor organization shall be deemed to be engaged in an industry affecting commerce if (1) it maintains or operates a hiring hall or hiring office which procures employees for an employer or procures for employees opportunities to work for an employer, or (2) the number of its members (or, where it is a labor organization composed of other labor organizations or their representatives, if the aggregate number of the members of such other labor organization) is (A) twenty-five or more during the first year after March 24, 1972 *[the date of enactment of the Equal Employment Opportunity Act of 1972]*, or (B) fifteen or more thereafter, and such labor organization-

   (1) is the certified representative of employees under the provisions of the National Labor Relations Act, as amended *[29 U.S.C. 151 et seq.]*, or the Railway Labor Act, as amended *[45 U.S.C. 151 et seq.]*;

   (2) although not certified, is a national or international labor organization or a local labor organization recognized or acting as the representative of employees of an employer or employers engaged in an industry affecting commerce; or

   (3) has chartered a local labor organization or subsidiary body which is representing or actively seeking to represent employees of employers within the meaning of paragraph (1) or (2); or

   (4) has been chartered by a labor organization representing or actively seeking to represent employees within the meaning of paragraph (1) or (2) as the local or subordinate body through which such employees may enjoy membership or become affiliated with such labor organization; or

   (5) is a conference, general committee, joint or system board, or joint council subordinate to a national or international labor organization, which includes a labor organization engaged in an industry affecting commerce within the meaning of any of the preceding paragraphs of this subsection.

(f) The term "employee" means an individual employed by an employer, except that the term "employee" shall not include any person elected to public office in any State or political subdivision of any State by the qualified voters thereof, or any person chosen by such officer to be on such officer's personal staff, or an appointee on the policy making level or an immediate adviser with respect to the exercise of the constitutional or legal powers of the office. The exemption set forth in the preceding sentence shall not include employees subject to the civil service laws of a State government, governmental agency or political subdivision. With respect to employment in a foreign country, such term includes an individual who is a citizen of the United States.

(g) The term "commerce" means trade, traffic, commerce, transportation, transmission, or communication among the several States; or between a State and any place outside thereof; or within the District of Columbia, or a possession of the United States; or between points in the same State but through a point outside thereof.

(h) The term "industry affecting commerce" means any activity, business, or industry in commerce or in which a labor dispute would hinder or obstruct commerce or the free flow of commerce and includes any activity or industry "affecting commerce" within the meaning of the Labor-Management Reporting and Disclosure Act of 1959 *[29 U.S.C. 401 et seq.]*, and further includes any governmental industry, business, or activity.

(i) The term "State" includes a State of the United States, the District of Columbia, Puerto Rico, the Virgin Islands, American Samoa, Guam, Wake Island, the Canal Zone, and Outer Continental Shelf lands defined in the Outer Continental Shelf Lands Act *[43 U.S.C. 1331 et seq.]*.

(j) The term "religion" includes all aspects of religious observance and practice, as well as belief, unless an employer demonstrates that he is unable to reasonably accommodate to an employee's or prospective employee's religious observance or practice without undue hardship on the conduct of the employer's business.

(k) The terms "because of sex" or "on the basis of sex" include, but are not limited to, because of or on the basis of pregnancy, childbirth, or related medical conditions; and women affected by pregnancy, childbirth, or related medical conditions shall be treated the same for all employment-related purposes, including receipt of benefits under fringe benefit programs, as other persons not so affected but similar in their ability or inability to work, and nothing in section 2000e-2(h) of this title *[section 703(h)]* shall be interpreted to permit otherwise. This subsection shall not require an employer to pay for health insurance benefits for abortion, except where the life of the mother would be endangered if the fetus were carried to term, or except where medical complications have arisen from an abortion: Provided, that nothing herein shall preclude an employer from providing abortion benefits or otherwise affect bargaining agreements in regard to abortion.

(l) The term "complaining party" means the Commission, the Attorney General, or a person who may bring an action or proceeding under this subchapter.

(m) The term "demonstrates" means meets the burdens of production and persuasion.

(n) The term "respondent" means an employer, employment agency, labor organization, joint labor management committee controlling apprenticeship or other training or retraining program, including an on-the-job training program, or Federal entity subject to section 2000e-16 of this title.

APPLICABILITY TO FOREIGN AND RELIGIOUS EMPLOYMENT

SEC. 2000e-1. *[Section 702]*

(a) Inapplicability of subchapter to certain aliens and employees of religious entities

  This subchapter shall not apply to an employer with respect to the employment of aliens outside any State, or to a religious corporation, association, educational institution, or society with respect to the employment of individuals of a particular religion to perform work connected with the carrying on by such corporation, association, educational institution, or society of its activities.

(b) Compliance with statute as violative of foreign law

  It shall not be unlawful under section 2000e-2 or 2000e-3 of this title *[section 703 or 704]* for an employer (or a corporation controlled by an employer), labor organization, employment agency, or joint labor-management committee controlling apprenticeship or other training or retraining (including on-the-job training programs) to take any action otherwise prohibited by such section, with respect to an employee in a workplace in a foreign country if compliance with such section would cause such employer (or such corporation), such organization, such agency, or such committee to violate the law of the foreign country in which such workplace is located.

(c) Control of corporation incorporated in foreign country

  (1) If an employer controls a corporation whose place of incorporation is a foreign country, any practice prohibited by section 2000e-2 or 2000e-3 of this title *[section 703 or 704]* engaged in by such corporation shall be presumed to be engaged in by such employer.

  (2) Sections 2000e-2 and 2000e-3 of this title *[sections 703 and 704]* shall not apply with respect to the foreign operations of an employer that is a foreign person not controlled by an American employer.

  (3) For purposes of this subsection, the determination of whether an employer controls a corporation shall be based on-

    (A) the interrelation of operations;

    (B) the common management;

    (C) the centralized control of labor relations; and

(D) the common ownership or financial control, of the employer and the corporation.

UNLAWFUL EMPLOYMENT PRACTICES

SEC. 2000e-2. *[Section 703]*

(a) Employer practices

  It shall be an unlawful employment practice for an employer-

  (1) to fail or refuse to hire or to discharge any individual, or otherwise to discriminate against any individual with respect to his compensation,

terms, conditions, or privileges of employment, because of such individual's race, color, religion, sex, or national origin; or

(2) to limit, segregate, or classify his employees or applicants for employment in any way which would deprive or tend to deprive any individual of employment opportunities or otherwise adversely affect his status as an employee, because of such individual's race, color, religion, sex, or national origin.

(b) Employment agency practices

It shall be an unlawful employment practice for an employment agency to fail or refuse to refer for employment, or otherwise to discriminate against, any individual because of his race, color, religion, sex, or national origin, or to classify or refer for employment any individual on the basis of his race, color, religion, sex, or national origin.

(c) Labor organization practices

It shall be an unlawful employment practice for a labor organization-

(1) to exclude or to expel from its membership, or otherwise to discriminate against, any individual because of his race, color, religion, sex, or national origin;

(2) to limit, segregate, or classify its membership or applicants for membership, or to classify or fail or refuse to refer for employment any individual, in any way which would deprive or tend to deprive any individual of employment opportunities, or would limit such employment opportunities or otherwise adversely affect his status as an employee or as an applicant for employment, because of such individual's race, color, religion, sex, or national origin; or

(3) to cause or attempt to cause an employer to discriminate against an individual in violation of this section.

(d) Training programs

It shall be an unlawful employment practice for any employer, labor organization, or joint labor–management committee controlling apprenticeship or other training or retraining, including on-the-job training programs to discriminate against any individual because of his race, color, religion, sex, or national origin in admission to, or employment in, any program established to provide apprenticeship or other training.

(e) Businesses or enterprises with personnel qualified on basis of religion, sex, or national origin; educational institutions with personnel of particular religion

Notwithstanding any other provision of this subchapter, (1) it shall not be an unlawful employment practice for an employer to hire and employ employees, for an employment agency to classify, or refer for employment any individual, for a labor organization to classify its membership or to classify or refer for employment any individual, or for an employer, labor organization, or joint labor–management committee controlling apprenticeship or other training or retraining programs to admit or employ any individual in any such program, on the basis of his religion, sex, or national origin in those certain instances where religion, sex, or national origin is a bona fide occupational qualification reasonably necessary to the normal operation of that particular business or enterprise, and (2) it shall not be an unlawful employment practice for a school, college, university, or other educational institution or institution of learning to hire and employ employees of a particular religion if such school, college, university, or other educational institution or institution of learning is, in whole or in substantial part, owned, supported, controlled, or managed by a particular religion or by a particular religious corporation, association, or society, or if the curriculum of such school, college, university, or other educational institution or institution of learning is directed toward the propagation of a particular religion.

(f) Members of Communist Party or Communist-action or Communist-front organizations

As used in this subchapter, the phrase "unlawful employment practice" shall not be deemed to include any action or measure taken by an employer, labor organization, joint labor–management committee, or employment agency with respect to an individual who is a member of the Communist Party of the United States or of any other organization required to register as a Communist-action or Communist-front organization by final order of the Subversive Activities Control Board pursuant to the Subversive Activities Control Act of 1950 *[50 U.S.C. 781 et seq.]*.

(g) National security

Notwithstanding any other provision of this subchapter, it shall not be an unlawful employment practice for an employer to fail or refuse to hire and employ any individual for any position, for an employer to discharge any individual from any position, or for an employment agency to fail or refuse to refer any individual for employment in any position, or for a labor organization to fail or refuse to refer any individual for employment in any position, if-

(1) the occupancy of such position, or access to the premises in or upon which any part of the duties of such position is performed or is to be performed, is subject to any requirement imposed in the interest of the national security of the United States under any security program in effect pursuant to or administered under any statute of the United States or any Executive order of the President; and

(2) such individual has not fulfilled or has ceased to fulfill that requirement.

(h) Seniority or merit system; quantity or quality of production; ability tests; compensation based on sex and authorized by minimum wage provisions

Notwithstanding any other provision of this subchapter, it shall not be an unlawful employment practice for an employer to apply different standards of compensation, or different terms, conditions, or privileges of employment pursuant to a bona fide seniority or merit system, or a system which measures earnings by quantity or quality of production or to employees who work in different locations, provided that such differences are not the result of an intention to discriminate because of race, color, religion, sex, or national origin, nor shall it be an unlawful employment practice for an employer to give and to act upon the results of any professionally developed ability test provided that such test, its administration or action upon the results is not designed, intended or used to discriminate because of race, color, religion, sex or national origin. It shall not be an unlawful employment practice under this subchapter for any employer to differentiate upon the basis of sex in determining the amount of the wages or compensation paid or to be paid to employees of such employer if such differentiation is authorized by the provisions of section 206(d) of Title 29 *[section 6(d) of the Labor Standards Act of 1938, as amended]*.

(i) Businesses or enterprises extending preferential treatment to Indians

Nothing contained in this subchapter shall apply to any business or enterprise on or near an Indian reservation with respect to any publicly announced employment practice of such business or enterprise under which a preferential treatment is given to any individual because he is an Indian living on or near a reservation.

(j) Preferential treatment not to be granted on account of existing number or percentage imbalance

Nothing contained in this subchapter shall be interpreted to require any employer, employment agency, labor organization, or joint labor-management committee subject to this subchapter to grant preferential treatment to any individual or to any group because of the race, color, religion, sex, or national origin of such individual or group on account of an imbalance which may exist with respect to the total number or percentage of persons of any race, color, religion, sex, or national origin employed by any employer, referred or classified for employment by any employment agency or labor organization, admitted to membership or classified by any labor organization, or admitted to, or employed in, any apprenticeship or other training program, in comparison with the total number or percentage of persons of such race, color, religion, sex, or national origin in any community, State, section, or other area, or in the available work force in any community, State, section, or other area.

(k) Burden of proof in disparate impact cases

(1) (A)   An unlawful employment practice based on disparate impact is established under this subchapter only if-

(i)   a complaining party demonstrates that a respondent uses a particular employment practice that causes a disparate impact on the basis of race, color, religion, sex, or national origin and the respondent fails to demonstrate that the challenged practice is job related for the position in question and consistent with business necessity; or

(ii)   the complaining party makes the demonstration described in subparagraph (C) with respect to an alternative employment practice and the respondent refuses to adopt such alternative employment practice.

(B) (i) With respect to demonstrating that a particular employment practice causes a disparate impact as described in subparagraph (A)(i), the complaining party shall demonstrate that each particular challenged employment practice causes a disparate impact, except that if the complaining party can demonstrate to the court that the elements of a respondent's decisionmaking process are not capable of separation for analysis, the decisionmaking process may be analyzed as one employment practice.

(ii) If the respondent demonstrates that a specific employment practice does not cause the disparate impact, the respondent shall not be required to demonstrate that such practice is required by business necessity.

(C) The demonstration referred to by subparagraph (A)(ii) shall be in accordance with the law as it existed on June 4, 1989, with respect to the concept of "alternative employment practice".

(2) A demonstration that an employment practice is required by business necessity may not be used as a defense against a claim of intentional discrimination under this subchapter.

(3) Notwithstanding any other provision of this subchapter, a rule barring the employment of an individual who currently and knowingly uses or possesses a controlled substance, as defined in schedules I and II of section 102(6) of the Controlled Substances Act (21 U.S.C. 802(6)), other than the use or possession of a drug taken under the supervision of a licensed health care professional, or any other use or possession authorized by the Controlled Substances Act *[21 U.S.C. 801 et seq.]* or any other provision of Federal law, shall be considered an unlawful employment practice under this subchapter only if such rule is adopted or applied with an intent to discriminate because of race, color, religion, sex, or national origin.

(l) Prohibition of discriminatory use of test scores

It shall be an unlawful employment practice for a respondent, in connection with the selection or referral of applicants or candidates for employment or promotion, to adjust the scores of, use different cutoff scores for, or otherwise alter the results of, employment related tests on the basis of race, color, religion, sex, or national origin.

(m) Impermissible consideration of race, color, religion, sex, or national origin in employment practices

Except as otherwise provided in this subchapter, an unlawful employment practice is established when the complaining party demonstrates that race, color, religion, sex, or national origin was a motivating factor for any employment practice, even though other factors also motivated the practice.

(n) Resolution of challenges to employment practices implementing litigated or consent judgments or orders

   (1) (A) Notwithstanding any other provision of law, and except as provided in paragraph (2), an employment practice that implements and is within the scope of a litigated or consent judgment or order that resolves a claim of employment discrimination under the Constitution or Federal civil rights laws may not be challenged under the circumstances described in subparagraph (B).

    (B) A practice described in subparagraph (A) may not be challenged in a claim under the Constitution or Federal civil rights laws-

      (i) by a person who, prior to the entry of the judgment or order described in subparagraph (A), had-

        (I) actual notice of the proposed judgment or order sufficient to apprise such person that such judgment or order might adversely affect the interests and legal rights of such person and that an opportunity was available to present objections to such judgment or order by a future date certain; and

        (II) a reasonable opportunity to present objections to such judgment or order; or

      (ii) by a person whose interests were adequately represented by another person who had previously challenged the judgment or order on the same legal grounds and with a similar factual situation, unless there has been an intervening change in law or fact.

   (2) Nothing in this subsection shall be construed to-

    (A) alter the standards for intervention under rule 24 of the Federal Rules of Civil Procedure or apply to the rights of parties who have successfully intervened pursuant to such rule in the proceeding in which the parties intervened;

    (B) apply to the rights of parties to the action in which a litigated or consent judgment or order was entered, or of members of a class represented or sought to be represented in such action, or of members of a group on whose behalf relief was sought in such action by the Federal Government;

    (C) prevent challenges to a litigated or consent judgment or order on the ground that such judgment or order was obtained through collusion or fraud, or is transparently invalid or was entered by a court lacking subject matter jurisdiction; or

    (D) authorize or permit the denial to any person of the due process of law required by the Constitution.

   (3) Any action not precluded under this subsection that challenges an employment consent judgment or order described in paragraph (1) shall be brought in the court, and if possible before the judge, that entered such judgment or order. Nothing in this subsection shall preclude a transfer of such action pursuant to section 1404 of Title 28 [*United States Code*].

OTHER UNLAWFUL EMPLOYMENT PRACTICES

SEC. 2000e-3. *[Section 704]*

(a) Discrimination for making charges, testifying, assisting, or participating in enforcement proceedings

It shall be an unlawful employment practice for an employer to discriminate against any of his employees or applicants for employment, for an employment agency, or joint labor–management committee controlling apprenticeship or other training or retraining, including on-the-job training programs, to discriminate against any individual, or for a labor organization to discriminate against any member thereof or applicant for membership, because he has opposed any practice made an unlawful employment practice by this subchapter, or because he has made a charge, testified, assisted, or participated in any manner in an investigation, proceeding, or hearing under this subchapter.

(b) Printing or publication of notices or advertisements indicating prohibited preference, limitation, specification, or discrimination; occupational qualification exception

It shall be an unlawful employment practice for an employer, labor organization, employment agency, or joint labor–management committee controlling apprenticeship or other training or retraining, including on-the-job training programs, to print or publish or cause to be printed or published any notice or advertisement relating to employment by such an employer or membership in or any classification or referral for employment by such a labor organization, or relating to any classification or referral for employment by such an employment agency, or relating to admission to, or employment in, any program established to provide apprenticeship or other training by such a joint labor–management committee, indicating any preference, limitation, specification, or discrimination, based on race, color, religion, sex, or national origin, except that such a notice or advertisement may indicate a preference, limitation, specification, or discrimination based on religion, sex, or national origin when religion, sex, or national origin is a bona fide occupational qualification for employment.

EQUAL EMPLOYMENT OPPORTUNITY COMMISSION

SEC. 2000e-4. *[Section 705]*

(a) Creation; composition; political representation; appointment; term; vacancies; Chairman and Vice Chairman; duties of Chairman; appointment of personnel; compensation of personnel

There is hereby created a Commission to be known as the Equal Employment Opportunity Commission, which shall be composed of five members, not more than three of whom shall be members of the same political party. Members of the Commission shall be appointed by the President by and with the advice and consent of the Senate for a term of five years. Any individual chosen to fill a vacancy shall be appointed only for the unexpired term of the member whom he shall succeed, and all members of the Commission shall continue to serve until their successors are appointed and qualified, except that no such member of the Commission shall continue to serve (1) for more than sixty days when the Congress is in session unless a nomination to fill such vacancy shall have been submitted to the Senate, or (2) after the adjournment sine die of the session of the Senate in which such nomination was submitted. The President shall designate one member to serve as Chairman of the Commission, and one member to serve as Vice Chairman. The Chairman shall be responsible on behalf of the Commission for the administrative operations of the Commission, and, except as provided in subsection (b) of this section, shall appoint, in accordance with the provisions of Title 5 *[United States Code]* governing appointments in the competitive service, such officers, agents, attorneys, administrative law judges *[originally, hearing examiners]*, and employees as he deems necessary to assist it in the performance of its functions and to fix their compensation in accordance with the provisions of chapter 51 and subchapter III of chapter 53 of Title 5 *[United States Code]*, relating to classification and General Schedule pay rates: Provided, That assignment, removal, and compensation of administrative law judges *[originally, hearing examiners]* shall be in accordance with sections 3105, 3344, 5372, and 7521 of Title 5 *[United States Code]*.

(b) General Counsel; appointment; term; duties; representation by attorneys and Attorney General

   (1) There shall be a General Counsel of the Commission appointed by the President, by and with the advice and consent of the Senate, for a term of four years. The General Counsel shall have responsibility for the conduct of litigation as provided in sections 2000e-5 and 2000e-6 of this title *[sections 706 and 707]*. The General Counsel shall have such other duties as the Commission may prescribe or as may be provided by law and shall concur with the Chairman of the Commission on the appointment and supervision of regional attorneys. The General Counsel of the Commission on the effective date of this Act shall continue in such position and perform the functions specified in this subsection until a successor is appointed and qualified.

   (2) Attorneys appointed under this section may, at the direction of the Commission, appear for and represent the Commission in any case in court, provided that the Attorney General shall conduct all litigation to which the Commission is a party in the Supreme Court pursuant to this subchapter.

(c) Exercise of powers during vacancy; quorum

   A vacancy in the Commission shall not impair the right of the remaining members to exercise all the powers of the Commission and three members thereof shall constitute a quorum.

(d) Seal; judicial notice

The Commission shall have an official seal which shall be judicially noticed.

(e) Reports to Congress and the President

The Commission shall at the close of each fiscal year report to the Congress and to the President concerning the action it has taken *[originally, the names, salaries, and duties of all individuals in its employ]* and the moneys it has disbursed. It shall make such further reports on the cause of and means of eliminating discrimination and such recommendations for further legislation as may appear desirable.

(f) Principal and other offices

The principal office of the Commission shall be in or near the District of Columbia, but it may meet or exercise any or all its powers at any other place. The Commission may establish such regional or State offices as it deems necessary to accomplish the purpose of this subchapter.

(g) Powers of Commission

The Commission shall have power-

(1) to cooperate with and, with their consent, utilize regional, State, local, and other agencies, both public and private, and individuals;

(2) to pay to witnesses whose depositions are taken or who are summoned before the Commission or any of its agents the same witness and mileage fees as are paid to witnesses in the courts of the United States;

(3) to furnish to persons subject to this subchapter such technical assistance as they may request to further their compliance with this subchapter or an order issued thereunder;

(4) upon the request of (i) any employer, whose employees or some of them, or (ii) any labor organization, whose members or some of them, refuse or threaten to refuse to cooperate in effectuating the provisions of this subchapter, to assist in such effectuation by conciliation or such other remedial action as is provided by this subchapter;

(5) to make such technical studies as are appropriate to effectuate the purposes and policies of this subchapter and to make the results of such studies available to the public;

(6) to intervene in a civil action brought under section 2000e-5 of this title *[section 706]* by an aggrieved party against a respondent other than a government, governmental agency or political subdivision.

(h) Cooperation with other departments and agencies in performance of educational or promotional activities; outreach activities

(1) The Commission shall, in any of its educational or promotional activities, cooperate with other departments and agencies in the performance of such educational and promotional activities.

(2) In exercising its powers under this subchapter, the Commission shall carry out educational and outreach activities (including dissemination of information in languages other than English) targeted to-

(A) individuals who historically have been victims of employment discrimination and have not been equitably served by the Commission; and

(B)   individuals on whose behalf the Commission has authority to enforce any other law prohibiting employment discrimination, concerning rights and obligations under this subchapter or such law, as the case may be.

(i) Personnel subject to political activity restrictions

All officers, agents, attorneys, and employees of the Commission shall be subject to the provisions of section 7324 of Title 5 *[originally, section 9 of the Act of August 2, 1939, as amended (the Hatch Act)]*, notwithstanding any exemption contained in such section.

(j) Technical Assistance Training Institute

(1)   The Commission shall establish a Technical Assistance Training Institute, through which the Commission shall provide technical assistance and training regarding the laws and regulations enforced by the Commission.

(2)   An employer or other entity covered under this subchapter shall not be excused from compliance with the requirements of this subchapter because of any failure to receive technical assistance under this subsection.

(3)   There are authorized to be appropriated to carry out this subsection such sums as may be necessary for fiscal year 1992.

(k) EEOC Education, Technical Assistance, and Training Revolving Fund

(1)   There is hereby established in the Treasury of the United States a revolving fund to be known as the "EEOC Education, Technical Assistance, and Training Revolving Fund" (hereinafter in this subsection referred to as the "Fund") and to pay the cost (including administrative and personnel expenses) of providing education, technical assistance, and training relating to laws administered by the Commission. Monies in the Fund shall be available without fiscal year limitation to the Commission for such purposes.

(2)   (A)   The Commission shall charge fees in accordance with the provisions of this paragraph to offset the costs of education, technical assistance, and training provided with monies in the Fund. Such fees for any education, technical assistance, or training--

(i)   shall be imposed on a uniform basis on persons and entities receiving such education, assistance, or training,

(ii)   shall not exceed the cost of providing such education, assistance, and training, and

(iii)   with respect to each person or entity receiving such education, assistance, or training, shall bear a reasonable relationship to the cost of providing such education, assistance, or training to such person or entity.

(B)   Fees received under subparagraph (A) shall be deposited in the Fund by the Commission.

(C)   The Commission shall include in each report made under subsection (e) of this section information with respect to the operation of the Fund, including information, presented in the aggregate, relating to--

(i)   the number of persons and entities to which the Commission provided education, technical assistance, or training with monies in the Fund, in the fiscal year for which such report is prepared,

(ii) the cost to the Commission to provide such education, technical assistance, or training to such persons and entities, and

(iii) the amount of any fees received by the Commission from such persons and entities for such education, technical assistance, or training.

(3) The Secretary of the Treasury shall invest the portion of the Fund not required to satisfy current expenditures from the Fund, as determined by the Commission, in obligations of the United States or obligations guaranteed as to principal by the United States. Investment proceeds shall be deposited in the Fund.

(4) There is hereby transferred to the Fund $1,000,000 from the Salaries and Expenses appropriation of the Commission.

ENFORCEMENT PROVISIONS

SEC. 2000e-5. *[Section 706]*

(a) Power of Commission to prevent unlawful employment practices

The Commission is empowered, as hereinafter provided, to prevent any person from engaging in any unlawful employment practice as set forth in section 2000e-2 or 2000e-3 of this title *[section 703 or 704]*.

(b) Charges by persons aggrieved or member of Commission of unlawful employment practices by employers, etc.; filing; allegations; notice to respondent; contents of notice; investigation by Commission; contents of charges; prohibition on disclosure of charges; determination of reasonable cause; conference, conciliation, and persuasion for elimination of unlawful practices; prohibition on disclosure of informal endeavors to end unlawful practices; use of evidence in subsequent proceedings; penalties for disclosure of information; time for determination of reasonable cause

Whenever a charge is filed by or on behalf of a person claiming to be aggrieved, or by a member of the Commission, alleging that an employer, employment agency, labor organization, or joint labor-management committee controlling apprenticeship or other training or retraining, including on-the-job training programs, has engaged in an unlawful employment practice, the Commission shall serve a notice of the charge (including the date, place and circumstances of the alleged unlawful employment practice) on such employer, employment agency, labor organization, or joint labor-management committee (hereinafter referred to as the "respondent") within ten days, and shall make an investigation thereof. Charges shall be in writing under oath or affirmation and shall contain such information and be in such form as the Commission requires. Charges shall not be made public by the Commission. If the Commission determines after such investigation that there is not reasonable cause to believe that the charge is true, it shall dismiss the charge and promptly notify the person claiming to be aggrieved and the respondent of its action. In determining whether reasonable cause exists, the Commission shall accord substantial weight to final findings and orders made by State or local authorities in proceedings commenced under State or local law pursuant to the requirements of subsections (c) and (d) of this section. If the Commission determines after such investigation that there is reasonable cause to believe that the charge is true, the Commission shall endeavor to eliminate any such alleged unlawful employment practice by informal methods of conference, conciliation, and persuasion. Nothing said or done during and as a part of such informal endeavors may be made public by the Commission, its officers or employees, or used as evidence in a subsequent proceeding without the written consent of the persons concerned. Any person who makes public information in violation of this subsection shall be fined not more than $1,000 or imprisoned for not more than one year, or both. The Commission shall make its determination on reasonable cause as promptly as possible and, so far as practicable, not later than one hundred and twenty days from the filing of the charge or, where applicable under subsection (c) or (d) of this section, from the date upon which the Commission is authorized to take action with respect to the charge.

(c) State or local enforcement proceedings; notification of State or local authority; time for filing charges with Commission; commencement of proceedings

In the case of an alleged unlawful employment practice occurring in a State, or political subdivision of a State, which has a State or local law prohibiting the unlawful employment practice alleged and establishing or authorizing a State or local authority to grant or seek relief from such practice or to institute criminal proceedings with respect thereto upon receiving notice thereof, no charge may be filed under subsection (a) of this section by the person aggrieved before the expiration of sixty days after proceedings have been commenced under the State or local law, unless such proceedings have been earlier terminated, provided that such sixty-day period shall be extended to one hundred and twenty days during the first year after the effective date of such State or local law. If any requirement for the commencement of such proceedings is imposed by a State or local authority other than a requirement of the filing of a written and signed statement of the facts upon which the proceeding is based, the proceeding shall be deemed to have been commenced for the purposes of this subsection at the time such statement is sent by registered mail to the appropriate State or local authority.

(d) State or local enforcement proceedings; notification of State or local authority; time for action on charges by Commission

In the case of any charge filed by a member of the Commission alleging an unlawful employment practice occurring in a State or political subdivision of a State which has a State or local law prohibiting the practice alleged and establishing or authorizing a State or local authority to grant or seek relief from such practice or to institute criminal proceedings with respect thereto upon receiving notice thereof, the Commission shall, before taking any action with respect to such charge, notify the appropriate State or local officials and, upon request, afford them a reasonable time, but not less than sixty days (provided that such sixty-day period shall be extended to one hundred and twenty days during the first year after the effective day of such State or local law), unless a shorter period is requested, to act under such State or local law to remedy the practice alleged.

(e) Time for filing charges; time for service of notice of charge on respondent; filing of charge by Commission with State or local agency; seniority system

(1) A charge under this section shall be filed within one hundred and eighty days after the alleged unlawful employment practice occurred and notice of the charge (including the date, place and circumstances of the alleged unlawful employment practice) shall be served upon the person against whom such charge is made within ten days thereafter, except that in a case of an unlawful employment practice with respect to which the person aggrieved has initially instituted proceedings with a State or local agency with authority to grant or seek relief from such practice or to institute criminal proceedings with respect thereto upon receiving notice thereof, such charge shall be filed by or on behalf of the person aggrieved within three hundred days after the alleged unlawful employment practice occurred, or within thirty days after receiving notice that the State or local agency has terminated the proceedings under the State or local law, whichever is earlier, and a copy of such charge shall be filed by the Commission with the State or local agency.

(2) For purposes of this section, an unlawful employment practice occurs, with respect to a seniority system that has been adopted for an intentionally discriminatory purpose in violation of this subchapter (whether or not that discriminatory purpose is apparent on the face of the seniority provision), when the seniority system is adopted, when an individual becomes subject to the seniority system, or when a person aggrieved is injured by the application of the seniority system or provision of the system.

(3) (A) For purposes of this section, an unlawful employment practice occurs, with respect to discrimination in compensation in violation of this title, when a discriminatory compensation decision or other practice is adopted, when an individual becomes subject to a discriminatory compensation decision or other practice, or when an individual is affected by application of a discriminatory compensation decision or other practice, including each time wages, benefits, or other compensation is paid, resulting in whole or in part from such a decision or other practice.

(B) In addition to any relief authorized by section 1977A of the Revised Statutes (42 U.S.C. 1981a), liability may accrue and an aggrieved person may obtain relief as provided in subsection (g)(1), including recovery of back pay for up to two years preceding the filing of the charge, where the unlawful employment practices that have occurred during the charge filing period are similar or related to unlawful employment practices with regard to discrimination in compensation that occurred outside the time for filing a charge.

(f) Civil action by Commission, Attorney General, or person aggrieved; preconditions; procedure; appointment of attorney; payment of fees, costs, or security; intervention; stay of Federal proceedings; action for appropriate temporary or preliminary relief pending final disposition of charge; jurisdiction and venue of United States courts; designation of judge to hear and determine case; assignment of case for hearing; expedition of case; appointment of master

(1) If within thirty days after a charge is filed with the Commission or within thirty days after expiration of any period of reference under subsection (c) or (d) of this section, the Commission has been unable to secure from the respondent a conciliation agreement acceptable to the Commission, the Commission may bring a civil action against any respondent not a government, governmental agency, or political subdivision named in the charge. In the case of a respondent which is a government, governmental agency, or political subdivision, if the Commission has been unable to secure from the respondent a conciliation agreement acceptable to the Commission, the Commission shall take no further action and shall refer the case to the Attorney General who may bring a civil action against such respondent in the appropriate United States district court. The person or persons aggrieved shall have the right to intervene in a civil action brought by the Commission or the Attorney General in a case involving a government, governmental agency, or political subdivision. If a charge filed with the Commission pursuant to subsection (b) of this section is dismissed by the Commission, or if within one hundred and eighty days from the filing of such charge or the expiration of any period of reference under subsection (c) or (d) of this section, whichever is later, the Commission has not filed a civil action under this section or the Attorney General has not filed a civil action in a case involving a government, governmental agency, or political subdivision, or the Commission has not entered into a conciliation agreement to which the person aggrieved is a party, the Commission, or the Attorney General in a case involving a government, governmental agency, or political subdivision, shall so notify the person aggrieved and within ninety days after the giving of such notice a civil action may be brought against the respondent named in the charge (A) by the person claiming to be aggrieved or (B) if such charge was filed by a member of the Commission, by any person whom the charge alleges was aggrieved by the alleged unlawful employment practice. Upon application by the complainant and in such circumstances as the court may deem just, the court may appoint an attorney for such complainant and may authorize the commencement of the action without the payment of fees, costs, or security. Upon timely

application, the court may, in its discretion, permit the Commission, or the Attorney General in a case involving a government, governmental agency, or political subdivision, to intervene in such civil action upon certification that the case is of general public importance. Upon request, the court may, in its discretion, stay further proceedings for not more than sixty days pending the termination of State or local proceedings described in subsection (c) or (d) of this section or further efforts of the Commission to obtain voluntary compliance.

(2) Whenever a charge is filed with the Commission and the Commission concludes on the basis of a preliminary investigation that prompt judicial action is necessary to carry out the purposes of this Act, the Commission, or the Attorney General in a case involving a government, governmental agency, or political subdivision, may bring an action for appropriate temporary or preliminary relief pending final disposition of such charge. Any temporary restraining order or other order granting preliminary or temporary relief shall be issued in accordance with rule 65 of the Federal Rules of Civil Procedure. It shall be the duty of a court having jurisdiction over proceedings under this section to assign cases for hearing at the earliest practicable date and to cause such cases to be in every way expedited.

(3) Each United States district court and each United States court of a place subject to the jurisdiction of the United States shall have jurisdiction of actions brought under this subchapter. Such an action may be brought in any judicial district in the State in which the unlawful employment practice is alleged to have been committed, in the judicial district in which the employment records relevant to such practice are maintained and administered, or in the judicial district in which the aggrieved person would have worked but for the alleged unlawful employment practice, but if the respondent is not found within any such district, such an action may be brought within the judicial district in which the respondent has his principal office. For purposes of sections 1404 and 1406 of Title 28 [United States Code], the judicial district in which the respondent has his principal office shall in all cases be considered a district in which the action might have been brought.

(4) It shall be the duty of the chief judge of the district (or in his absence, the acting chief judge) in which the case is pending immediately to designate a judge in such district to hear and determine the case. In the event that no judge in the district is available to hear and determine the case, the chief judge of the district, or the acting chief judge, as the case may be, shall certify this fact to the chief judge of the circuit (or in his absence, the acting chief judge) who shall then designate a district or circuit judge of the circuit to hear and determine the case.

(5) It shall be the duty of the judge designated pursuant to this subsection to assign the case for hearing at the earliest practicable date and to cause the case to be in every way expedited. If such judge has not scheduled

the case for trial within one hundred and twenty days after issue has been joined, that judge may appoint a master pursuant to rule 53 of the Federal Rules of Civil Procedure.

(g) Injunctions; appropriate affirmative action; equitable relief; accrual of back pay; reduction of back pay; limitations on judicial orders

(1) If the court finds that the respondent has intentionally engaged in or is intentionally engaging in an unlawful employment practice charged in the complaint, the court may enjoin the respondent from engaging in such unlawful employment practice, and order such affirmative action as may be appropriate, which may include, but is not limited to, reinstatement or hiring of employees, with or without back pay (payable by the employer, employment agency, or labor organization, as the case may be, responsible for the unlawful employment practice), or any other equitable relief as the court deems appropriate. Back pay liability shall not accrue from a date more than two years prior to the filing of a charge with the Commission. Interim earnings or amounts earnable with reasonable diligence by the person or persons discriminated against shall operate to reduce the back pay otherwise allowable.

(2) (A) No order of the court shall require the admission or reinstatement of an individual as a member of a union, or the hiring, reinstatement, or promotion of an individual as an employee, or the payment to him of any back pay, if such individual was refused admission, suspended, or expelled, or was refused employment or advancement or was suspended or discharged for any reason other than discrimination on account of race, color, religion, sex, or national origin or in violation of section 2000e-3(a) of this Title *[section 704(a)]*.

(B) On a claim in which an individual proves a violation under section 2000e-2(m) of this title *[section 703(m)]* and a respondent demonstrates that the respondent would have taken the same action in the absence of the impermissible motivating factor, the court—

(i) may grant declaratory relief, injunctive relief (except as provided in clause (ii)), and attorney's fees and costs demonstrated to be directly attributable only to the pursuit of a claim under section 2000e-2(m) of this title *[section 703(m)]*; and

(ii) shall not award damages or issue an order requiring any admission, reinstatement, hiring, promotion, or payment, described in subparagraph (A).

(h) Provisions of chapter 6 of Title 29 not applicable to civil actions for prevention of unlawful practices

The provisions of chapter 6 of title 29 [the Act entitled "An Act to amend the Judicial Code and to define and limit the jurisdiction of courts sitting in equity, and for other purposes," approved March 23, 1932 (29 U.S.C. 105-115)] shall not apply with respect to civil actions brought under this section.

(i) Proceedings by Commission to compel compliance with judicial orders
In any case in which an employer, employment agency, or labor organization fails to comply with an order of a court issued in a civil action
brought under this section, the Commission may commence proceedings
to compel compliance with such order.

(j) Appeals
Any civil action brought under this section and any proceedings brought
under subsection (i) of this section shall be subject to appeal as provided in sections 1291 and 1292, Title 28 *[United States Code]*.

(k) Attorney's fee; liability of Commission and United States for costs
In any action or proceeding under this subchapter the court, in its discretion, may allow the prevailing party, other than the Commission or the
United States, a reasonable attorney's fee (including expert fees) as part
of the costs, and the Commission and the United States shall be liable
for costs the same as a private person.

CIVIL ACTIONS BY THE ATTORNEY GENERAL

SEC. 2000e-6. *[Section 707]*

(a) Complaint
Whenever the Attorney General has reasonable cause to believe that any
person or group of persons is engaged in a pattern or practice of resis-
tance to the full enjoyment of any of the rights secured by this subchapter, and that the pattern or practice is of such a nature and is intended
to deny the full exercise of the rights herein described, the Attorney
General may bring a civil action in the appropriate district court of the
United States by filing with it a complaint (1) signed by him (or in his
absence the Acting Attorney General), (2) setting forth facts pertaining
to such pattern or practice, and (3) requesting such relief, including an
application for a permanent or temporary injunction, restraining order
or other order against the person or persons responsible for such pattern
or practice, as he deems necessary to insure the full enjoyment of the
rights herein described.

(b) Jurisdiction; three-judge district court for cases of general public importance: hearing, determination, expedition of action, review by Supreme
Court; single judge district court: hearing, determination, expedition of
action
The district courts of the United States shall have and shall exercise jurisdiction of proceedings instituted pursuant to this section, and in any such
proceeding the Attorney General may file with the clerk of such court a
request that a court of three judges be convened to hear and determine
the case. Such request by the Attorney General shall be accompanied
by a certificate that, in his opinion, the case is of general public importance. A copy of the certificate and request for a three-judge court shall
be immediately furnished by such clerk to the chief judge of the circuit

(or in his absence, the presiding circuit judge of the circuit) in which the case is pending. Upon receipt of such request it shall be the duty of the chief judge of the circuit or the presiding circuit judge, as the case may be, to designate immediately three judges in such circuit, of whom at least one shall be a circuit judge and another of whom shall be a district judge of the court in which the proceeding was instituted, to hear and determine such case, and it shall be the duty of the judges so designated to assign the case for hearing at the earliest practicable date, to partici- pate in the hearing and determination thereof, and to cause the case to be in every way expedited. An appeal from the final judgment of such court will lie to the Supreme Court.

In the event the Attorney General fails to file such a request in any such proceeding, it shall be the duty of the chief judge of the district (or in his absence, the acting chief judge) in which the case is pending immedi- ately to designate a judge in such district to hear and determine the case. In the event that no judge in the district is available to hear and deter- mine the case, the chief judge of the district, or the acting chief judge, as the case may be, shall certify this fact to the chief judge of the circuit (or in his absence, the acting chief judge) who shall then designate a district or circuit judge of the circuit to hear and determine the case.

It shall be the duty of the judge designated pursuant to this section to assign the case for hearing at the earliest practicable date and to cause the case to be in every way expedited.

(c) Transfer offunctions, etc., to Commission; effective date; prerequisite to transfer; execution of functions by Commission

Effective two years after March 24, 1972 *[the date of enactment of the Equal Employment Opportunity Act of 1972]*, the functions of the Attorney General under this section shall be transferred to the Commission, together with such personnel, property, records, and unexpended bal- ances of appropriations, allocations, and other funds employed, used, held, available, or to be made available in connection with such func- tions unless the President submits, and neither House of Congress vetoes, a reorganization plan pursuant to chapter 9 of Title 5 *[United States Code]*, inconsistent with the provisions of this subsection. The Commission shall carry out such functions in accordance with subsec- tions (d) and (e) of this section.

(d) Transfer of functions, etc., not to affect suits commenced pursuant to this section prior to date of transfer

Upon the transfer of functions provided for in subsection (c) of this section, in all suits commenced pursuant to this section prior to the date of such transfer, proceedings shall continue without abatement, all court orders and decrees shall remain in effect, and the Commission shall be substi- tuted as a party for the United States of America, the Attorney General, or the Acting Attorney General, as appropriate.

(e) Investigation and action by Commission pursuant to filing of charge of discrimination; procedure

Subsequent to March 24, 1972 *[the date of enactment of the Equal
Employment Opportunity Act of 1972]*, the Commission shall have
authority to investigate and act on a charge of a pattern or practice of
discrimination, whether filed by or on behalf of a person claiming to
be aggrieved or by a member of the Commission. All such actions shall
be conducted in accordance with the procedures set forth in section
2000e-5of this title *[section 706]*.

EFFECT ON STATE LAWS

SEC. 2000e-7. *[Section 708]*

Nothing in this subchapter shall be deemed to exempt or relieve any person from
any liability, duty, penalty, or punishment provided by any present or future law of
any State or political subdivision of a State, other than any such law which purports
to require or permit the doing of any act which would be an unlawful employment
practice under this subchapter.

INVESTIGATIONS

SEC. 2000e-8. *[Section 709]*

(a) Examination and copying of evidence related to unlawful employment
practices

In connection with any investigation of a charge filed under section 2000e-5
of this title *[section 706]*, the Commission or its designated representative
shall at all reasonable times have access to, for the purposes of examination, and the right to copy any evidence of any person being investigated
or proceeded against that relates to unlawful employment practices covered by this subchapter and is relevant to the charge under investigation.

(b) Cooperation with State and local agencies administering State fair employment practices laws; participation in and contribution to research and other
projects; utilization of services; payment in advance or reimbursement;
agreements and rescission of agreements

The Commission may cooperate with State and local agencies charged with
the administration of State fair employment practices laws and, with the
consent of such agencies, may, for the purpose of carrying out its functions and duties under this subchapter and within the limitation of funds
appropriated specifically for such purpose, engage in and contribute to
the cost of research and other projects of mutual interest undertaken
by such agencies, and utilize the services of such agencies and their
employees, and, notwithstanding any other provision of law, pay by
advance or reimbursement such agencies and their employees for services rendered to assist the Commission in carrying out this subchapter. In furtherance of such cooperative efforts, the Commission may

enter into written agreements with such State or local agencies and such agreements may include provisions under which the Commission shall refrain from processing a charge in any cases or class of cases specified in such agreements or under which the Commission shall relieve any person or class of persons in such State or locality from requirements imposed under this section. The Commission shall rescind any such agreement whenever it determines that the agreement no longer serves the interest of effective enforcement of this subchapter.

(c) Execution, retention, and preservation of records; reports to Commission; training program records; appropriate relief from regulation or order for undue hardship; procedure for exemption; judicial action to compel compliance

Every employer, employment agency, and labor organization subject to this subchapter shall (1) make and keep such records relevant to the determinations of whether unlawful employment practices have been or are being committed, (2) preserve such records for such periods, and (3) make such reports therefrom as the Commission shall prescribe by regulation or order, after public hearing, as reasonable, necessary, or appropriate for the enforcement of this subchapter or the regulations or orders thereunder. The Commission shall, by regulation, require each employer, labor organization, and joint labor-management committee subject to this subchapter which controls an apprenticeship or other training program to maintain such records as are reasonably necessary to carry out the purposes of this subchapter, including, but not limited to, a list of applicants who wish to participate in such program, including the chronological order in which applications were received, and to furnish to the Commission upon request, a detailed description of the manner in which persons are selected to participate in the apprenticeship or other training program. Any employer, employment agency, labor organization, or joint labor-management committee which believes that the application to it of any regulation or order issued under this section would result in undue hardship may apply to the Commission for an exemption from the application of such regulation or order, and, if such application for an exemption is denied, bring a civil action in the United States district court for the district where such records are kept. If the Commission or the court, as the case may be, finds that the application of the regulation or order to the employer, employment agency, or labor organization in question would impose an undue hardship, the Commission or the court, as the case may be, may grant appropriate relief. If any person required to comply with the provisions of this subsection fails or refuses to do so, the United States district court for the district in which such person is found, resides, or transacts business, shall, upon application of the Commission, or the Attorney General in a case involving a government, governmental agency or political subdivision, have jurisdiction to issue to such person an order requiring him to comply.

(d) Consultation and coordination between Commission and interested State and Federal agencies in prescribing recordkeeping and reporting requirements; availability of information furnished pursuant to recordkeeping and reporting requirements; conditions on availability

In prescribing requirements pursuant to subsection (c) of this section, the Commission shall consult with other interested State and Federal agencies and shall endeavor to coordinate its requirements with those adopted by such agencies. The Commission shall furnish upon request and without cost to any State or local agency charged with the administration of a fair employment practice law information obtained pursuant to subsection (c) of this section from any employer, employment agency, labor organization, or joint labor-management committee subject to the jurisdiction of such agency. Such information shall be furnished on condition that it not be made public by the recipient agency prior to the institution of a proceeding under State or local law involving such information. If this condition is violated by a recipient agency, the Commission may decline to honor subsequent requests pursuant to this subsection.

(e) Prohibited disclosures; penalties

It shall be unlawful for any officer or employee of the Commission to make public in any manner whatever any information obtained by the Commission pursuant to its authority under this section prior to the institution of any proceeding under this subchapter involving such information. Any officer or employee of the Commission who shall make public in any manner whatever any information in violation of this subsection shall be guilty of a misdemeanor and upon conviction thereof, shall be fined not more than $1,000, or imprisoned not more than one year.

CONDUCT OF HEARINGS AND INVESTIGATIONS PURSUANT TO SECTION 161 OF TITLE 29

SEC. 2000e-9. *[Section 710]*

For the purpose of all hearings and investigations conducted by the Commission or its duly authorized agents or agencies, section 161 of Title 29 *[section 11 of the National Labor Relations Act]* shall apply.

POSTING OF NOTICES; PENALTIES

SEC. 2000e-10. *[Section 711]*

(a) Every employer, employment agency, and labor organization, as the case may be, shall post and keep posted in conspicuous places upon its premises where notices to employees, applicants for employment, and members are customarily posted a notice to be prepared or approved by the Commission setting forth excerpts from or, summaries of, the pertinent provisions of this subchapter and information pertinent to the filing of a complaint.

(b) A willful violation of this section shall be punishable by a fine of not more than $100 for each separate offense.

VETERANS' SPECIAL RIGHTS OR PREFERENCE

SEC. 2000e-11. *[Section 712]*

Nothing contained in this subchapter shall be construed to repeal or modify any Federal, State, territorial, or local law creating special rights or preference for veterans.

REGULATIONS; CONFORMITY OF REGULATIONS WITH ADMINISTRATIVE PROCEDURE PROVISIONS; RELIANCE ON INTERPRETATIONS AND INSTRUCTIONS OF COMMISSION

SEC. 2000e-12. *[Section 713]*

(a) The Commission shall have authority from time to time to issue, amend, or rescind suitable procedural regulations to carry out the provisions of this subchapter. Regulations issued under this section shall be in conformity with the standards and limitations of subchapter II of chapter 5 of Title 5 *[originally, the Administrative Procedure Act]*.

(b) In any action or proceeding based on any alleged unlawful employment practice, no person shall be subject to any liability or punishment for or on account of (1) the commission by such person of an unlawful employment practice if he pleads and proves that the act or omission complained of was in good faith, in conformity with, and in reliance on any written interpretation or opinion of the Commission, or (2) the failure of such person to publish and file any information required by any provision of this subchapter if he pleads and proves that he failed to publish and file such information in good faith, in conformity with the instructions of the Commission issued under this subchapter regarding the filing of such information. Such a defense, if established, shall be a bar to the action or proceeding, notwithstanding that (A) after such act or omission, such interpretation or opinion is modified or rescinded or is determined by judicial authority to be invalid or of no legal effect, or (B) after publishing or filing the description and annual reports, such publication or filing is determined by judicial authority not to be in conformity with the requirements of this subchapter.

APPLICATION TO PERSONNEL OF COMMISSION OF SECTIONS 111 AND 1114 OF TITLE 18; PUNISHMENT FOR VIOLATION OF SECTION 1114 OF TITLE 18

SEC. 2000e-13. *[Section 714]*

The provisions of sections 111 and 1114, Title 18 *[United States Code]*, shall apply to officers, agents, and employees of the Commission in the performance of their official duties. Notwithstanding the provisions of sections 111 and 1114 of Title 18 *[United States Code]*, whoever in violation of the provisions of section 1114 of such title kills a person while engaged in or on account of the performance of his official functions under this Act shall be punished by imprisonment for any term of years or for life.

TRANSFER OF AUTHORITY

[Administration of the duties of the Equal Employment Opportunity Coordinating Council was transferred to the Equal Employment Opportunity Commission effective July 1, 1978, under the President's Reorganization Plan of 1978.]

EQUAL EMPLOYMENT OPPORTUNITY COORDINATING COUNCIL; ESTABLISHMENT; COMPOSITION; DUTIES; REPORT TO PRESIDENT AND CONGRESS

SEC. 2000e-14. *[Section 715]*

[Original introductory text: There shall be established an Equal Employment Opportunity Coordinating Council (hereinafter referred to in this section as the Council) composed of the Secretary of Labor, the Chairman of the Equal Employment Opportunity Commission, the Attorney General, the Chairman of the United States Civil Service Commission, and the Chairman of the United States Civil Rights Commission, or their respective delegates.]

The Equal Employment Opportunity Commission *[originally, Council]* shall have the responsibility for developing and implementing agreements, policies and practices designed to maximize effort, promote efficiency, and eliminate conflict, competition, duplication and inconsistency among the operations, functions and jurisdictions of the various departments, agencies and branches of the Federal Government responsible for the implementation and enforcement of equal employment opportunity legislation, orders, and policies. On or before October 1 *[originally, July 1]* of each year, the Equal Employment Opportunity Commission *[originally, Council]* shall transmit to the President and to the Congress a report of its activities, together with such recommendations for legislative or administrative changes as it concludes are desirable to further promote the purposes of this section.

PRESIDENTIAL CONFERENCES; ACQUAINTANCE OF LEADERSHIP WITH PROVISIONS FOR EMPLOYMENT RIGHTS AND OBLIGATIONS; PLANS FOR FAIR ADMINISTRATION; MEMBERSHIP

SEC. 2000e-15. *[Section 716]*

[Original text: (a) This title shall become effective one year after the date of its enactment.

(b) Notwithstanding subsection (a), sections of this title other than sections 703, 704, 706, and 707 shall become effective immediately.

(c)] The President shall, as soon as feasible after July 2, 1964 *[the date of enactment of this title]*, convene one or more conferences for the purpose of enabling the leaders of groups whose members will be affected by this subchapter to become familiar with the rights afforded and obligations imposed by its provisions, and for the purpose of making plans which will result in the fair and effective administration of this subchapter when all of its provisions become effective. The President shall invite the participation in such conference or conferences of (1) the members of the President's Committee on Equal Employment Opportunity, (2) the members

of the Commission on Civil Rights, (3) representatives of State and local agencies engaged in furthering equal employment opportunity, (4) representatives of private agencies engaged in furthering equal employment opportunity, and (5) representatives of employers, labor organizations, and employment agencies who will be subject to this subchapter.

TRANSFER OF AUTHORITY

[Enforcement of Section 717 was transferred to the Equal Employment Opportunity Commission from the Civil Service Commission (Office of Personnel Management) effective January 1, 1979 under the President's Reorganization Plan No. 1 of 1978.]

EMPLOYMENT BY FEDERAL GOVERNMENT

SEC. 2000e-16. *[Section 717]*

(a) Discriminatory practices prohibited; employees or applicants for employment subject to coverage

All personnel actions affecting employees or applicants for employment (except with regard to aliens employed outside the limits of the United States) in military departments as defined in section 102 of Title 5 *[United States Code]*, in executive agencies *[originally, other than the General Accounting Office]* as defined in section 105 of Title 5 *[United States Code]* (including employees and applicants for employment who are paid from nonappropriated funds), in the United States Postal Service and the Postal Regulatory Commission, in those units of the Government of the District of Columbia having positions in the competitive service, and in those units of the judicial branch of the Federal Government having positions in the competitive service, in the Smithsonian Institution, and in the Government Printing Office, the Government Accountability Office, and the Library of Congress shall be made free from any discrimination based on race, color, religion, sex, or national origin.

(b) Equal Employment Opportunity Commission; enforcement powers; issuance of rules, regulations, etc.; annual review and approval of national and regional equal employment opportunity plans; review and evaluation of equal employment opportunity programs and publication of progress reports; consultations with interested parties; compliance with rules, regulations, etc.; contents of national and regional equal employment opportunity plans; authority of Librarian of Congress

Except as otherwise provided in this subsection, the Equal Employment Opportunity Commission *[originally, Civil Service Commission]* shall have authority to enforce the provisions of subsection (a) of this section through appropriate remedies, including reinstatement or hiring of employees with or without back pay, as will effectuate the policies of this section, and shall issue such rules, regulations, orders and instructions as it deems necessary and appropriate to carry out its responsibilities under this section. The Equal Employment Opportunity Commission *[originally, Civil Service Commission]* shall-

(1) be responsible for the annual review and approval of a national and regional equal employment opportunity plan which each department and agency and each appropriate unit referred to in subsection (a) of this section shall submit in order to maintain an affirmative program of equal employment opportunity for all such employees and applicants for employment;

(2) be responsible for the review and evaluation of the operation of all agency equal employment opportunity programs, periodically obtaining and publishing (on at least a semiannual basis) progress reports from each such department, agency, or unit; and

(3) consult with and solicit the recommendations of interested individuals, groups, and organizations relating to equal employment opportunity.

The head of each such department, agency, or unit shall comply with such rules, regulations, orders, and instructions which shall include a provision that an employee or applicant for employment shall be notified of any final action taken on any complaint of discrimination filed by him thereunder. The plan submitted by each department, agency, and unit shall include, but not be limited to-

(1) provision for the establishment of training and education programs designed to provide a maximum opportunity for employees to advance so as to perform at their highest potential; and

(2) a description of the qualifications in terms of training and experience relating to equal employment opportunity for the principal and operating officials of each such department, agency, or unit responsible for carrying out the equal employment opportunity program and of the allocation of personnel and resources proposed by such department, agency, or unit to carry out its equal employment opportunity program.

With respect to employment in the Library of Congress, authorities granted in this subsection to the Equal Employment Opportunity Commission *[originally, Civil Service Commission]* shall be exercised by the Librarian of Congress.

(c) Civil action by employee or applicant for employment for redress of grievances; time for bringing of action; head of department, agency, or unit as defendant

Within 90 days of receipt of notice of final action taken by a department, agency, or unit referred to in subsection (a) of this section, or by the Equal Employment Opportunity Commission *[originally, Civil Service Commission]* upon an appeal from a decision or order of such department, agency, or unit on a complaint of discrimination based on race, color, religion, sex or national origin, brought pursuant to subsection (a) of this section, Executive Order 11478 or any succeeding Executive orders, or after one hundred and eighty days from the filing of the initial charge with the department, agency, or unit or with the Equal Employment Opportunity Commission *[originally, Civil Service Commission]* on appeal from a decision or order of such department, agency, or unit until such time as final action may be taken by a department, agency, or unit, an employee or applicant for employment, if aggrieved by the final disposition of his complaint, or by the failure to take final action on his complaint, may file a civil action as provided in section 2000e-5 of this title *[section 706]*, in which civil action the head of the department, agency, or unit, as appropriate, shall be the defendant.

(d) Section 2000e-5(f) through (k) of this title applicable to civil actions
The provisions of section 2000e-5(f) through (k) of this title *[section 706(f) through (k)]*, as applicable, shall govern civil actions brought hereunder, and the same interest to compensate for delay in payment shall be available as in cases involving nonpublic parties.

(e) Government agency or official not relieved of responsibility to assure nondiscrimination in employment or equal employment opportunity
Nothing contained in this Act shall relieve any Government agency or official of its or his primary responsibility to assure nondiscrimination in employment as required by the Constitution and statutes or of its or his responsibilities under Executive Order 11478 relating to equal employment opportunity in the Federal Government.

(f) Section 2000e-5(e)(3) *[Section 706(e)(3)]* shall apply to complaints of discrimination in compensation under this section.

PROCEDURE FOR DENIAL, WITHHOLDING, TERMINATION, OR SUSPENSION OF GOVERNMENT CONTRACT SUBSEQUENT TO ACCEPTANCE BY GOVERNMENT OF AFFIRMATIVE ACTION PLAN OF EMPLOYER; TIME OF ACCEPTANCE OF PLAN

SEC. 2000e-17. *[Section 718]*

No Government contract, or portion thereof, with any employer, shall be denied, withheld, terminated, or suspended, by any agency or officer of the United States under any equal employment opportunity law or order, where such employer has an affirmative action plan which has previously been accepted by the Government for the same facility within the past twelve months without first according such employer full hearing and adjudication under the provisions of section 554 of Title 5 *[United States Code]*, and the following pertinent sections: Provided, That if such employer has deviated substantially from such previously agreed to affirmative action plan, this section shall not apply: Provided further, That for the purposes of this section an affirmative action plan shall be deemed to have been accepted by the Government at the time the appropriate compliance agency has accepted such plan unless within forty-five days thereafter the Office of Federal Contract Compliance has disapproved such plans.

Americans with Disabilities Act

To establish a clear and comprehensive prohibition of discrimination on the basis of disability.

Be it enacted by the Senate and House of Representatives of the United States of America in Congress assembled,

SECTION 1. SHORT TITLE; TABLE OF CONTENTS.

(a) Short Title—This Act may be cited as the 'Americans with Disabilities Act of 1990'.
(b) Table of Contents—The table of contents is as follows:

TITLE V—MISCELLANEOUS PROVISIONS

Sec. 501. Construction.
Sec. 502. State immunity.
Sec. 503. Prohibition against retaliation and coercion.
Sec. 504. Regulations by the Architectural and Transportation Barriers Compliance
  Board.
Sec. 505. Attorney's fees.
Sec. 506. Technical assistance.
Sec. 507. Federal wilderness areas.
Sec. 508. Transvestites.
Sec. 509. Coverage of Congress and the agencies of the legislative branch.
Sec. 510. Illegal use of drugs.
Sec. 511. Definitions.
Sec. 512. Amendments to the Rehabilitation Act.
Sec. 513. Alternative means of dispute resolution.
Sec. 514. Severability.

SEC. 2. FINDINGS AND PURPOSES.

(a) FINDINGS—The Congress finds that—

(1) some 43,000,000 Americans have one or more physical or mental
     disabilities, and this number is increasing as the population as a whole
     is growing older;

(2) historically, society has tended to isolate and segregate individuals with
     disabilities, and, despite some improvements, such forms of discrimi-
     nation against individuals with disabilities continue to be a serious and
     pervasive social problem;

(3) discrimination against individuals with disabilities persists in such
     critical areas as employment, housing, public accommodations, edu-
     cation, transportation, communication, recreation, institutionalization,
     health services, voting, and access to public services;

(4) unlike individuals who have experienced discrimination on the basis of
     race, color, sex, national origin, religion, or age, individuals who have
     experienced discrimination on the basis of disability have often had no
     legal recourse to redress such discrimination;

(5) individuals with disabilities continually encounter various forms of
     discrimination, including outright intentional exclusion, the discrimi-
     natory effects of architectural, transportation, and communication bar-
     riers, overprotective rules and policies, failure to make modifications
     to existing facilities and practices, exclusionary qualification standards
     and criteria, segregation, and relegation to lesser services, programs,
     activities, benefits, jobs, or other opportunities;

(6) census data, national polls, and other studies have documented that
     people with disabilities, as a group, occupy an inferior status in our

society, and are severely disadvantaged socially, vocationally, economi-
cally, and educationally;

(7) individuals with disabilities are a discrete and insular minority who
have been faced with restrictions and limitations, subjected to a history
of purposeful unequal treatment, and relegated to a position of political
powerlessness in our society, based on characteristics that are beyond
the control of such individuals and resulting from stereotypic assumptions not truly indicative of the individual ability of such individuals to
participate in, and contribute to, society;

(8) the Nation's proper goals regarding individuals with disabilities are to
assure equality of opportunity, full participation, independent living,
and economic self-sufficiency for such individuals; and

(9) the continuing existence of unfair and unnecessary discrimination
and prejudice denies people with disabilities the opportunity to compete on an equal basis and to pursue those opportunities for which our
free society is justifiably famous, and costs the United States billions
of dollars in unnecessary expenses resulting from dependency and
nonproductivity.

(b) PURPOSE—It is the purpose of this Act—

(1) to provide a clear and comprehensive national mandate for the elimination of discrimination against individuals with disabilities;

(2) to provide clear, strong, consistent, enforceable standards addressing
discrimination against individuals with disabilities;

(3) to ensure that the Federal Government plays a central role in enforcing the standards established in this Act on behalf of individuals with
disabilities; and

(4) to invoke the sweep of congressional authority, including the power to
enforce the Fourteenth Amendment and to regulate commerce, in order
to address the major areas of discrimination faced day-to-day by people
with disabilities.

SEC. 3. DEFINITIONS.

As used in this Act:

(1) AUXILIARY AIDS AND SERVICES—The term 'auxiliary aids and
services' includes—

(A) qualified interpreters or other effective methods of making
aurally delivered materials available to individuals with hearing
impairments;

(B) qualified readers, taped texts, or other effective methods of making visually delivered materials available to individuals with visual
impairments;

(C) acquisition or modification of equipment or devices; and

(D) other similar services and actions.

(2) DISABILITY—The term 'disability' means, with respect to an individual—

(A) a physical or mental impairment that substantially limits one or more of the major life activities of such individual;

(B) a record of such an impairment; or

(C) being regarded as having such an impairment.

(3) STATE—The term 'State' means each of the several States, the District of Columbia, the Commonwealth of Puerto Rico, Guam, American Samoa, the Virgin Islands, the Trust Territory of the Pacific Islands, and the Commonwealth of the Northern Mariana Islands.

TITLE I—EMPLOYMENT

Sec. 101. Definitions.

As used in this title:

(1) Commission—The term 'Commission' means the Equal Employment Opportunity Commission established by section 705 of the Civil Rights Act of 1964 (42 U.S.C. 2000e-4).

(2) Covered ENTITY—The term 'covered entity' means an employer, employment agency, labor organization, or joint labor-management committee.

(3) DIRECT THREAT—The term 'direct threat' means a significant risk to the health or safety of others that cannot be eliminated by reasonable accommodation.

(4) EMPLOYEE—The term 'employee' means an individual employed by an employer.

(5) EMPLOYER—

(A) IN GENERAL—The term 'employer' means a person engaged in an industry affecting commerce who has 15 or more employees for each working day in each of 20 or more calendar weeks in the current or preceding calendar year, and any agent of such person, except that, for two years following the effective date of this title, an employer means a person engaged in an industry affecting commerce who has 25 or more employees for each working day in each of 20 or more calendar weeks in the current or preceding year, and any agent of such person.

(B) EXCEPTIONS—The term 'employer' does not include—

(i) the United States, a corporation wholly owned by the government of the United States, or an Indian tribe; or

(ii) a bona fide private membership club (other than a labor organization) that is exempt from taxation under section 501(c) of the Internal Revenue Code of 1986.

(6) ILLEGAL USE OF DRUGS—

(A) IN GENERAL—The term 'illegal use of drugs' means the use of drugs, the possession or distribution of which is unlawful under the Controlled Substances Act (21 U.S.C. 812). Such term does not include the use of a drug taken under supervision by a licensed health care professional, or other uses authorized by the Controlled Substances Act or other provisions of Federal law.

　(B)　DRUGS—The term 'drug' means a controlled substance, as defined in schedules I through V of section 202 of the Controlled Substances Act.

(7) PERSON, ETC.—The terms 'person', 'labor organization', 'employment agency', 'commerce', and 'industry affecting commerce', shall have the same meaning given such terms in section 701 of the Civil Rights Act of 1964 (42 U.S.C. 2000e).

(8) QUALIFIED INDIVIDUAL WITH A DISABILITY—The term 'qualified individual with a disability' means an individual with a disability who, with or without reasonable accommodation, can perform the essential functions of the employment position that such individual holds or desires. For the purposes of this title, consideration shall be given to the employer's judgment as to what functions of a job are essential, and if an employer has prepared a written description before advertising or interviewing applicants for the job, this description shall be considered evidence of the essential functions of the job.

(9) REASONABLE ACCOMMODATION—The term 'reasonable accommodation' may include—

　(A)　making existing facilities used by employees readily accessible to and usable by individuals with disabilities; and

　(B)　job restructuring, part-time or modified work schedules, reassignment to a vacant position, acquisition or modification of equipment or devices, appropriate adjustment or modifications of examinations, training materials or policies, the provision of qualified readers or interpreters, and other similar accommodations for individuals with disabilities.

(10) UNDUE HARDSHIP—

　(A)　IN GENERAL—The term 'undue hardship' means an action requiring significant difficulty or expense, when considered in light of the factors set forth in subparagraph (B).

　(B)　FACTORS TO BE CONSIDERED—In determining whether an accommodation would impose an undue hardship on a covered entity, factors to be considered include—

　　(i)　the nature and cost of the accommodation needed under this Act;

　　(ii)　the overall financial resources of the facility or facilities involved in the provision of the reasonable accommodation; the number of persons employed at such facility; the effect on expenses and resources, or the impact otherwise of such accommodation upon the operation of the facility;

　　(iii)　the overall financial resources of the covered entity; the overall size of the business of a covered entity with respect to the number of its employees; the number, type, and location of its facilities; and

(iv) the type of operation or operations of the covered entity, including the composition, structure, and functions of the workforce of such entity; the geographic separateness, administrative, or fiscal relationship of the facility or facilities in question to the covered entity.

Sec. 102. Discrimination.

(a) GENERAL RULE—No covered entity shall discriminate against a qualified individual with a disability because of the disability of such individual in regard to job application procedures, the hiring, advancement, or discharge of employees, employee compensation, job training, and other terms, conditions, and privileges of employment.

(b) CONSTRUCTION—As used in subsection (a), the term 'discriminate' includes—

    (1) limiting, segregating, or classifying a job applicant or employee in a way that adversely affects the opportunities or status of such applicant or employee because of the disability of such applicant or employee;

    (2) participating in a contractual or other arrangement or relationship that has the effect of subjecting a covered entity's qualified applicant or employee with a disability to the discrimination prohibited by this title (such relationship includes a relationship with an employment or referral agency, labor union, an organization providing fringe benefits to an employee of the covered entity, or an organization providing training and apprenticeship programs);

    (3) utilizing standards, criteria, or methods of administration—

        (A) that have the effect of discrimination on the basis of disability; or

        (B) that perpetuate the discrimination of others who are subject to common administrative control;

    (4) excluding or otherwise denying equal jobs or benefits to a qualified individual because of the known disability of an individual with whom the qualified individual is known to have a relationship or association;

    (5)

        (A) not making reasonable accommodations to the known physical or mental limitations of an otherwise qualified individual with a disability who is an applicant or employee, unless such covered entity can demonstrate that the accommodation would impose an undue hardship on the operation of the business of such covered entity; or

        (B) denying employment opportunities to a job applicant or employee who is an otherwise qualified individual with a disability, if such denial is based on the need of such covered entity to make reasonable accommodation to the physical or mental impairments of the employee or applicant;

    (6) using qualification standards, employment tests or other selection criteria that screen out or tend to screen out an individual with a disability or a class of individuals with disabilities unless the standard, test or other selection criteria, as used by the covered entity, is shown to be

job-related for the position in question and is consistent with business necessity; and

(7) failing to select and administer tests concerning employment in the most effective manner to ensure that, when such test is administered to a job applicant or employee who has a disability that impairs sensory, manual, or speaking skills, such test results accurately reflect the skills, aptitude, or whatever other factor of such applicant or employee that such test purports to measure, rather than reflecting the impaired sensory, manual, or speaking skills of such employee or applicant (except where such skills are the factors that the test purports to measure).

(c) MEDICAL EXAMINATIONS AND INQUIRIES—

(1) IN GENERAL—The prohibition against discrimination as referred to in subsection (a) shall include medical examinations and inquiries.

(2) PREEMPLOYMENT—

(A) PROHIBITED EXAMINATION OR INQUIRY—Except as provided in paragraph (3), a covered entity shall not conduct a medical examination or make inquiries of a job applicant as to whether such applicant is an individual with a disability or as to the nature or severity of such disability.

(B) ACCEPTABLE INQUIRY—A covered entity may make preemployment inquiries into the ability of an applicant to perform job-related functions.

(3) EMPLOYMENT ENTRANCE EXAMINATION—A covered entity may require a medical examination after an offer of employment has been made to a job applicant and prior to the commencement of the employment duties of such applicant, and may condition an offer of employment on the results of such examination, if—

(A) all entering employees are subjected to such an examination regardless of disability;

(B) information obtained regarding the medical condition or history of the applicant is collected and maintained on separate forms and in separate medical files and is treated as a confidential medical record, except that—

(i) supervisors and managers may be informed regarding necessary restrictions on the work or duties of the employee and necessary accommodations;

(ii) first aid and safety personnel may be informed, when appropriate, if the disability might require emergency treatment; and

(iii) government officials investigating compliance with this Act shall be provided relevant information on request; and

(C) the results of such examination are used only in accordance with this title.

(4) EXAMINATION AND INQUIRY—

(A) PROHIBITED EXAMINATIONS AND INQUIRIES—A covered entity shall not require a medical examination and shall not make inquiries of an employee as to whether such employee is an individual with a disability or as to the nature or severity of the disability, unless such examination or inquiry is shown to be job-related and consistent with business necessity.

(B) ACCEPTABLE EXAMINATIONS AND INQUIRIES—A covered entity may conduct voluntary medical examinations, including voluntary medical histories, which are part of an employee health program available to employees at that work site. A covered entity may make inquiries into the ability of an employee to perform job-related functions.

(C) REQUIREMENT—Information obtained under subparagraph (B) regarding the medical condition or history of any employee are subject to the requirements of subparagraphs (B) and (C) of paragraph (3).

Sec. 103. Defenses.

(a) IN GENERAL—It may be a defense to a charge of discrimination under this Act that an alleged application of qualification standards, tests, or selection criteria that screen out or tend to screen out or otherwise deny a job or benefit to an individual with a disability has been shown to be job-related and consistent with business necessity, and such performance cannot be accomplished by reasonable accommodation, as required under this title.

(b) QUALIFICATION STANDARDS—The term 'qualification standards' may include a requirement that an individual shall not pose a direct threat to the health or safety of other individuals in the workplace.

(c) RELIGIOUS ENTITIES—

(1) IN GENERAL—This title shall not prohibit a religious corporation, association, educational institution, or society from giving preference in employment to individuals of a particular religion to perform work connected with the carrying on by such corporation, association, educational institution, or society of its activities.

(2) RELIGIOUS TENETS REQUIREMENT—Under this title, a religious organization may require that all applicants and employees conform to the religious tenets of such organization.

(d) List of Infectious and Communicable Diseases—

(1) IN GENERAL—The Secretary of Health and Human Services, not later than 6 months after the date of enactment of this Act, shall—

(A) review all infectious and communicable diseases which may be transmitted through handling the food supply;

(B) publish a list of infectious and communicable diseases which are transmitted through handling the food supply;

(C) publish the methods by which such diseases are transmitted; and

(D) widely disseminate such information regarding the list of diseases and their modes of transmissability to the general public.

Such list shall be updated annually.

(2) APPLICATIONS—In any case in which an individual has an infectious or communicable disease that is transmitted to others through the handling of food, that is included on the list developed by the Secretary

of Health and Human Services under paragraph (1), and which cannot be eliminated by reasonable accommodation, a covered entity may refuse to assign or continue to assign such individual to a job involving food handling.

(3) CONSTRUCTION—Nothing in this Act shall be construed to preempt, modify, or amend any State, county, or local law, ordinance, or regulation applicable to food handling which is designed to protect the public health from individuals who pose a significant risk to the health or safety of others, which cannot be eliminated by reasonable accommodation, pursuant to the list of infectious or communicable diseases and the modes of transmissability published by the Secretary of Health and Human Services.

Sec. 104. Illegal Use of Drugs and Alcohol.

(a) QUALIFIED INDIVIDUAL WITH A DISABILITY—For purposes of this title, the term 'qualified individual with a disability' shall not include any employee or applicant who is currently engaging in the illegal use of drugs, when the covered entity acts on the basis of such use.

(b) RULES OF CONSTRUCTION—Nothing in subsection (a) shall be construed to exclude as a qualified individual with a disability an individual who—

(1) has successfully completed a supervised drug rehabilitation program and is no longer engaging in the illegal use of drugs, or has otherwise been rehabilitated successfully and is no longer engaging in such use;

(2) is participating in a supervised rehabilitation program and is no longer engaging in such use; or

(3) is erroneously regarded as engaging in such use, but is not engaging in such use; except that it shall not be a violation of this Act for a covered entity to adopt or administer reasonable policies or procedures, including but not limited to drug testing, designed to ensure that an individual described in paragraph (1) or (2) is no longer engaging in the illegal use of drugs.

(c) AUTHORITY OF COVERED ENTITY—A covered entity—

(1) may prohibit the illegal use of drugs and the use of alcohol at the workplace by all employees;

(2) may require that employees shall not be under the influence of alcohol or be engaging in the illegal use of drugs at the workplace;

(3) may require that employees behave in conformance with the requirements established under the Drug-Free Workplace Act of 1988 (41 U.S.C. 701 et seq.);

(4) may hold an employee who engages in the illegal use of drugs or who is an alcoholic to the same qualification standards for employment or job performance and behavior that such entity holds other employees, even if any unsatisfactory performance or behavior is related to the drug use or alcoholism of such employee; and

(5) may, with respect to Federal regulations regarding alcohol and the illegal use of drugs, require that—

    (A) employees comply with the standards established in such regulations of the Department of Defense, if the employees of the covered entity are employed in an industry subject to such regulations, including complying with regulations (if any) that apply to employment in sensitive positions in such an industry, in the case of employees of the covered entity who are employed in such positions (as defined in the regulations of the Department of Defense);

    (B) employees comply with the standards established in such regulations of the Nuclear Regulatory Commission, if the employees of the covered entity are employed in an industry subject to such regulations, including complying with regulations (if any) that apply to employment in sensitive positions in such an industry, in the case of employees of the covered entity who are employed in such positions (as defined in the regulations of the Nuclear Regulatory Commission); and

    (C) employees comply with the standards established in such regulations of the Department of Transportation, if the employees of the covered entity are employed in a transportation industry subject to such regulations, including complying with such regulations (if any) that apply to employment in sensitive positions in such an industry, in the case of employees of the covered entity who are employed in such positions (as defined in the regulations of the Department of Transportation).

(d) DRUG TESTING—

    (1) IN GENERAL—For purposes of this title, a test to determine the illegal use of drugs shall not be considered a medical examination.

    (2) CONSTRUCTION—Nothing in this title shall be construed to encourage, prohibit, or authorize the conducting of drug testing for the illegal use of drugs by job applicants or employees or making employment decisions based on such test results.

(e) TRANSPORTATION EMPLOYEES—Nothing in this title shall be construed to encourage, prohibit, restrict, or authorize the otherwise lawful exercise by entities subject to the jurisdiction of the Department of Transportation of authority to—

    (1) test employees of such entities in, and applicants for, positions involving safety-sensitive duties for the illegal use of drugs and for on-duty impairment by alcohol; and

    (2) remove such persons who test positive for illegal use of drugs and on-duty impairment by alcohol pursuant to paragraph (1) from safety-sensitive duties in implementing subsection (c).

Sec. 105. Posting Notices.

Every employer, employment agency, labor organization, or joint labor-management committee covered under this title shall post notices in an accessible format to applicants, employees, and members describing the applicable provisions of this Act, in the manner prescribed by section 711 of the Civil Rights Act of 1964 (42 U.S.C. 2000e-10).

Sec. 106. Regulations.

Not later than 1 year after the date of enactment of this Act, the Commission shall issue regulations in an accessible format to carry out this title in accordance with subchapter II of chapter 5 of title 5, United States Code.

Sec. 107. Enforcement.

(a) POWERS, REMEDIES, AND PROCEDURES—The powers, remedies, and procedures set forth in sections 705, 706, 707, 709, and 710 of the Civil Rights Act of 1964 (42 U.S.C. 2000e-4, 2000e-5, 2000e-6, 2000e-8, and 2000e-9) shall be the powers, remedies, and procedures this title provides to the Commission, to the Attorney General, or to any person alleging discrimination on the basis of disability in violation of any provision of this Act, or regulations promulgated under section 106, concerning employment.

(b) COORDINATION—The agencies with enforcement authority for actions which allege employment discrimination under this title and under the Rehabilitation Act of 1973 shall develop procedures to ensure that administrative complaints filed under this title and under the Rehabilitation Act of 1973 are dealt with in a manner that avoids duplication of effort and prevents imposition of inconsistent or conflicting standards for the same requirements under this title and the Rehabilitation Act of 1973. The Commission, the Attorney General, and the Office of Federal Contract Compliance Programs shall establish such coordinating mechanisms (similar to provisions contained in the joint regulations promulgated by the Commission and the Attorney General at part 42 of title 28 and part 1691 of title 29, Code of Federal Regulations, and the Memorandum of Understanding between the Commission and the Office of Federal Contract Compliance Programs dated January 16, 1981 (46 Fed. Reg. 7435, January 23, 1981)) in regulations implementing this title and Rehabilitation Act of 1973 not later than 18 months after the date of enactment of this Act.

Sec. 108. Effective Date.

This title shall become effective 24 months after the date of enactment.

Title II—Public Services

Subtitle A—Prohibition Against Discrimination and Other Generally Applicable Provisions

Sec. 201. Definition.

As used in this title:

(1) PUBLIC ENTITY—The term 'public entity' means—
  (A) any State or local government;
  (B) any department, agency, special purpose district, or other instrumental-
       ity of a State or States or local government; and
  (C) the National Railroad Passenger Corporation, and any commuter
       authority (as defined in section 103(8) of the Rail Passenger Service
       Act).
(2) QUALIFIED INDIVIDUAL WITH A DISABILITY—The term 'quali-
    fied individual with a disability' means an individual with a disability who,
    with or without reasonable modifications to rules, policies, or practices,
    the removal of architectural, communication, or transportation barriers, or
    the provision of auxiliary aids and services, meets the essential eligibility
    requirements for the receipt of services or the participation in programs or
    activities provided by a public entity.

Sec. 202. Discrimination.

Subject to the provisions of this title, no qualified individual with a disability shall,
by reason of such disability, be excluded from participation in or be denied the benefits of the services, programs, or activities of a public entity, or be subjected to
discrimination by any such entity.

Sec. 203. Enforcement.

The remedies, procedures, and rights set forth in section 505 of the Rehabilitation
Act of 1973 (29 U.S.C. 794a) shall be the remedies, procedures, and rights this title
provides to any person alleging discrimination on the basis of disability in violation
of section 202.

Sec. 204. Regulations.

  (a) IN GENERAL—Not later than 1 year after the date of enactment of this
       Act, the Attorney General shall promulgate regulations in an accessible
       format that implement this subtitle. Such regulations shall not include any
       matter within the scope of the authority of the Secretary of Transportation
       under section 223, 229, or 244.
  (b) RELATIONSHIP TO OTHER REGULATIONS—Except for 'program
       accessibility, existing facilities', and 'communications', regulations under
       subsection (a) shall be consistent with this Act and with the coordination
       regulations under part 41 of title 28, Code of Federal Regulations (as pro-
       mulgated by the Department of Health, Education, and Welfare on January
       13, 1978), applicable to recipients of Federal financial assistance under sec-
       tion 504 of the Rehabilitation Act of 1973 (29 U.S.C. 794). With respect
       to 'program accessibility, existing facilities', and 'communications', such
       regulations shall be consistent with regulations and analysis as in part 39
       of title 28 of the Code of Federal Regulations, applicable to federally con-
       ducted activities under such section 504.
  (c) STANDARDS—Regulations under subsection (a) shall include stan-
       dards applicable to facilities and vehicles covered by this subtitle, other

than facilities, stations, rail passenger cars, and vehicles covered by subtitle B. Such standards shall be consistent with the minimum guidelines
and requirements issued by the Architectural and Transportation Barriers
Compliance Board in accordance with section 504(a) of this Act.

Sec. 205. Effective Date.

(a) GENERAL RULE—Except as provided in subsection (b), this subtitle shall
become effective 18 months after the date of enactment of this Act.
(b) EXCEPTION—Section 204 shall become effective on the date of enactment
of this Act.

Subtitle B—Actions Applicable to Public Transportation Provided by Public Entities
Considered Discriminatory

PART I—PUBLIC TRANSPORTATION OTHER THAN BY AIRCRAFT OR CERTAIN RAIL OPERATIONS

Sec. 221. Definitions.

As used in this part:

(1) DEMAND RESPONSIVE SYSTEM—The term 'demand responsive system' means any system of providing designated public transportation which
is not a fixed route system.
(2) DESIGNATED PUBLIC TRANSPORTATION—The term 'designated
public transportation' means transportation (other than public school transportation) by bus, rail, or any other conveyance (other than transportation
by aircraft or intercity or commuter rail transportation (as defined in section 241)) that provides the general public with general or special service
(including charter service) on a regular and continuing basis.
(3) FIXED ROUTE SYSTEM—The term 'fixed route system' means a system
of providing designated public transportation on which a vehicle is operated
along a prescribed route according to a fixed schedule.
(4) OPERATES—The term 'operates', as used with respect to a fixed route
system or demand responsive system, includes operation of such system by
a person under a contractual or other arrangement or relationship with a
public entity.
(5) PUBLIC SCHOOL TRANSPORTATION—The term 'public school transportation' means transportation by school bus vehicles of schoolchildren,
personnel, and equipment to and from a public elementary or secondary
school and school-related activities.
(6) SECRETARY—The term 'Secretary' means the Secretary of
Transportation.

Sec. 222. Public Entities Operating Fixed Route Systems.

(a) PURCHASE AND LEASE OF NEW VEHICLES—It shall be considered
discrimination for purposes of section 202 of this Act and section 504 of

the Rehabilitation Act of 1973 (29 U.S.C. 794) for a public entity which operates a fixed route system to purchase or lease a new bus, a new rapid rail vehicle, a new light rail vehicle, or any other new vehicle to be used on such system, if the solicitation for such purchase or lease is made after the 30th day following the effective date of this subsection and if such bus, rail vehicle, or other vehicle is not readily accessible to and usable by individuals with disabilities, including individuals who use wheelchairs.

(b) PURCHASE AND LEASE OF USED VEHICLES—Subject to subsection (c)(1), it shall be considered discrimination for purposes of section 202 of this Act and section 504 of the Rehabilitation Act of 1973 (29 U.S.C. 794) for a public entity which operates a fixed route system to purchase or lease, after the 30th day following the effective date of this subsection, a used vehicle for use on such system unless such entity makes demonstrated good faith efforts to purchase or lease a used vehicle for use on such system that is readily accessible to and usable by individuals with disabilities, including individuals who use wheelchairs.

(c) REMANUFACTURED VEHICLES—

(1) GENERAL RULE—Except as provided in paragraph (2), it shall be considered discrimination for purposes of section 202 of this Act and section 504 of the Rehabilitation Act of 1973 (29 U.S.C. 794) for a public entity which operates a fixed route system—

(A) to remanufacture a vehicle for use on such system so as to extend its usable life for 5 years or more, which remanufacture begins (or for which the solicitation is made) after the 30th day following the effective date of this subsection; or

(B) to purchase or lease for use on such system a remanufactured vehicle which has been remanufactured so as to extend its usable life for 5 years or more, which purchase or lease occurs after such 30th day and during the period in which the usable life is extended;

unless, after remanufacture, the vehicle is, to the maximum extent feasible, readily accessible to and usable by individuals with disabilities, including individuals who use wheelchairs.

(2) EXCEPTION FOR HISTORIC VEHICLES—

(A) GENERAL RULE—If a public entity operates a fixed route system any segment of which is included on the National Register of Historic Places and if making a vehicle of historic character to be used solely on such segment readily accessible to and usable by individuals with disabilities would significantly alter the historic character of such vehicle, the public entity only has to make (or to purchase or lease a remanufactured vehicle with) those modifications which are necessary to meet the requirements of paragraph (1) and which do not significantly alter the historic character of such vehicle.

(B) VEHICLES OF HISTORIC CHARACTER DEFINED BY REGULATIONS—For purposes of this paragraph and section 228(b), a vehicle of historic character shall be defined by the regulations issued by the Secretary to carry out this subsection.

Sec. 223. Paratransit as a Complement to Fixed Route Service.

(a) GENERAL RULE—It shall be considered discrimination for purposes of section 202 of this Act and section 504 of the Rehabilitation Act of 1973 (29 U.S.C. 794) for a public entity which operates a fixed route system (other than a system which provides solely commuter bus service) to fail to provide with respect to the operations of its fixed route system, in accordance with this section, paratransit and other special transportation services to individuals with disabilities, including individuals who use wheelchairs, that are sufficient to provide to such individuals a level of service (1) which is comparable to the level of designated public transportation services provided to individuals without disabilities using such system; or (2) in the case of response time, which is comparable, to the extent practicable, to the level of designated public transportation services provided to individuals without disabilities using such system.

(b) ISSUANCE OF REGULATIONS—Not later than 1 year after the effective date of this subsection, the Secretary shall issue final regulations to carry out this section.

(c) REQUIRED CONTENTS OF REGULATIONS—

    (1) ELIGIBLE RECIPIENTS OF SERVICE—The regulations issued under this section shall require each public entity which operates a fixed route system to provide the paratransit and other special transportation services required under this section—

        (A)

            (i) to any individual with a disability who is unable, as a result of a physical or mental impairment (including a vision impairment) and without the assistance of another individual (except an operator of a wheelchair lift or other boarding assistance device), to board, ride, or disembark from any vehicle on the system which is readily accessible to and usable by individuals with disabilities;

            (ii) to any individual with a disability who needs the assistance of a wheelchair lift or other boarding assistance device (and is able with such assistance) to board, ride, and disembark from any vehicle which is readily accessible to and usable by individuals with disabilities if the individual wants to travel on a route on the system during the hours of operation of the system at a time (or within a reasonable period of such time) when such a vehicle is not being used to provide designated public transportation on the route; and

            (iii) to any individual with a disability who has a specific impairment-related condition which prevents such individual from traveling to a boarding location or from a disembarking location on such system;

        (B) to one other individual accompanying the individual with the disability; and

(C) to other individuals, in addition to the one individual described in subparagraph (B), accompanying the individual with a disability provided that space for these additional individuals is available on the paratransit vehicle carrying the individual with a disability and that the transportation of such additional individuals will not result in a denial of service to individuals with disabilities.

For purposes of clauses (i) and (ii) of subparagraph (A), boarding or disembarking from a vehicle does not include travel to the boarding location or from the disembarking location.

(2) SERVICE AREA—The regulations issued under this section shall require the provision of paratransit and special transportation services required under this section in the service area of each public entity which operates a fixed route system, other than any portion of the service area in which the public entity solely provides commuter bus service.

(3) SERVICE CRITERIA—Subject to paragraphs (1) and (2), the regulations issued under this section shall establish minimum service criteria for determining the level of services to be required under this section.

(4) UNDUE FINANCIAL BURDEN LIMITATION—The regulations issued under this section shall provide that, if the public entity is able to demonstrate to the satisfaction of the Secretary that the provision of paratransit and other special transportation services otherwise required under this section would impose an undue financial burden on the public entity, the public entity, notwithstanding any other provision of this section (other than paragraph (5)), shall only be required to provide such services to the extent that providing such services would not impose such a burden.

(5) ADDITIONAL SERVICES—The regulations issued under this section shall establish circumstances under which the Secretary may require a public entity to provide, notwithstanding paragraph (4), paratransit and other special transportation services under this section beyond the level of paratransit and other special transportation services which would otherwise be required under paragraph (4).

(6) PUBLIC PARTICIPATION—The regulations issued under this section shall require that each public entity which operates a fixed route system hold a public hearing, provide an opportunity for public comment, and consult with individuals with disabilities in preparing its plan under paragraph (7).

(7) PLANS—The regulations issued under this section shall require that each public entity which operates a fixed route system—

(A) within 18 months after the effective date of this subsection, submit to the Secretary, and commence implementation of, a plan for providing paratransit and other special transportation services which meets the requirements of this section; and

     (B) on an annual basis thereafter, submit to the Secretary, and commence implementation of, a plan for providing such services.

  (8) PROVISION OF SERVICES BY OTHERS—The regulations issued under this section shall—

     (A) require that a public entity submitting a plan to the Secretary under this section identify in the plan any person or other public entity which is providing a paratransit or other special transportation service for individuals with disabilities in the service area to which the plan applies; and

     (B) provide that the public entity submitting the plan does not have to provide under the plan such service for individuals with disabilities.

  (9) OTHER PROVISIONS—The regulations issued under this section shall include such other provisions and requirements as the Secretary determines are necessary to carry out the objectives of this section.

(d) REVIEW OF PLAN—

  (1) GENERAL RULE—The Secretary shall review a plan submitted under this section for the purpose of determining whether or not such plan meets the requirements of this section, including the regulations issued under this section.

  (2) DISAPPROVAL—If the Secretary determines that a plan reviewed under this subsection fails to meet the requirements of this section, the Secretary shall disapprove the plan and notify the public entity which submitted the plan of such disapproval and the reasons therefore.

  (3) MODIFICATION OF DISAPPROVED PLAN—Not later than 90 days after the date of disapproval of a plan under this subsection, the public entity which submitted the plan shall modify the plan to meet the requirements of this section and shall submit to the Secretary, and commence implementation of, such modified plan.

(e) DISCRIMINATION DEFINED—As used in subsection (a), the term 'discrimination' includes—

  (1) a failure of a public entity to which the regulations issued under this section apply to submit, or commence implementation of, a plan in accordance with subsections (c)(6) and (c)(7);

  (2) a failure of such entity to submit, or commence implementation of, a modified plan in accordance with subsection (d)(3);

  (3) submission to the Secretary of a modified plan under subsection (d)(3) which does not meet the requirements of this section; or

  (4) a failure of such entity to provide paratransit or other special transportation services in accordance with the plan or modified plan the public entity submitted to the Secretary under this section.

(f) STATUTORY CONSTRUCTION—Nothing in this section shall be construed as preventing a public entity—

  (1) from providing paratransit or other special transportation services at a level which is greater than the level of such services which are required by this section,

(2) from providing paratransit or other special transportation services in addition to those paratransit and special transportation services required by this section, or

(3) from providing such services to individuals in addition to those individuals to whom such services are required to be provided by this section.

Sec. 224. Public Entity Operating a Demand-Responsive System.

If a public entity operates a demand responsive system, it shall be considered discrimination, for purposes of section 202 of this Act and section 504 of the Rehabilitation Act of 1973 (29 U.S.C. 794), for such entity to purchase or lease a new vehicle for use on such system, for which a solicitation is made after the 30th day following the effective date of this section, that is not readily accessible to and usable by individuals with disabilities, including individuals who use wheelchairs, unless such system, when viewed in its entirety, provides a level of service to such individuals equivalent to the level of service such system provides to individuals without disabilities.

Sec. 225. Temporary Relief Where Lifts Are Unavailable.

(a) GRANTING—With respect to the purchase of new buses, a public entity may apply for, and the Secretary may temporarily relieve such public entity from the obligation under section 222(a) or 224 to purchase new buses that are readily accessible to and usable by individuals with disabilities if such public entity demonstrates to the satisfaction of the Secretary—

(1) that the initial solicitation for new buses made by the public entity specified that all new buses were to be lift-equipped and were to be otherwise accessible to and usable by individuals with disabilities;

(2) the unavailability from any qualified manufacturer of hydraulic, electromechanical, or other lifts for such new buses;

(3) that the public entity seeking temporary relief has made good faith efforts to locate a qualified manufacturer to supply the lifts to the manufacturer of such buses in sufficient time to comply with such solicitation; and

(4) that any further delay in purchasing new buses necessary to obtain such lifts would significantly impair transportation services in the community served by the public entity.

(b) DURATION AND NOTICE TO CONGRESS—Any relief granted under subsection (a) shall be limited in duration by a specified date, and the appropriate committees of Congress shall be notified of any such relief granted.

(c) FRAUDULENT APPLICATION—If, at any time, the Secretary has reasonable cause to believe that any relief granted under subsection (a) was fraudulently applied for, the Secretary shall—

(1) cancel such relief if such relief is still in effect; and

(2) take such other action as the Secretary considers appropriate.

Sec. 226. New Facilities.

For purposes of section 202 of this Act and section 504 of the Rehabilitation Act of 1973 (29 U.S.C. 794), it shall be considered discrimination for a public entity to construct a new facility to be used in the provision of designated public transportation services unless such facility is readily accessible to and usable by individuals with disabilities, including individuals who use wheelchairs.

Sec. 227. Alterations of Existing Facilities.

(a) GENERAL RULE—With respect to alterations of an existing facility or part thereof used in the provision of designated public transportation services that affect or could affect the usability of the facility or part thereof, it shall be considered discrimination, for purposes of section 202 of this Act and section 504 of the Rehabilitation Act of 1973 (29 U.S.C. 794), for a public entity to fail to make such alterations (or to ensure that the alterations are made) in such a manner that, to the maximum extent feasible, the altered portions of the facility are readily accessible to and usable by individuals with disabilities, including individuals who use wheelchairs, upon the completion of such alterations. Where the public entity is undertaking an alteration that affects or could affect usability of or access to an area of the facility containing a primary function, the entity shall also make the alterations in such a manner that, to the maximum extent feasible, the path of travel to the altered area and the bathrooms, telephones, and drinking fountains serving the altered area, are readily accessible to and usable by individuals with disabilities, including individuals who use wheelchairs, upon completion of such alterations, where such alterations to the path of travel or the bathrooms, telephones, and drinking fountains serving the altered area are not disproportionate to the overall alterations in terms of cost and scope (as determined under criteria established by the Attorney General).

(b) Special Rule for Stations—

  (1) GENERAL RULE—For purposes of section 202 of this Act and section 504 of the Rehabilitation Act of 1973 (29 U.S.C. 794), it shall be considered discrimination for a public entity that provides designated public transportation to fail, in accordance with the provisions of this subsection, to make key stations (as determined under criteria established by the Secretary by regulation) in rapid rail and light rail systems readily accessible to and usable by individuals with disabilities, including individuals who use wheelchairs.

  (2) Rapid rail and light rail key stations—

    (A) ACCESSIBILITY—Except as otherwise provided in this paragraph, all key stations (as determined under criteria established by the Secretary by regulation) in rapid rail and light rail systems shall be made readily accessible to and usable by individuals with disabilities, including individuals who use wheelchairs, as soon as practicable but in no event later than the last day of the 3-year period beginning on the effective date of this paragraph.

(B) EXTENSION FOR EXTRAORDINARILY EXPENSIVE STRUCTURAL CHANGES—The Secretary may extend the 3-year period under subparagraph (A) up to a 30-year period for key stations in a rapid rail or light rail system which stations need extraordinarily expensive structural changes to, or replacement of, existing facilities; except that by the last day of the 20th year following the date of the enactment of this Act at least 2/3 of such key stations must be readily accessible to and usable by individuals with disabilities.

(3) PLANS AND MILESTONES—The Secretary shall require the appropriate public entity to develop and submit to the Secretary a plan for compliance with this subsection—

(A) that reflects consultation with individuals with disabilities affected by such plan and the results of a public hearing and public comments on such plan, and

(B) that establishes milestones for achievement of the requirements of this subsection.

Sec. 228. Public Transportation Programs and Activities in Existing Facilities and One-Car-Per-Train Rule.

(a) PUBLIC TRANSPORTATION PROGRAMS AND ACTIVITIES IN EXISTING FACILITIES—

(1) IN GENERAL—With respect to existing facilities used in the provision of designated public transportation services, it shall be considered discrimination, for purposes of section 202 of this Act and section 504 of the Rehabilitation Act of 1973 (29 U.S.C. 794), for a public entity to fail to operate a designated public transportation program or activity conducted in such facilities so that, when viewed in the entirety, the program or activity is readily accessible to and usable by individuals with disabilities.

(2) EXCEPTION—Paragraph (1) shall not require a public entity to make structural changes to existing facilities in order to make such facilities accessible to individuals who use wheelchairs, unless and to the extent required by section 227(a) (relating to alterations) or section 227(b) (relating to key stations).

(3) UTILIZATION—Paragraph (1) shall not require a public entity to which paragraph (2) applies, to provide to individuals who use wheelchairs services made available to the general public at such facilities when such individuals could not utilize or benefit from such services provided at such facilities.

(b) ONE-CAR-PER-TRAIN RULE—

(1) GENERAL RULE—Subject to paragraph (2), with respect to 2 or more vehicles operated as a train by a light or rapid rail system, for purposes of section 202 of this Act and section 504 of the Rehabilitation Act of 1973 (29 U.S.C. 794), it shall be considered discrimination for a public entity to fail to have at least 1 vehicle per train that is accessible

to individuals with disabilities, including individuals who use wheelchairs, as soon as practicable but in no event later than the last day of
the 5-year period beginning on the effective date of this section.

(2) HISTORIC TRAINS—In order to comply with paragraph (1) with respect
to the remanufacture of a vehicle of historic character which is to be used on
a segment of a light or rapid rail system which is included on the National
Register of Historic Places, if making such vehicle readily accessible to
and usable by individuals with disabilities would significantly alter the historic character of such vehicle, the public entity which operates such system
only has to make (or to purchase or lease a remanufactured vehicle with)
those modifications which are necessary to meet the requirements of section 222(c)(1) and which do not significantly alter the historic character of
such vehicle.

Sec. 229. Regulations.

(a) IN GENERAL—Not later than 1 year after the date of enactment of this
Act, the Secretary of Transportation shall issue regulations, in an accessible
format, necessary for carrying out this part (other than section 223).

(b) STANDARDS—The regulations issued under this section and section 223
shall include standards applicable to facilities and vehicles covered by this
subtitle. The standards shall be consistent with the minimum guidelines
and requirements issued by the Architectural and Transportation Barriers
Compliance Board in accordance with section 504 of this Act.

Sec. 230. Interim Accessibility Requirements.

If final regulations have not been issued pursuant to section 229, for new construction or alterations for which a valid and appropriate State or local building permit
is obtained prior to the issuance of final regulations under such section, and for
which the construction or alteration authorized by such permit begins within one
year of the receipt of such permit and is completed under the terms of such permit,
compliance with the Uniform Federal Accessibility Standards in effect at the time
the building permit is issued shall suffice to satisfy the requirement that facilities
be readily accessible to and usable by persons with disabilities as required under
sections 226 and 227, except that, if such final regulations have not been issued
one year after the Architectural and Transportation Barriers Compliance Board has
issued the supplemental minimum guidelines required under section 504(a) of this
Act, compliance with such supplemental minimum guidelines shall be necessary to
satisfy the requirement that facilities be readily accessible to and usable by persons
with disabilities prior to issuance of the final regulations.

Sec. 231. Effective Date.

(a) GENERAL RULE—Except as provided in subsection (b), this part shall
become effective 18 months after the date of enactment of this Act.

(b) EXCEPTION—Sections 222, 223 (other than subsection (a)), 224, 225,
227(b), 228(b), and 229 shall become effective on the date of enactment of
this Act.

Part II—Public Transportation by Intercity and Commuter Rail

Sec. 241. Definitions.

As used in this part:

(1) COMMUTER AUTHORITY—The term 'commuter authority' has the meaning given such term in section 103(8) of the Rail Passenger Service Act (45 U.S.C. 502(8)).

(2) COMMUTER RAIL TRANSPORTATION—The term 'commuter rail transportation' has the meaning given the term 'commuter service' in section 103(9) of the Rail Passenger Service Act (45 U.S.C. 502(9)).

(3) INTERCITY RAIL TRANSPORTATION—The term 'intercity rail transportation' means transportation provided by the National Railroad Passenger Corporation.

(4) RAIL PASSENGER CAR—The term 'rail passenger car' means, with respect to intercity rail transportation, single-level and bi-level coach cars, single-level and bi-level dining cars, single-level and bi-level sleeping cars, single-level and bi-level lounge cars, and food service cars.

(5) RESPONSIBLE PERSON—The term 'responsible person' means—

  (A) in the case of a station more than 50% of which is owned by a public entity, such public entity;

  (B) in the case of a station more than 50% of which is owned by a private party, the persons providing intercity or commuter rail transportation to such station, as allocated on an equitable basis by regulation by the Secretary of Transportation; and

  (C) in a case where no party owns more than 50% of a station, the persons providing intercity or commuter rail transportation to such station and the owners of the station, other than private party owners, as allocated on an equitable basis by regulation by the Secretary of Transportation.

(6) STATION—The term 'station' means the portion of a property located appurtenant to a right-of-way on which intercity or commuter rail transportation is operated, where such portion is used by the general public and is related to the provision of such transportation, including passenger platforms, designated waiting areas, ticketing areas, restrooms, and, where a public entity providing rail transportation owns the property, concession areas, to the extent that such public entity exercises control over the selection, design, construction, or alteration of the property, but such term does not include flag stops.

Sec. 242. Intercity and Commuter Rail Actions Considered Discriminatory.

  (a) INTERCITY RAIL TRANSPORTATION—

    (1) ONE-CAR-PER-TRAIN RULE—It shall be considered discrimination for purposes of section 202 of this Act and section 504 of the Rehabilitation Act of 1973 (29 U.S.C. 794) for a person who provides intercity rail transportation to fail to have at least one passenger car per train that is readily accessible to and usable by individuals with

disabilities, including individuals who use wheelchairs, in accordance with regulations issued under section 244, as soon as practicable, but in no event later than 5 years after the date of enactment of this Act.

(2) NEW INTERCITY CARS—

   (A) GENERAL RULE—Except as otherwise provided in this subsection with respect to individuals who use wheelchairs, it shall be considered discrimination for purposes of section 202 of this Act and section 504 of the Rehabilitation Act of 1973 (29 U.S.C. 794) for a person to purchase or lease any new rail passenger cars for use in intercity rail transportation, and for which a solicitation is made later than 30 days after the effective date of this section, unless all such rail cars are readily accessible to and usable by individuals with disabilities, including individuals who use wheelchairs, as prescribed by the Secretary of Transportation in regulations issued under section 244.

   (B) SPECIAL RULE FOR SINGLE-LEVEL PASSENGER COACHES FOR INDIVIDUALS WHO USE WHEELCHAIRS—Single-level passenger coaches shall be required to—

      (i)   be able to be entered by an individual who uses a wheelchair;

      (ii) have space to park and secure a wheelchair;

      (iii) have a seat to which a passenger in a wheelchair can transfer, and a space to fold and store such passenger's wheelchair; and

      (iv) have a restroom usable by an individual who uses a wheelchair, only to the extent provided in paragraph (3).

   (C) SPECIAL RULE FOR SINGLE-LEVEL DINING CARS FOR INDIVIDUALS WHO USE WHEELCHAIRS—Single-level dining cars shall not be required to—

      (i)   be able to be entered from the station platform by an individual who uses a wheelchair; or

      (ii) have a restroom usable by an individual who uses a wheelchair if no restroom is provided in such car for any passenger.

   (D) SPECIAL RULE FOR BI-LEVEL DINING CARS FOR INDIVIDUALS WHO USE WHEELCHAIRS—Bi-level dining cars shall not be required to—

      (i)   be able to be entered by an individual who uses a wheelchair;

      (ii) have space to park and secure a wheelchair;

      (iii) have a seat to which a passenger in a wheelchair can transfer, or a space to fold and store such passenger's wheelchair; or

      (iv) have a restroom usable by an individual who uses a wheelchair.

(3) ACCESSIBILITY OF SINGLE-LEVEL COACHES—

   (A) GENERAL RULE—It shall be considered discrimination for purposes of section 202 of this Act and section 504 of the Rehabilitation Act of 1973 (29 U.S.C. 794) for a person who provides intercity rail transportation to fail to have on each train which includes one or more single-level rail passenger coaches—

      (i)   a number of spaces—

(I)   to park and secure wheelchairs (to accommodate individuals who wish to remain in their wheelchairs) equal to not less than one-half of the number of single-level rail passenger coaches in such train; and

(II)   to fold and store wheelchairs (to accommodate individuals who wish to transfer to coach seats) equal to not less than one-half of the number of single-level rail passenger coaches in such train, as soon as practicable, but in no event later than 5 years after the date of enactment of this Act; and

(ii)   a number of spaces—

(I)   to park and secure wheelchairs (to accommodate individuals who wish to remain in their wheelchairs) equal to not less than the total number of single-level rail passenger coaches in such train; and

(II)   to fold and store wheelchairs (to accommodate individuals who wish to transfer to coach seats) equal to not less than the total number of single-level rail passenger coaches in such train, as soon as practicable, but in no event later than 10 years after the date of enactment of this Act.

(B)   LOCATION—Spaces required by subparagraph (A) shall be located in single-level rail passenger coaches or food service cars.

(C)   LIMITATION—Of the number of spaces required on a train by subparagraph (A), not more than two spaces to park and secure wheelchairs nor more than two spaces to fold and store wheelchairs shall be located in any one coach or food service car.

(D)   OTHER ACCESSIBILITY FEATURES—Single-level rail passenger coaches and food service cars on which the spaces required by subparagraph (A) are located shall have a restroom usable by an individual who uses a wheelchair and shall be able to be entered from the station platform by an individual who uses a wheelchair.

(4)   FOOD SERVICE—

(A)   SINGLE-LEVEL DINING CARS—On any train in which a single-level dining car is used to provide food service—

(i)   if such single-level dining car was purchased after the date of enactment of this Act, table service in such car shall be provided to a passenger who uses a wheelchair if—

(I)   the car adjacent to the end of the dining car through which a wheelchair may enter is itself accessible to a wheelchair;

(II)   such passenger can exit to the platform from the car such passenger occupies, move down the platform, and enter the adjacent accessible car described in subclause (I) without the necessity of the train being moved within the station; and

(III)   space to park and secure a wheelchair is available in the dining car at the time such passenger wishes to eat (if such passenger wishes to remain in a wheelchair), or space to store and fold a wheelchair is available in the dining car at the time such

passenger wishes to eat (if such passenger wishes to transfer to a dining car seat); and
(ii) appropriate auxiliary aids and services, including a hard surface on which to eat, shall be provided to ensure that other equivalent food service is available to individuals with disabilities, including individuals who use wheelchairs, and to passengers traveling with such individuals.

> Unless not practicable, a person providing intercity rail transportation shall place an accessible car adjacent to the end of a dining car described in clause (i) through which an individual who uses a wheelchair may enter.

(B) BI-LEVEL DINING CARS—On any train in which a bi-level dining car is used to provide food service—
    (i)   if such train includes a bi-level lounge car purchased after the date of enactment of this Act, table service in such lounge car shall be provided to individuals who use wheelchairs and to other passengers; and
    (ii) appropriate auxiliary aids and services, including a hard surface on which to eat, shall be provided to ensure that other equivalent food service is available to individuals with disabilities, including individuals who use wheelchairs, and to passengers traveling with such individuals.
(b) COMMUTER RAIL TRANSPORTATION—
    (1) ONE-CAR-PER-TRAIN RULE—It shall be considered discrimination for purposes of section 202 of this Act and section 504 of the Rehabilitation Act of 1973 (29 U.S.C. 794) for a person who provides commuter rail transportation to fail to have at least one passenger car per train that is readily accessible to and usable by individuals with disabilities, including individuals who use wheelchairs, in accordance with regulations issued under section 244, as soon as practicable, but in no event later than 5 years after the date of enactment of this Act.
    (2) NEW COMMUTER RAIL CARS—
        (A) GENERAL RULE—It shall be considered discrimination for purposes of section 202 of this Act and section 504 of the Rehabilitation Act of 1973 (29 U.S.C. 794) for a person to purchase or lease any new rail passenger cars for use in commuter rail transportation, and for which a solicitation is made later than 30 days after the effective date of this section, unless all such rail cars are readily accessible to and usable by individuals with disabilities, including individuals who use wheelchairs, as prescribed by the Secretary of Transportation in regulations issued under section 244.
        (B) ACCESSIBILITY—For purposes of section 202 of this Act and section 504 of the Rehabilitation Act of 1973 (29 U.S.C. 794), a requirement that a rail passenger car used in commuter rail transportation be accessible to or readily accessible to and usable by

individuals with disabilities, including individuals who use wheel-chairs, shall not be construed to require—

(i) a restroom usable by an individual who uses a wheelchair if no restroom is provided in such car for any passenger;

(ii) space to fold and store a wheelchair; or

(iii) a seat to which a passenger who uses a wheelchair can transfer.

(c) USED RAIL CARS—It shall be considered discrimination for purposes of section 202 of this Act and section 504 of the Rehabilitation Act of 1973 (29 U.S.C. 794) for a person to purchase or lease a used rail passenger car for use in intercity or commuter rail transportation, unless such person makes demonstrated good faith efforts to purchase or lease a used rail car that is readily accessible to and usable by individuals with disabilities, including individuals who use wheelchairs, as prescribed by the Secretary of Transportation in regulations issued under section 244.

(d) REMANUFACTURED RAIL CARS—

(1) REMANUFACTURING—It shall be considered discrimination for purposes of section 202 of this Act and section 504 of the Rehabilitation Act of 1973 (29 U.S.C. 794) for a person to remanufacture a rail passenger car for use in intercity or commuter rail transportation so as to extend its usable life for 10 years or more, unless the rail car, to the maximum extent feasible, is made readily accessible to and usable by individuals with disabilities, including individuals who use wheelchairs, as prescribed by the Secretary of Transportation in regulations issued under section 244.

(2) PURCHASE OR LEASE—It shall be considered discrimination for purposes of section 202 of this Act and section 504 of the Rehabilitation Act of 1973 (29 U.S.C. 794) for a person to purchase or lease a remanufactured rail passenger car for use in intercity or commuter rail transportation unless such car was remanufactured in accordance with paragraph (1).

(e) STATIONS—

(1) NEW STATIONS—It shall be considered discrimination for purposes of section 202 of this Act and section 504 of the Rehabilitation Act of 1973 (29 U.S.C. 794) for a person to build a new station for use in intercity or commuter rail transportation that is not readily accessible to and usable by individuals with disabilities, including individuals who use wheelchairs, as prescribed by the Secretary of Transportation in regulations issued under section 244.

(2) EXISTING STATIONS—

(A) FAILURE TO MAKE READILY ACCESSIBLE—

(i) GENERAL RULE—It shall be considered discrimination for purposes of section 202 of this Act and section 504 of the Rehabilitation Act of 1973 (29 U.S.C. 794) for a responsible person to fail to make existing stations in the intercity rail transportation system, and existing key stations in commuter rail transportation systems, readily accessible to and usable by individuals with disabilities, including individuals who use wheelchairs, as prescribed by the Secretary of Transportation in regulations issued under section 244.

(ii) PERIOD FOR COMPLIANCE—

(I) INTERCITY RAIL—All stations in the intercity rail transportation system shall be made readily accessible to and usable by individuals with disabilities, including individuals who use wheelchairs, as soon as practicable, but in no event later than 20 years after the date of enactment of this Act.

(II) COMMUTER RAIL—Key stations in commuter rail transportation systems shall be made readily accessible to and usable by individuals with disabilities, including individuals who use wheelchairs, as soon as practicable but in no event later than 3 years after the date of enactment of this Act, except that the time limit may be extended by the Secretary of Transportation up to 20 years after the date of enactment of this Act in a case where the raising of the entire passenger platform is the only means available of attaining accessibility or where other extraordinarily expensive structural changes are necessary to attain accessibility.

(iii) DESIGNATION OF KEY STATIONS—Each commuter authority shall designate the key stations in its commuter rail transportation system, in consultation with individuals with disabilities and organizations representing such individuals, taking into consideration such factors as high ridership and whether such station serves as a transfer or feeder station. Before the final designation of key stations under this clause, a commuter authority shall hold a public hearing.

(iv) PLANS AND MILESTONES—The Secretary of Transportation shall require the appropriate person to develop a plan for carrying out this subparagraph that reflects consultation with individuals with disabilities affected by such plan and that establishes milestones for achievement of the requirements of this subparagraph.

(B) REQUIREMENT WHEN MAKING ALTERATIONS—

(i) GENERAL RULE—It shall be considered discrimination, for purposes of section 202 of this Act and section 504 of the Rehabilitation Act of 1973 (29 U.S.C. 794), with respect to alterations of an existing station or part thereof in the intercity or commuter rail transportation systems that affect or could affect the usability of the station or part thereof, for the responsible person, owner, or person in control of the station to fail to make the alterations in such a manner that, to the maximum extent feasible, the altered portions of the station are readily accessible to and usable by individuals with disabilities, including individuals who use wheelchairs, upon completion of such alterations.

   (ii) ALTERATIONS TO A PRIMARY FUNCTION AREA—It shall be considered discrimination, for purposes of section 202 of this Act and section 504 of the Rehabilitation Act of 1973 (29 U.S.C. 794), with respect to alterations that affect or could affect the usability of or access to an area of the station containing a primary function, for the responsible person, owner, or person in control of the station to fail to make the alterations in such a manner that, to the maximum extent feasible, the path of travel to the altered area, and the bathrooms, telephones, and drinking fountains serving the altered area, are readily accessible to and usable by individuals with disabilities, including individuals who use wheelchairs, upon completion of such alterations, where such alterations to the path of travel or the bathrooms, telephones, and drinking fountains serving the altered area are not disproportionate to the overall alterations in terms of cost and scope (as determined under criteria established by the Attorney General).

(C) REQUIRED COOPERATION—It shall be considered discrimination for purposes of section 202 of this Act and section 504 of the Rehabilitation Act of 1973 (29 U.S.C. 794) for an owner, or person in control, of a station governed by subparagraph (A) or (B) to fail to provide reasonable cooperation to a responsible person with respect to such station in that responsible person's efforts to comply with such subparagraph. An owner, or person in control, of a station shall be liable to a responsible person for any failure to provide reasonable cooperation as required by this subparagraph. Failure to receive reasonable cooperation required by this subparagraph shall not be a defense to a claim of discrimination under this Act.

Sec. 243. Conformance of Accessibility Standards.

Accessibility standards included in regulations issued under this part shall be consistent with the minimum guidelines issued by the Architectural and Transportation Barriers Compliance Board under section 504(a) of this Act.

Sec. 244. Regulations.

Not later than 1 year after the date of enactment of this Act, the Secretary of Transportation shall issue regulations, in an accessible format, necessary for carrying out this part.

Sec. 245. Interim Accessibility Requirements.

(a) STATIONS—If final regulations have not been issued pursuant to section 244, for new construction or alterations for which a valid and appropriate State or local building permit is obtained prior to the issuance of final regulations under such section, and for which the construction or alteration authorized by such permit begins within one year of the receipt of such permit and is completed under the terms of such permit, compliance with the Uniform Federal Accessibility Standards in effect at the time the building permit is issued shall suffice to satisfy the requirement that stations be readily accessible to and usable by persons with disabilities as required under section 242(e), except that, if such final regulations have not been issued one year after the Architectural and Transportation Barriers Compliance Board has issued the supplemental minimum guidelines required under section 504(a) of this Act, compliance with such supplemental minimum guidelines shall be necessary to satisfy the requirement that stations be readily accessible to and usable by persons with disabilities prior to issuance of the final regulations.

(b) RAIL PASSENGER CARS—If final regulations have not been issued pursuant to section 244, a person shall be considered to have complied with the requirements of section 242 (a) through (d) that a rail passenger car be readily accessible to and usable by individuals with disabilities, if the design for such car complies with the laws and regulations (including the Minimum Guidelines and Requirements for Accessible Design and such supplemental minimum guidelines as are issued under section 504(a) of this Act) governing accessibility of such cars, to the extent that such laws and regulations are not inconsistent with this part and are in effect at the time such design is substantially completed.

Sec. 246. Effective Date.

(a) GENERAL RULE—Except as provided in subsection (b), this part shall become effective 18 months after the date of enactment of this Act.

(b) EXCEPTION—Sections 242 and 244 shall become effective on the date of enactment of this Act.

Title III—Public Accommodations and Services Operated by Private Entities

Sec. 301. Definitions.

As used in this title:

(1) COMMERCE—The term 'commerce' means travel, trade, traffic, commerce, transportation, or communication—

(A) among the several States;

(B) between any foreign country or any territory or possession and any State; or

(C) between points in the same State but through another State or foreign country.

(2) COMMERCIAL FACILITIES—The term 'commercial facilities' means facilities—

(A) that are intended for nonresidential use; and

(B) whose operations will affect commerce.

Such term shall not include railroad locomotives, railroad freight cars, railroad cabooses, railroad cars described in section 242 or covered under this title, railroad rights-of-way, or facilities that are covered or expressly exempted from coverage under the Fair Housing Act of 1968 (42 U.S.C. 3601 et seq.).

(3) DEMAND RESPONSIVE SYSTEM—The term 'demand responsive system' means any system of providing transportation of individuals by a vehicle, other than a system which is a fixed route system.

(4) FIXED ROUTE SYSTEM—The term 'fixed route system' means a system of providing transportation of individuals (other than by aircraft) on which a vehicle is operated along a prescribed route according to a fixed schedule.

(5) OVER-THE-ROAD BUS—The term 'over-the-road bus' means a bus characterized by an elevated passenger deck located over a baggage compartment.

(6) PRIVATE ENTITY—The term 'private entity' means any entity other than a public entity (as defined in section 201(1)).

(7) PUBLIC ACCOMMODATION—The following private entities are considered public accommodations for purposes of this title, if the operations of such entities affect commerce—

 (A)   an inn, hotel, motel, or other place of lodging, except for an establishment located within a building that contains not more than five rooms for rent or hire and that is actually occupied by the proprietor of such establishment as the residence of such proprietor;

 (B)   a restaurant, bar, or other establishment serving food or drink;

 (C)   a motion picture house, theater, concert hall, stadium, or other place of exhibition or entertainment;

 (D)   an auditorium, convention center, lecture hall, or other place of public gathering;

 (E)   a bakery, grocery store, clothing store, hardware store, shopping center, or other sales or rental establishment;

 (F)   a laundromat, dry-cleaner, bank, barber shop, beauty shop, travel service, shoe repair service, funeral parlor, gas station, office of an accountant or lawyer, pharmacy, insurance office, professional office of a health care provider, hospital, or other service establishment;

 (G)   a terminal, depot, or other station used for specified public transportation;

 (H)   a museum, library, gallery, or other place of public display or collection;

 (I)   a park, zoo, amusement park, or other place of recreation;

 (J)   a nursery, elementary, secondary, undergraduate, or postgraduate private school, or other place of education;

 (K)   a day care center, senior citizen center, homeless shelter, food bank, adoption agency, or other social service center establishment; and

 (L)   a gymnasium, health spa, bowling alley, golf course, or other place of exercise or recreation.

(8) RAIL AND RAILROAD—The terms 'rail' and 'railroad' have the meaning given the term 'railroad' in section 202(e) of the Federal Railroad Safety Act of 1970 (45 U.S.C. 431(e)).

(9) READILY ACHIEVABLE—The term 'readily achievable' means easily accomplishable and able to be carried out without much difficulty or expense. In determining whether an action is readily achievable, factors to be considered include—

(A)   the nature and cost of the action needed under this Act;

(B)   the overall financial resources of the facility or facilities involved in the action; the number of persons employed at such facility; the effect on expenses and resources, or the impact otherwise of such action upon the operation of the facility;

(C)   the overall financial resources of the covered entity; the overall size of the business of a covered entity with respect to the number of its employees; the number, type, and location of its facilities; and

(D)   the type of operation or operations of the covered entity, including the composition, structure, and functions of the workforce of such entity; the geographic separateness, administrative or fiscal relationship of the facility or facilities in question to the covered entity.

(10) SPECIFIED PUBLIC TRANSPORTATION—The term 'specified public transportation' means transportation by bus, rail, or any other conveyance (other than by aircraft) that provides the general public with general or special service (including charter service) on a regular and continuing basis.

(11) VEHICLE—The term 'vehicle' does not include a rail passenger car, railroad locomotive, railroad freight car, railroad caboose, or a railroad car described in section 242 or covered under this title.

Sec. 302. Prohibition of Discrimination by Public Accommodations.

(a)   GENERAL RULE—No individual shall be discriminated against on the basis of disability in the full and equal enjoyment of the goods, services, facilities, privileges, advantages, or accommodations of any place of public accommodation by any person who owns, leases (or leases to), or operates a place of public accommodation.

(b)   CONSTRUCTION—

(1)   GENERAL PROHIBITION—

(A)   ACTIVITIES—

(i)   DENIAL OF PARTICIPATION—It shall be discriminatory to subject an individual or class of individuals on the basis of a disability or disabilities of such individual or class, directly, or through contractual, licensing, or other arrangements, to a denial of the opportunity of the individual or class to participate in or benefit from the goods, services, facilities, privileges, advantages, or accommodations of an entity.

(ii)   PARTICIPATION IN UNEQUAL BENEFIT—It shall be discriminatory to afford an individual or class of individuals, on the basis of a disability or disabilities of such individual or class, directly, or through contractual, licensing, or other arrangements with the opportunity to participate in or benefit from a good, service, facility, privilege, advantage, or accommodation that is not equal to that afforded to other individuals.

(iii) SEPARATE BENEFIT—It shall be discriminatory to provide an individual or class of individuals, on the basis of a disability or disabilities of such individual or class, directly, or through contractual, licensing, or other arrangements with a good, service, facility, privilege, advantage, or accommodation that is different or separate from that provided to other individuals, unless such action is necessary to provide the individual or class of individuals with a good, service, facility, privilege, advantage, or accommodation, or other opportunity that is as effective as that provided to others.

(iv) INDIVIDUAL OR CLASS OF INDIVIDUALS—For purposes of clauses (i) through (iii) of this subparagraph, the term 'individual or class of individuals' refers to the clients or customers of the covered public accommodation that enters into the contractual, licensing or other arrangement.

(B) INTEGRATED SETTINGS—Goods, services, facilities, privileges, advantages, and accommodations shall be afforded to an individual with a disability in the most integrated setting appropriate to the needs of the individual.

(C) OPPORTUNITY TO PARTICIPATE—Notwithstanding the existence of separate or different programs or activities provided in accordance with this section, an individual with a disability shall not be denied the opportunity to participate in such programs or activities that are not separate or different.

(D) ADMINISTRATIVE METHODS—An individual or entity shall not, directly or through contractual or other arrangements, utilize standards or criteria or methods of administration—

(i) that have the effect of discriminating on the basis of disability; or

(ii) that perpetuate the discrimination of others who are subject to common administrative control.

(E) ASSOCIATION—It shall be discriminatory to exclude or otherwise deny equal goods, services, facilities, privileges, advantages, accommodations, or other opportunities to an individual or entity because of the known disability of an individual with whom the individual or entity is known to have a relationship or association.

(2) SPECIFIC PROHIBITIONS—

(A) DISCRIMINATION—For purposes of subsection (a), discrimination includes—

(i) the imposition or application of eligibility criteria that screen out or tend to screen out an individual with a disability or any class of individuals with disabilities from fully and equally enjoying any goods, services, facilities, privileges, advantages, or accommodations, unless such criteria can be shown to be necessary for the provision of the goods, services, facilities, privileges, advantages, or accommodations being offered;

(ii) a failure to make reasonable modifications in policies, practices, or procedures, when such modifications are necessary to afford such goods, services, facilities, privileges, advantages, or accommodations to individuals with disabilities, unless the entity can demonstrate that making such modifications would fundamentally alter the nature of such goods, services, facilities, privileges, advantages, or accommodations;

(iii) a failure to take such steps as may be necessary to ensure that no individual with a disability is excluded, denied services, segregated or otherwise treated differently than other individuals because of the absence of auxiliary aids and services, unless the entity can demonstrate that taking such steps would fundamentally alter the nature of the good, service, facility, privilege, advantage, or accommodation being offered or would result in an undue burden;

(iv) a failure to remove architectural barriers, and communication barriers that are structural in nature, in existing facilities, and transportation barriers in existing vehicles and rail passenger cars used by an establishment for transporting individuals (not including barriers that can only be removed through the retrofitting of vehicles or rail passenger cars by the installation of a hydraulic or other lift), where such removal is readily achievable; and

(v) where an entity can demonstrate that the removal of a barrier under clause (iv) is not readily achievable, a failure to make such goods, services, facilities, privileges, advantages, or accommodations available through alternative methods if such methods are readily achievable.

(B) FIXED ROUTE SYSTEM—

(i) ACCESSIBILITY—It shall be considered discrimination for a private entity which operates a fixed route system and which is not subject to section 304 to purchase or lease a vehicle with a seating capacity in excess of 16 passengers (including the driver) for use on such system, for which a solicitation is made after the 30th day following the effective date of this subparagraph, that is not readily accessible to and usable by individuals with disabilities, including individuals who use wheelchairs.

(ii) EQUIVALENT SERVICE—If a private entity which operates a fixed route system and which is not subject to section 304 purchases or leases a vehicle with a seating capacity of 16 passengers or less (including the driver) for use on such system after the effective date of this subparagraph that is not readily accessible to or usable by individuals with disabilities, it shall be considered discrimination for such entity to fail to operate such system so that, when viewed in its entirety, such system ensures a level of service to individuals with disabilities, including individuals who use wheelchairs, equivalent to the level of service provided to individuals without disabilities.

(C) DEMAND RESPONSIVE SYSTEM—For purposes of subsection (a), discrimination includes—

   (i) a failure of a private entity which operates a demand responsive system and which is not subject to section 304 to operate such system so that, when viewed in its entirety, such system ensures a level of service to individuals with disabilities, including individuals who use wheelchairs, equivalent to the level of service provided to individuals without disabilities; and

   (ii) the purchase or lease by such entity for use on such system of a vehicle with a seating capacity in excess of 16 passengers (including the driver), for which solicitations are made after the 30th day following the effective date of this subparagraph, that is not readily accessible to and usable by individuals with disabilities (including individuals who use wheelchairs) unless such entity can demonstrate that such system, when viewed in its entirety, provides a level of service to individuals with disabilities equivalent to that provided to individuals without disabilities.

(D) OVER-THE-ROAD BUSES—

   (i) LIMITATION ON APPLICABILITY—Subparagraphs (B) and (C) do not apply to over-the-road buses.

   (ii) ACCESSIBILITY REQUIREMENTS—For purposes of subsection (a), discrimination includes (I) the purchase or lease of an over-the-road bus which does not comply with the regulations issued under section 306(a)(2) by a private entity which provides transportation of individuals and which is not primarily engaged in the business of transporting people, and (II) any other failure of such entity to comply with such regulations.

(3) SPECIFIC CONSTRUCTION—Nothing in this title shall require an entity to permit an individual to participate in or benefit from the goods, services, facilities, privileges, advantages and accommodations of such entity where such individual poses a direct threat to the health or safety of others. The term 'direct threat' means a significant risk to the health or safety of others that cannot be eliminated by a modification of policies, practices, or procedures or by the provision of auxiliary aids or services.

Sec. 303. New Construction and Alterations in Public Accommodations and Commercial Facilities.

(a) APPLICATION OF TERM—Except as provided in subsection (b), as applied to public accommodations and commercial facilities, discrimination for purposes of section 302(a) includes—

(1) a failure to design and construct facilities for first occupancy later than 30 months after the date of enactment of this Act that are readily accessible to and usable by individuals with disabilities, except where an entity can demonstrate that it is structurally impracticable to meet the requirements of such subsection in accordance with standards set forth or incorporated by reference in regulations issued under this title; and

(2) with respect to a facility or part thereof that is altered by, on behalf of, or for the use of an establishment in a manner that affects or could affect the usability of the facility or part thereof, a failure to make alterations in such a manner that, to the maximum extent feasible, the altered portions of the facility are readily accessible to and usable by individuals with disabilities, including individuals who use wheelchairs. Where the entity is undertaking an alteration that affects or could affect usability of or access to an area of the facility containing a primary function, the entity shall also make the alterations in such a manner that, to the maximum extent feasible, the path of travel to the altered area and the bathrooms, telephones, and drinking fountains serving the altered area, are readily accessible to and usable by individuals with disabilities where such alterations to the path of travel or the bathrooms, telephones, and drinking fountains serving the altered area are not disproportionate to the overall alterations in terms of cost and scope (as determined under criteria established by the Attorney General).

(b) ELEVATOR—Subsection (a) shall not be construed to require the installation of an elevator for facilities that are less than three stories or have less than 3,000 square feet per story unless the building is a shopping center, a shopping mall, or the professional office of a health care provider or unless the Attorney General determines that a particular category of such facilities requires the installation of elevators based on the usage of such facilities.

Sec. 304. Prohibition of Discrimination in Specified Public Transportation Services Provided by Private Entities.

(a) GENERAL RULE—No individual shall be discriminated against on the basis of disability in the full and equal enjoyment of specified public transportation services provided by a private entity that is primarily engaged in the business of transporting people and whose operations affect commerce.

(b) CONSTRUCTION—For purposes of subsection (a), discrimination includes—

(1) the imposition or application by a entity described in subsection (a) of eligibility criteria that screen out or tend to screen out an individual with a disability or any class of individuals with disabilities from fully enjoying the specified public transportation services provided by the entity, unless such criteria can be shown to be necessary for the provision of the services being offered;

(2) the failure of such entity to—

    (A) make reasonable modifications consistent with those required under section 302(b)(2)(A)(ii);

    (B) provide auxiliary aids and services consistent with the requirements of section 302(b)(2)(A)(iii); and

    (C) remove barriers consistent with the requirements of section 302(b)(2)(A) and with the requirements of section 303(a)(2);

(3) the purchase or lease by such entity of a new vehicle (other than an automobile, a van with a seating capacity of less than 8 passengers, including the driver, or an over-the-road bus) which is to be used to provide specified public transportation and for which a solicitation is made after the 30th day following the effective date of this section, that is not readily accessible to and usable by individuals with disabilities, including individuals who use wheelchairs; except that the new vehicle need not be readily accessible to and usable by such individuals if the new vehicle is to be used solely in a demand responsive system and if the entity can demonstrate that such system, when viewed in its entirety, provides a level of service to such individuals equivalent to the level of service provided to the general public;

(4)

    (A) the purchase or lease by such entity of an over-the-road bus which does not comply with the regulations issued under section 306(a)(2); and

    (B) any other failure of such entity to comply with such regulations; and

(5) the purchase or lease by such entity of a new van with a seating capacity of less than 8 passengers, including the driver, which is to be used to provide specified public transportation and for which a solicitation is made after the 30th day following the effective date of this section that is not readily accessible to or usable by individuals with disabilities, including individuals who use wheelchairs; except that the new van need not be readily accessible to and usable by such individuals if the entity can demonstrate that the system for which the van is being purchased or leased, when viewed in its entirety, provides a level of service to such individuals equivalent to the level of service provided to the general public;

(6) the purchase or lease by such entity of a new rail passenger car that is to be used to provide specified public transportation, and for which a solicitation is made later than 30 days after the effective date of this paragraph, that is not readily accessible to and usable by individuals with disabilities, including individuals who use wheelchairs; and

(7) the remanufacture by such entity of a rail passenger car that is to be used to provide specified public transportation so as to extend its usable life for 10 years or more, or the purchase or lease by such entity of such a rail car, unless the rail car, to the maximum extent feasible, is made

readily accessible to and usable by individuals with disabilities, including individuals who use wheelchairs.

(c) HISTORICAL OR ANTIQUATED CARS—

(1) EXCEPTION—To the extent that compliance with subsection (b)(2)(C) or (b)(7) would significantly alter the historic or antiquated character of a historical or antiquated rail passenger car, or a rail station served exclusively by such cars, or would result in violation of any rule, regulation, standard, or order issued by the Secretary of Transportation under the Federal Railroad Safety Act of 1970, such compliance shall not be required.

(2) DEFINITION—As used in this subsection, the term 'historical or antiquated rail passenger car' means a rail passenger car—

(A) which is not less than 30 years old at the time of its use for transporting individuals;

(B) the manufacturer of which is no longer in the business of manufacturing rail passenger cars; and

(C) which—

(i) has a consequential association with events or persons significant to the past; or

(ii) embodies, or is being restored to embody, the distinctive characteristics of a type of rail passenger car used in the past, or to represent a time period which has passed.

Sec. 305. Study.

(a) PURPOSES—The Office of Technology Assessment shall undertake a study to determine—

(1) the access needs of individuals with disabilities to over-the-road buses and over-the-road bus service; and

(2) the most cost-effective methods for providing access to over-the-road buses and over-the-road bus service to individuals with disabilities, particularly individuals who use wheelchairs, through all forms of boarding options.

(b) CONTENTS—The study shall include, at a minimum, an analysis of the following:

(1) The anticipated demand by individuals with disabilities for accessible over-the-road buses and over-the-road bus service.

(2) The degree to which such buses and service, including any service required under sections 304(b)(4) and 306(a)(2), are readily accessible to and usable by individuals with disabilities.

(3) The effectiveness of various methods of providing accessibility to such buses and service to individuals with disabilities.

(4) The cost of providing accessible over-the-road buses and bus service to individuals with disabilities, including consideration of recent technological and cost saving developments in equipment and devices.

(5) Possible design changes in over-the-road buses that could enhance accessibility, including the installation of accessible restrooms which do not result in a loss of seating capacity.

(6) The impact of accessibility requirements on the continuation of over-the-road bus service, with particular consideration of the impact of such requirements on such service to rural communities.

(c) ADVISORY COMMITTEE—In conducting the study required by subsection (a), the Office of Technology Assessment shall establish an advisory committee, which shall consist of—

(1) members selected from among private operators and manufacturers of over-the-road buses;

(2) members selected from among individuals with disabilities, particularly individuals who use wheelchairs, who are potential riders of such buses; and

(3) members selected for their technical expertise on issues included in the study, including manufacturers of boarding assistance equipment and devices.

The number of members selected under each of paragraphs (1) and (2) shall be equal, and the total number of members selected under paragraphs (1) and (2) shall exceed the number of members selected under paragraph (3).

(d) DEADLINE—The study required by subsection (a), along with recommendations by the Office of Technology Assessment, including any policy options for legislative action, shall be submitted to the President and Congress within 36 months after the date of the enactment of this Act. If the President determines that compliance with the regulations issued pursuant to section 306(a)(2)(B) on or before the applicable deadlines specified in section 306(a)(2)(B) will result in a significant reduction in intercity over-the-road bus service, the President shall extend each such deadline by 1 year.

(e) REVIEW—In developing the study required by subsection (a), the Office of Technology Assessment shall provide a preliminary draft of such study to the Architectural and Transportation Barriers Compliance Board established under section 502 of the Rehabilitation Act of 1973 (29 U.S.C. 792). The Board shall have an opportunity to comment on such draft study, and any such comments by the Board made in writing within 120 days after the Board's receipt of the draft study shall be incorporated as part of the final study required to be submitted under subsection (d).

Sec. 306. Regulations.

(a) TRANSPORTATION PROVISIONS—

(1) GENERAL RULE—Not later than 1 year after the date of the enactment of this Act, the Secretary of Transportation shall issue regulations in an accessible format to carry out sections 302(b)(2) (B) and (C) and to carry out section 304 (other than subsection (b)(4)).

(2) SPECIAL RULES FOR PROVIDING ACCESS TO OVER-THE-ROAD BUSES-

   (A) INTERIM REQUIREMENTS—

      (i) ISSUANCE—Not later than 1 year after the date of the enactment of this Act, the Secretary of Transportation shall issue regulations in an accessible format to carry out sections 304(b)(4) and 302(b)(2)(D)(ii) that require each private entity which uses an over-the-road bus to provide transportation of individuals to provide accessibility to such bus; except that such regulations shall not require any structural changes in over-the-road buses in order to provide access to individuals who use wheelchairs during the effective period of such regulations and shall not require the purchase of boarding assistance devices to provide access to such individuals.

      (ii) EFFECTIVE PERIOD—The regulations issued pursuant to this subparagraph shall be effective until the effective date of the regulations issued under subparagraph (B).

   (B) FINAL REQUIREMENT—

      (i) REVIEW OF STUDY AND INTERIM REQUIREMENTS—The Secretary shall review the study submitted under section 305 and the regulations issued pursuant to subparagraph (A).

      (ii) ISSUANCE—Not later than 1 year after the date of the submission of the study under section 305, the Secretary shall issue in an accessible format new regulations to carry out sections 304(b)(4) and 302(b)(2)(D)(ii) that require, taking into account the purposes of the study under section 305 and any recommendations resulting from such study, each private entity which uses an over-the-road bus to provide transportation to individuals to provide accessibility to such bus to individuals with disabilities, including individuals who use wheelchairs.

      (iii) EFFECTIVE PERIOD—Subject to section 305(d), the regulations issued pursuant to this subparagraph shall take effect—

         (I) with respect to small providers of transportation (as defined by the Secretary), 7 years after the date of the enactment of this Act; and

         (II) with respect to other providers of transportation, 6 years after such date of enactment.

   (C) LIMITATION ON REQUIRING INSTALLATION OF ACCESSIBLE RESTROOMS—The regulations issued pursuant to this paragraph shall not require the installation of accessible restrooms in over-the-road buses if such installation would result in a loss of seating capacity.

(3) STANDARDS—The regulations issued pursuant to this subsection shall include standards applicable to facilities and vehicles covered by sections 302(b)(2) and 304.

(b) OTHER PROVISIONS—Not later than 1 year after the date of the enactment of this Act, the Attorney General shall issue regulations in an accessible format to carry out the provisions of this title not referred to in subsection (a) that include standards applicable to facilities and vehicles covered under section 302.

(c) CONSISTENCY WITH ATBCB GUIDELINES—Standards included in regulations issued under subsections (a) and (b) shall be consistent with the minimum guidelines and requirements issued by the Architectural and Transportation Barriers Compliance Board in accordance with section 504 of this Act.

(d) INTERIM ACCESSIBILITY STANDARDS—

(1) FACILITIES—If final regulations have not been issued pursuant to this section, for new construction or alterations for which a valid and appropriate State or local building permit is obtained prior to the issuance of final regulations under this section, and for which the construction or alteration authorized by such permit begins within one year of the receipt of such permit and is completed under the terms of such permit, compliance with the Uniform Federal Accessibility Standards in effect at the time the building permit is issued shall suffice to satisfy the requirement that facilities be readily accessible to and usable by persons with disabilities as required under section 303, except that, if such final regulations have not been issued one year after the Architectural and Transportation Barriers Compliance Board has issued the supplemental minimum guidelines required under section 504(a) of this Act, compliance with such supplemental minimum guidelines shall be necessary to satisfy the requirement that facilities be readily accessible to and usable by persons with disabilities prior to issuance of the final regulations.

(2) VEHICLES AND RAIL PASSENGER CARS—If final regulations have not been issued pursuant to this section, a private entity shall be considered to have complied with the requirements of this title, if any, that a vehicle or rail passenger car be readily accessible to and usable by individuals with disabilities, if the design for such vehicle or car complies with the laws and regulations (including the Minimum Guidelines and Requirements for Accessible Design and such supplemental minimum guidelines as are issued under section 504(a) of this Act) governing accessibility of such vehicles or cars, to the extent that such laws and regulations are not inconsistent with this title and are in effect at the time such design is substantially completed.

Sec. 307. Exemptions for Private Clubs and Religious Organizations.

The provisions of this title shall not apply to private clubs or establishments exempted from coverage under title II of the Civil Rights Act of 1964 (42 U.S.C. 2000-a(e)) or to religious organizations or entities controlled by religious organizations, including places of worship.

Sec. 308. Enforcement.

(a) IN GENERAL—

(1) AVAILABILITY OF REMEDIES AND PROCEDURES—The remedies and procedures set forth in section 204(a) of the Civil Rights Act of 1964 (42 U.S.C. 2000a-3(a)) are the remedies and procedures this title provides to any person who is being subjected to discrimination on the basis of disability in violation of this title or who has reasonable grounds for believing that such person is about to be subjected to discrimination in violation of section 303. Nothing in this section shall require a person with a disability to engage in a futile gesture if such person has actual notice that a person or organization covered by this title does not intend to comply with its provisions.

(2) INJUNCTIVE RELIEF—In the case of violations of sections 302(b) (2)(A)(iv) and section 303(a), injunctive relief shall include an order to alter facilities to make such facilities readily accessible to and usable by individuals with disabilities to the extent required by this title. Where appropriate, injunctive relief shall also include requiring the provision of an auxiliary aid or service, modification of a policy, or provision of alternative methods, to the extent required by this title.

(b) ENFORCEMENT BY THE ATTORNEY GENERAL—

(1) DENIAL OF RIGHTS—

(A) DUTY TO INVESTIGATE—

(i)   IN GENERAL—The Attorney General shall investigate alleged violations of this title, and shall undertake periodic reviews of compliance of covered entities under this title.

(ii) ATTORNEY GENERAL CERTIFICATION—On the application of a State or local government, the Attorney General may, in consultation with the Architectural and Transportation Barriers Compliance Board, and after prior notice and a public hearing at which persons, including individuals with disabilities, are provided an opportunity to testify against such certification, certify that a State law or local building code or similar ordinance that establishes accessibility requirements meets or exceeds the minimum requirements of this Act for the accessibility and usability of covered facilities under this title. At any enforcement proceeding under this section, such certification by the Attorney General shall be rebuttable evidence that such State law or local ordinance does meet or exceed the minimum requirements of this Act.

(B) POTENTIAL VIOLATION—If the Attorney General has reasonable cause to believe that—

(i)   any person or group of persons is engaged in a pattern or practice of discrimination under this title; or

(ii) any person or group of persons has been discriminated against under this title and such discrimination raises an issue of general public importance, the Attorney General may commence a civil action in any appropriate United States district court.

(2) AUTHORITY OF COURT—In a civil action under paragraph (1)(B), the court—

   (A) may grant any equitable relief that such court considers to be appropriate, including, to the extent required by this title—

      (i) granting temporary, preliminary, or permanent relief;

      (ii) providing an auxiliary aid or service, modification of policy, practice, or procedure, or alternative method; and

      (iii) making facilities readily accessible to and usable by individuals with disabilities;

   (B) may award such other relief as the court considers to be appropriate, including monetary damages to persons aggrieved when requested by the Attorney General; and

   (C) may, to vindicate the public interest, assess a civil penalty against the entity in an amount—

      (i) not exceeding $50,000 for a first violation; and

      (ii) not exceeding $100,000 for any subsequent violation.

(3) SINGLE VIOLATION—For purposes of paragraph (2)(C), in determining whether a first or subsequent violation has occurred, a determination in a single action, by judgment or settlement, that the covered entity has engaged in more than one discriminatory act shall be counted as a single violation.

(4) PUNITIVE DAMAGES—For purposes of subsection (b)(2)(B), the term 'monetary damages' and 'such other relief' does not include punitive damages.

(5) JUDICIAL CONSIDERATION—In a civil action under paragraph (1)(B), the court, when considering what amount of civil penalty, if any, is appropriate, shall give consideration to any good faith effort or attempt to comply with this Act by the entity. In evaluating good faith, the court shall consider, among other factors it deems relevant, whether the entity could have reasonably anticipated the need for an appropriate type of auxiliary aid needed to accommodate the unique needs of a particular individual with a disability.

Sec. 309. Examinations and Courses.

Any person that offers examinations or courses related to applications, licensing, certification, or credentialing for secondary or postsecondary education, professional, or trade purposes shall offer such examinations or courses in a place and manner accessible to persons with disabilities or offer alternative accessible arrangements for such individuals.

Sec. 310. Effective Date.

(a) GENERAL RULE—Except as provided in subsections (b) and (c), this title shall become effective 18 months after the date of the enactment of this Act.

(b) CIVIL ACTIONS—Except for any civil action brought for a violation of section 303, no civil action shall be brought for any act or omission described in section 302 which occurs—

    (1) during the first 6 months after the effective date, against businesses that employ 25 or fewer employees and have gross receipts of $1,000,000 or less; and

    (2) during the first year after the effective date, against businesses that employ 10 or fewer employees and have gross receipts of $500,000 or less.

(c) EXCEPTION—Sections 302(a) for purposes of section 302(b)(2) (B) and (C) only, 304(a) for purposes of section 304(b)(3) only, 304(b)(3), 305, and 306 shall take effect on the date of the enactment of this Act.

TITLE IV—TELECOMMUNICATIONS

Sec. 401. Telecommunications Relay Services for Hearing-Impaired and Speech-Impaired Individuals.

(a) TELECOMMUNICATIONS—Title II of the Communications Act of 1934 (47 U.S.C. 201 et seq.) is amended by adding at the end thereof the following new section:

'SEC. 225. TELECOMMUNICATIONS SERVICES FOR HEARING-IMPAIRED AND SPEECH-IMPAIRED INDIVIDUALS.

'(a) DEFINITIONS—As used in this section—

    '(1) COMMON CARRIER OR CARRIER—The term 'common carrier' or 'carrier' includes any common carrier engaged in interstate communication by wire or radio as defined in section 3(h) and any common carrier engaged in intrastate communication by wire or radio, notwithstanding sections 2(b) and 221(b).

    '(2) TDD—The term 'TDD' means a Telecommunications Device for the Deaf, which is a machine that employs graphic communication in the transmission of coded signals through a wire or radio communication system.

    '(3) TELECOMMUNICATIONS RELAY SERVICES—The term 'telecommunications relay services' means telephone transmission services that provide the ability for an individual who has a hearing impairment or speech impairment to engage in communication by wire or radio with a hearing individual in a manner that is functionally equivalent to the ability of an individual who does not have a hearing impairment or speech impairment to communicate using voice communication services by wire or radio. Such term includes services that enable two-way communication between an individual who uses a TDD or other non-voice terminal device and an individual who does not use such a device.

    '(b) AVAILABILITY OF TELECOMMUNICATIONS RELAY SERVICES—

'(1) IN GENERAL—In order to carry out the purposes established under section 1, to make available to all individuals in the United States a rapid, efficient nationwide communication service, and to increase the utility of the telephone system of the Nation, the Commission shall ensure that interstate and intrastate telecommunications relay services are available, to the extent possible and in the most efficient manner, to hearing-impaired and speech-impaired individuals in the United States.

'(2) USE OF GENERAL AUTHORITY AND REMEDIES—For the purposes of administering and enforcing the provisions of this section and the regulations prescribed thereunder, the Commission shall have the same authority, power, and functions with respect to common carriers engaged in intrastate communication as the Commission has in administering and enforcing the provisions of this title with respect to any common carrier engaged in interstate communication. Any violation of this section by any common carrier engaged in intrastate communication shall be subject to the same remedies, penalties, and procedures as are applicable to a violation of this Act by a common carrier engaged in interstate communication.

'(c) PROVISION OF SERVICES—Each common carrier providing telephone voice transmission services shall, not later than 3 years after the date of enactment of this section, provide in compliance with the regulations prescribed under this section, throughout the area in which it offers service, telecommunications relay services, individually, through designees, through a competitively selected vendor, or in concert with other carriers. A common carrier shall be considered to be in compliance with such regulations—

'(1) with respect to intrastate telecommunications relay services in any State that does not have a certified program under subsection (f) and with respect to interstate telecommunications relay services, if such common carrier (or other entity through which the carrier is providing such relay services) is in compliance with the Commission's regulations under subsection (d); or

'(2) with respect to intrastate telecommunications relay services in any State that has a certified program under subsection (f) for such State, if such common carrier (or other entity through which the carrier is providing such relay services) is in compliance with the program certified under subsection (f) for such State.

'(d) REGULATIONS-

'(1) IN GENERAL—The Commission shall, not later than 1 year after the date of enactment of this section, prescribe regulations to implement this section, including regulations that—

'(A) establish functional requirements, guidelines, and operations procedures for telecommunications relay services;

'(B) establish minimum standards that shall be met in carrying out subsection (c);

'(C) require that telecommunications relay services operate every day for 24 hours per day;

'(D) require that users of telecommunications relay services pay rates no greater than the rates paid for functionally equivalent voice communication services with respect to such factors as the duration of the call, the time of day, and the distance from point of origination to point of termination;

'(E) prohibit relay operators from failing to fulfill the obligations of common carriers by refusing calls or limiting the length of calls that use telecommunications relay services;

'(F) prohibit relay operators from disclosing the content of any relayed conversation and from keeping records of the content of any such conversation beyond the duration of the call; and

'(G) prohibit relay operators from intentionally altering a relayed conversation.

'(2) TECHNOLOGY—The Commission shall ensure that regulations prescribed to implement this section encourage, consistent with section 7(a) of this Act, the use of existing technology and do not discourage or impair the development of improved technology.

'(3) JURISDICTIONAL SEPARATION OF COSTS—

'(A) IN GENERAL—Consistent with the provisions of section 410 of this Act, the Commission shall prescribe regulations governing the jurisdictional separation of costs for the services provided pursuant to this section.

'(B) RECOVERING COSTS—Such regulations shall generally provide that costs caused by interstate telecommunications relay services shall be recovered from all subscribers for every interstate service and costs caused by intrastate telecommunications relay services shall be recovered from the intrastate jurisdiction. In a State that has a certified program under subsection (f), a State commission shall permit a common carrier to recover the costs incurred in providing intrastate telecommunications relay services by a method consistent with the requirements of this section.

'(e) ENFORCEMENT—

'(1) IN GENERAL—Subject to subsections (f) and (g), the Commission shall enforce this section.

'(2) COMPLAINT—The Commission shall resolve, by final order, a complaint alleging a violation of this section within 180 days after the date such complaint is filed.

'(f) CERTIFICATION—

'(1)   STATE DOCUMENTATION—Any State desiring to establish a State program under this section shall submit documentation to the Commission that describes the program of such State for implementing intrastate telecommunications relay services and the procedures and remedies available for enforcing any requirements imposed by the State program.

'(2)   REQUIREMENTS FOR CERTIFICATION—After review of such documentation, the Commission shall certify the State program if the Commission determines that—

'(A)   the program makes available to hearing-impaired and speech-impaired individuals, either directly, through designees, through a competitively selected vendor, or through regulation of intrastate common carriers, intrastate telecommunications relay services in such State in a manner that meets or exceeds the requirements of regulations prescribed by the Commission under subsection (d); and

'(B)   the program makes available adequate procedures and remedies for enforcing the requirements of the State program.

'(3)   METHOD OF FUNDING—Except as provided in subsection (d), the Commission shall not refuse to certify a State program based solely on the method such State will implement for funding intrastate telecommunication relay services.

'(4)   SUSPENSION OR REVOCATION OF CERTIFICATION—The Commission may suspend or revoke such certification if, after notice and opportunity for hearing, the Commission determines that such certification is no longer warranted. In a State whose program has been suspended or revoked, the Commission shall take such steps as may be necessary, consistent with this section, to ensure continuity of telecommunications relay services.

'(g)   COMPLAINT—

'(1)   REFERRAL OF COMPLAINT—If a complaint to the Commission alleges a violation of this section with respect to intrastate telecommunications relay services within a State and certification of the program of such State under subsection (f) is in effect, the Commission shall refer such complaint to such State.

'(2)   JURISDICTION OF COMMISSION—After referring a complaint to a State under paragraph (1), the Commission shall exercise jurisdiction over such complaint only if—

'(A)   final action under such State program has not been taken on such complaint by such State—

'(i)   within 180 days after the complaint is filed with such State; or

'(ii)   within a shorter period as prescribed by the regulations of such State; or

'(B)   the Commission determines that such State program is no longer qualified for certification under subsection (f).'.

(b) CONFORMING AMENDMENTS—The Communications Act of 1934 (47 U.S.C. 151 et seq.) is amended—

    (1) in section 2(b) (47 U.S.C. 152(b)), by striking 'section 224' and inserting 'sections 224 and 225'; and

    (2) in section 221(b) (47 U.S.C. 221(b)), by striking 'section 301' and inserting 'sections 225 and 301'.

Sec. 402. Closed-Captioning of Public Service Announcements.

Section 711 of the Communications Act of 1934 is amended to read as follows:

    'SEC. 711. CLOSED-CAPTIONING OF PUBLIC SERVICE ANNOUNCEMENTS.

    'Any television public service announcement that is produced or funded in whole or in part by any agency or instrumentality of Federal Government shall include closed captioning of the verbal content of such announcement. A television broadcast station licensee—

        '(1) shall not be required to supply closed captioning for any such announcement that fails to include it; and

        '(2) shall not be liable for broadcasting any such announcement without transmitting a closed caption unless the licensee intentionally fails to transmit the closed caption that was included with the announcement.'.

TITLE V—MISCELLANEOUS PROVISIONS

Sec. 501. Construction.

(a) IN GENERAL—Except as otherwise provided in this Act, nothing in this Act shall be construed to apply a lesser standard than the standards applied under title V of the Rehabilitation Act of 1973 (29 U.S.C. 790 et seq.) or the regulations issued by Federal agencies pursuant to such title.

(b) RELATIONSHIP TO OTHER LAWS—Nothing in this Act shall be construed to invalidate or limit the remedies, rights, and procedures of any Federal law or law of any State or political subdivision of any State or jurisdiction that provides greater or equal protection for the rights of individuals with disabilities than are afforded by this Act. Nothing in this Act shall be construed to preclude the prohibition of, or the imposition of restrictions on, smoking in places of employment covered by title I, in transportation covered by title II or III, or in places of public accommodation covered by title III.

(c) INSURANCE—Titles I through IV of this Act shall not be construed to prohibit or restrict—

    (1) an insurer, hospital or medical service company, health maintenance organization, or any agent, or entity that administers benefit plans, or similar organizations from underwriting risks, classifying risks, or administering such risks that are based on or not inconsistent with State law; or

(2) a person or organization covered by this Act from establishing, sponsoring, observing or administering the terms of a bona fide benefit plan that are based on underwriting risks, classifying risks, or administering such risks that are based on or not inconsistent with State law; or

(3) a person or organization covered by this Act from establishing, sponsoring, observing or administering the terms of a bona fide benefit plan that is not subject to State laws that regulate insurance.

Paragraphs (1), (2), and (3) shall not be used as a subterfuge to evade the purposes of title I and III.

(d) ACCOMMODATIONS AND SERVICES—Nothing in this Act shall be construed to require an individual with a disability to accept an accommodation, aid, service, opportunity, or benefit which such individual chooses not to accept.

Sec. 502. State Immunity.

A State shall not be immune under the eleventh amendment to the Constitution of the United States from an action in Federal or State court of competent jurisdiction for a violation of this Act. In any action against a State for a violation of the requirements of this Act, remedies (including remedies both at law and in equity) are available for such a violation to the same extent as such remedies are available for such a violation in an action against any public or private entity other than a State.

Sec. 503. Prohibition Against Retaliation and Coercion.

(a) RETALIATION—No person shall discriminate against any individual because such individual has opposed any act or practice made unlawful by this Act or because such individual made a charge, testified, assisted, or participated in any manner in an investigation, proceeding, or hearing under this Act.

(b) INTERFERENCE, COERCION, OR INTIMIDATION—It shall be unlawful to coerce, intimidate, threaten, or interfere with any individual in the exercise or enjoyment of, or on account of his or her having exercised or enjoyed, or on account of his or her having aided or encouraged any other individual in the exercise or enjoyment of, any right granted or protected by this Act.

(c) REMEDIES AND PROCEDURES—The remedies and procedures available under sections 107, 203, and 308 of this Act shall be available to aggrieved persons for violations of subsections (a) and (b), with respect to title I, title II and title III, respectively.

Sec. 504. Regulations by the Architectural and Transportation Barriers Compliance Board.

(a) ISSUANCE OF GUIDELINES—Not later than 9 months after the date of enactment of this Act, the Architectural and Transportation Barriers Compliance Board shall issue minimum guidelines that shall supplement the existing Minimum Guidelines and Requirements for Accessible Design for purposes of titles II and III of this Act.

(b) CONTENTS OF GUIDELINES—The supplemental guidelines issued under subsection (a) shall establish additional requirements, consistent with this Act, to ensure that buildings, facilities, rail passenger cars, and vehicles are accessible, in terms of architecture and design, transportation, and communication, to individuals with disabilities.

(c) QUALIFIED HISTORIC PROPERTIES—

    (1) IN GENERAL—The supplemental guidelines issued under subsection (a) shall include procedures and requirements for alterations that will threaten or destroy the historic significance of qualified historic buildings and facilities as defined in 4.1.7(1)(a) of the Uniform Federal Accessibility Standards.

    (2) SITES ELIGIBLE FOR LISTING IN NATIONAL REGISTER—With respect to alterations of buildings or facilities that are eligible for listing in the National Register of Historic Places under the National Historic Preservation Act (16 U.S.C. 470 et seq.), the guidelines described in paragraph (1) shall, at a minimum, maintain the procedures and requirements established in 4.1.7 (1) and (2) of the Uniform Federal Accessibility Standards.

    (3) OTHER SITES—With respect to alterations of buildings or facilities designated as historic under State or local law, the guidelines described in paragraph (1) shall establish procedures equivalent to those established by 4.1.7(1) (b) and (c) of the Uniform Federal Accessibility Standards, and shall require, at a minimum, compliance with the requirements established in 4.1.7(2) of such standards.

Sec. 505. Attorney's Fees.

In any action or administrative proceeding commenced pursuant to this Act, the court or agency, in its discretion, may allow the prevailing party, other than the United States, a reasonable attorney's fee, including litigation expenses, and costs, and the United States shall be liable for the foregoing the same as a private individual.

Sec. 506. Technical Assistance.

(a) PLAN FOR ASSISTANCE—

    (1) IN GENERAL—Not later than 180 days after the date of enactment of this Act, the Attorney General, in consultation with the Chair of the Equal Employment Opportunity Commission, the Secretary of Transportation, the Chair of the Architectural and Transportation Barriers Compliance Board, and the Chairman of the Federal Communications Commission, shall develop a plan to assist entities covered under this Act, and other Federal agencies, in understanding the responsibility of such entities and agencies under this Act.

    (2) PUBLICATION OF PLAN—The Attorney General shall publish the plan referred to in paragraph (1) for public comment in accordance with

subchapter II of chapter 5 of title 5, United States Code (commonly known as the Administrative Procedure Act).

(b) AGENCY AND PUBLIC ASSISTANCE—The Attorney General may obtain the assistance of other Federal agencies in carrying out subsection (a), including the National Council on Disability, the President's Committee on Employment of People with Disabilities, the Small Business Administration, and the Department of Commerce.

(c) IMPLEMENTATION-

(1) RENDERING ASSISTANCE—Each Federal agency that has responsibility under paragraph (2) for implementing this Act may render technical assistance to individuals and institutions that have rights or duties under the respective title or titles for which such agency has responsibility.

(2) IMPLEMENTATION OF TITLES—

(A) TITLE I—The Equal Employment Opportunity Commission and the Attorney General shall implement the plan for assistance developed under subsection (a), for title I.

(B) TITLE II-

(i) SUBTITLE A—The Attorney General shall implement such plan for assistance for subtitle A of title II.

(ii) SUBTITLE B—The Secretary of Transportation shall implement such plan for assistance for subtitle B of title II.

(C) TITLE III—The Attorney General, in coordination with the Secretary of Transportation and the Chair of the Architectural Transportation Barriers Compliance Board, shall implement such plan for assistance for title III, except for section 304, the plan for assistance for which shall be implemented by the Secretary of Transportation.

(D) TITLE IV—The Chairman of the Federal Communications Commission, in coordination with the Attorney General, shall implement such plan for assistance for title IV.

(3) TECHNICAL ASSISTANCE MANUALS—Each Federal agency that has responsibility under paragraph (2) for implementing this Act shall, as part of its implementation responsibilities, ensure the availability and provision of appropriate technical assistance manuals to individuals or entities with rights or duties under this Act no later than six months after applicable final regulations are published under titles I, II, III, and IV.

(d) GRANTS AND CONTRACTS—

(1) IN GENERAL—Each Federal agency that has responsibility under subsection (c)(2) for implementing this Act may make grants or award contracts to effectuate the purposes of this section, subject to the availability of appropriations. Such grants and contracts may be awarded to individuals, institutions not organized for profit and no part of the net earnings of which inures to the benefit of any private shareholder or individual (including educational institutions), and

associations representing individuals who have rights or duties under this Act. Contracts may be awarded to entities organized for profit, but such entities may not be the recipients or grants described in this paragraph.

(2) DISSEMINATION OF INFORMATION—Such grants and contracts, among other uses, may be designed to ensure wide dissemination of information about the rights and duties established by this Act and to provide information and technical assistance about techniques for effective compliance with this Act.

(e) FAILURE TO RECEIVE ASSISTANCE—An employer, public accommodation, or other entity covered under this Act shall not be excused from compliance with the requirements of this Act because of any failure to receive technical assistance under this section, including any failure in the development or dissemination of any technical assistance manual authorized by this section.

Sec. 507. Federal Wilderness Areas.

(a) STUDY—The National Council on Disability shall conduct a study and report on the effect that wilderness designations and wilderness land management practices have on the ability of individuals with disabilities to use and enjoy the National Wilderness Preservation System as established under the Wilderness Act (16 U.S.C. 1131 et seq.).

(b) SUBMISSION OF REPORT—Not later than 1 year after the enactment of this Act, the National Council on Disability shall submit the report required under subsection (a) to Congress.

(c) SPECIFIC WILDERNESS ACCESS—

(1) IN GENERAL—Congress reaffirms that nothing in the Wilderness Act is to be construed as prohibiting the use of a wheelchair in a wilderness area by an individual whose disability requires use of a wheelchair, and consistent with the Wilderness Act no agency is required to provide any form of special treatment or accommodation, or to construct any facilities or modify any conditions of lands within a wilderness area in order to facilitate such use.

(2) DEFINITION—For purposes of paragraph (1), the term 'wheelchair' means a device designed solely for use by a mobility-impaired person for locomotion, that is suitable for use in an indoor pedestrian area.

Sec. 508. Transvestites.

For the purposes of this Act, the term 'disabled' or 'disability' shall not apply to an individual solely because that individual is a transvestite.

Sec. 509. Coverage Of Congress and the Agencies of the Legislative Branch.

(a) COVERAGE OF THE SENATE—

(1) COMMITMENT TO RULE XLII—The Senate reaffirms its commitment to Rule XLII of the Standing Rules of the Senate which provides as follows:

'No member, officer, or employee of the Senate shall, with respect to employment by the Senate or any office thereof—

'(a) fail or refuse to hire an individual;

> '(b) discharge an individual; or
>
> '(c) otherwise discriminate against an individual with respect to promotion, compensation, or terms, conditions, or privileges of employment on the basis of such individual's race, color, religion, sex, national origin, age, or state of physical handicap.'.

(2) APPLICATION TO SENATE EMPLOYMENT—The rights and protections provided pursuant to this Act, the Civil Rights Act of 1990 (S. 2104, 101st Congress), the Civil Rights Act of 1964, the Age Discrimination in Employment Act of 1967, and the Rehabilitation Act of 1973 shall apply with respect to employment by the United States Senate.

(3) INVESTIGATION AND ADJUDICATION OF CLAIMS—All claims raised by any individual with respect to Senate employment, pursuant to the Acts referred to in paragraph (2), shall be investigated and adjudicated by the Select Committee on Ethics, pursuant to S. Res. 338, 88th Congress, as amended, or such other entity as the Senate may designate.

(4) RIGHTS OF EMPLOYEES—The Committee on Rules and Administration shall ensure that Senate employees are informed of their rights under the Acts referred to in paragraph (2).

(5) APPLICABLE REMEDIES—When assigning remedies to individuals found to have a valid claim under the Acts referred to in paragraph (2), the Select Committee on Ethics, or such other entity as the Senate may designate, should to the extent practicable apply the same remedies applicable to all other employees covered by the Acts referred to in paragraph (2). Such remedies shall apply exclusively.

(6) MATTERS OTHER THAN EMPLOYMENT—

> (A) IN GENERAL—The rights and protections under this Act shall, subject to subparagraph (B), apply with respect to the conduct of the Senate regarding matters other than employment.
>
> (B) REMEDIES—The Architect of the Capitol shall establish remedies and procedures to be utilized with respect to the rights and protections provided pursuant to subparagraph (A). Such remedies and procedures shall apply exclusively, after approval in accordance with subparagraph (C).
>
> (C) PROPOSED REMEDIES AND PROCEDURES—For purposes of subparagraph (B), the Architect of the Capitol shall submit proposed remedies and procedures to the Senate Committee on Rules and Administration. The remedies and procedures shall be effective upon the approval of the Committee on Rules and Administration.

(7) EXERCISE OF RULEMAKING POWER—Notwithstanding any other provision of law, enforcement and adjudication of the rights and protections referred to in paragraph (2) and (6)(A) shall be within the

exclusive jurisdiction of the United States Senate. The provisions of paragraph (1), (3), (4), (5), (6)(B), and (6)(C) are enacted by the Senate as an exercise of the rulemaking power of the Senate, with full recognition of the right of the Senate to change its rules, in the same manner, and to the same extent, as in the case of any other rule of the Senate.

(b) COVERAGE OF THE HOUSE OF REPRESENTATIVES—

    (1) IN GENERAL—Notwithstanding any other provision of this Act or of law, the purposes of this Act shall, subject to paragraphs (2) and (3), apply in their entirety to the House of Representatives.

    (2) EMPLOYMENT IN THE HOUSE—

        (A) APPLICATION—The rights and protections under this Act shall, subject to subparagraph (B), apply with respect to any employee in an employment position in the House of Representatives and any employing authority of the House of Representatives.

        (B) ADMINISTRATION—

            (i) IN GENERAL—In the administration of this paragraph, the remedies and procedures made applicable pursuant to the resolution described in clause (ii) shall apply exclusively.

            (ii) RESOLUTION—The resolution referred to in clause (i) is House Resolution 15 of the One Hundred First Congress, as agreed to January 3, 1989, or any other provision that continues in effect the provisions of, or is a successor to, the Fair Employment Practices Resolution (House Resolution 558 of the One Hundredth Congress, as agreed to October 4, 1988).

        (C) EXERCISE OF RULEMAKING POWER—The provisions of subparagraph (B) are enacted by the House of Representatives as an exercise of the rulemaking power of the House of Representatives, with full recognition of the right of the House to change its rules, in the same manner, and to the same extent as in the case of any other rule of the House.

    (3) MATTERS OTHER THAN EMPLOYMENT—

        (A) IN GENERAL—The rights and protections under this Act shall, subject to subparagraph (B), apply with respect to the conduct of the House of Representatives regarding matters other than employment.

        (B) REMEDIES—The Architect of the Capitol shall establish remedies and procedures to be utilized with respect to the rights and protections provided pursuant to subparagraph (A). Such remedies and procedures shall apply exclusively, after approval in accordance with subparagraph (C).

        (C) APPROVAL—For purposes of subparagraph (B), the Architect of the Capitol shall submit proposed remedies and procedures to the Speaker of the House of Representatives. The remedies and procedures shall be effective upon the approval of the Speaker, after consultation with the House Office Building Commission.

(c) INSTRUMENTALITIES OF CONGRESS-

   (1) IN GENERAL—The rights and protections under this Act shall, subject to paragraph (2), apply with respect to the conduct of each instrumentality of the Congress.

   (2) ESTABLISHMENT OF REMEDIES AND PROCEDURES BY INSTRUMENTALITIES—The chief official of each instrumentality of the Congress shall establish remedies and procedures to be utilized with respect to the rights and protections provided pursuant to paragraph (1). Such remedies and procedures shall apply exclusively.

   (3) REPORT TO CONGRESS—The chief official of each instrumentality of the Congress shall, after establishing remedies and procedures for purposes of paragraph (2), submit to the Congress a report describing the remedies and procedures.

   (4) DEFINITION OF INSTRUMENTALITIES—For purposes of this section, instrumentalities of the Congress include the following: the Architect of the Capitol, the Congressional Budget Office, the General Accounting Office, the Government Printing Office, the Library of Congress, the Office of Technology Assessment, and the United States Botanic Garden.

  (5) CONSTRUCTION—Nothing in this section shall alter the enforcement procedures for individuals with disabilities provided in the General Accounting Office Personnel Act of 1980 and regulations promulgated pursuant to that Act.

Sec. 510. Illegal Use of Drugs.

(a) IN GENERAL—For purposes of this Act, the term 'individual with a disability' does not include an individual who is currently engaging in the illegal use of drugs, when the covered entity acts on the basis of such use.

(b) RULES OF CONSTRUCTION—Nothing in subsection (a) shall be construed to exclude as an individual with a disability an individual who—

   (1) has successfully completed a supervised drug rehabilitation program and is no longer engaging in the illegal use of drugs, or has otherwise been rehabilitated successfully and is no longer engaging in such use;

   (2) is participating in a supervised rehabilitation program and is no longer engaging in such use; or

   (3) is erroneously regarded as engaging in such use, but is not engaging in such use; except that it shall not be a violation of this Act for a covered entity to adopt or administer reasonable policies or procedures, including but not limited to drug testing, designed to ensure that an individual described in paragraph (1) or (2) is no longer engaging in the illegal use of drugs; however, nothing in this section shall be construed to encourage, prohibit, restrict, or authorize the conducting of testing for the illegal use of drugs.

(c) HEALTH AND OTHER SERVICES—Notwithstanding subsection (a) and section 511(b)(3), an individual shall not be denied health services, or services provided in connection with drug rehabilitation, on the basis of the current illegal use of drugs if the individual is otherwise entitled to such services.

(d) DEFINITION OF ILLEGAL USE OF DRUGS—

(1) IN GENERAL—The term 'illegal use of drugs' means the use of drugs, the possession or distribution of which is unlawful under the Controlled Substances Act (21 U.S.C. 812). Such term does not include the use of a drug taken under supervision by a licensed health care professional, or other uses authorized by the Controlled Substances Act or other provisions of Federal law.

(2) DRUGS—The term 'drug' means a controlled substance, as defined in schedules I through V of section 202 of the Controlled Substances Act.

Sec. 511. Definitions.

(a) HOMOSEXUALITY AND BISEXUALITY—For purposes of the definition of 'disability' in section 3(2), homosexuality and bisexuality are not impairments and as such are not disabilities under this Act.

(b) CERTAIN CONDITIONS—Under this Act, the term 'disability' shall not include—

(1) transvestism, transsexualism, pedophilia, exhibitionism, voyeurism, gender identity disorders not resulting from physical impairments, or other sexual behavior disorders;

(2) compulsive gambling, kleptomania, or pyromania; or

(3) psychoactive substance use disorders resulting from current illegal use of drugs.

Sec. 512. Amendments to the Rehabilitation Act.

(a) DEFINITION OF HANDICAPPED INDIVIDUAL—Section 7(8) of the Rehabilitation Act of 1973 (29 U.S.C. 706(8)) is amended by redesignating subparagraph (C) as subparagraph (D), and by inserting after subparagraph (B) the following subparagraph:

'(C) (i)   For purposes of title V, the term 'individual with handicaps' does not include an individual who is currently engaging in the illegal use of drugs, when a covered entity acts on the basis of such use.

'(ii)   Nothing in clause (i) shall be construed to exclude as an individual with handicaps an individual who—

'(I)   has successfully completed a supervised drug rehabilitation program and is no longer engaging in the illegal use of drugs, or has otherwise been rehabilitated successfully and is no longer engaging in such use;

'(II)   is participating in a supervised rehabilitation program and is no longer engaging in such use; or

'(III)   is erroneously regarded as engaging in such use, but is not engaging in such use; except that it shall not be a violation of this Act for a covered entity to adopt or administer reasonable policies or procedures, including but not limited to drug testing, designed to ensure that an individual described in subclause (I) or (II) is no longer engaging in the illegal use of drugs.

'(iii) Notwithstanding clause (i), for purposes of programs and activities providing health services and services provided under titles I, II and III, an individual shall not be excluded from the benefits of such programs or activities on the basis of his or her current illegal use of drugs if he or she is otherwise entitled to such services.

'(iv) For purposes of programs and activities providing educational services, local educational agencies may take disciplinary action pertaining to the use or possession of illegal drugs or alcohol against any handicapped student who currently is engaging in the illegal use of drugs or in the use of alcohol to the same extent that such disciplinary action is taken against nonhandicapped students. Furthermore, the due process procedures at 34 CFR 104.36 shall not apply to such disciplinary actions.

'(v) For purposes of sections 503 and 504 as such sections relate to employment, the term 'individual with handicaps' does not include any individual who is an alcoholic whose current use of alcohol prevents such individual from performing the duties of the job in question or whose employment, by reason of such current alcohol abuse, would constitute a direct threat to property or the safety of others.'.

(b) DEFINITION OF ILLEGAL DRUGS—Section 7 of the Rehabilitation Act of 1973 (29 U.S.C. 706) is amended by adding at the end the following new paragraph:

'(22)(A) The term 'drug' means a controlled substance, as defined in schedules I through V of section 202 of the Controlled Substances Act (21 U.S.C. 812).

'(B) The term 'illegal use of drugs' means the use of drugs, the possession or distribution of which is unlawful under the Controlled Substances Act. Such term does not include the use of a drug taken under supervision by a licensed health care professional, or other uses authorized by the Controlled Substances Act or other provisions of Federal law.'.

(c) CONFORMING   AMENDMENTS—Section   7(8)(B)   of   the Rehabilitation Act of 1973 (29 U.S.C. 706(8)(B)) is amended—

(1) in the first sentence, by striking 'Subject to the second sentence of this subparagraph,' and inserting 'Subject to subparagraphs (C) and (D),'; and

(2) by striking the second sentence.

Sec. 513. Alternative Means of Dispute Resolution.

Where appropriate and to the extent authorized by law, the use of alternative means of dispute resolution, including settlement negotiations, conciliation, facilitation, mediation, fact-finding, mini trials, and arbitration, is encouraged to resolve disputes arising under this Act.

Sec. 514. Severability.

Should any provision in this Act be found to be unconstitutional by a court of law, such provision shall be severed from the remainder of the Act, and such action shall not affect the enforceability of the remaining provisions of the Act.

ADA Amendments Act of 2008

PL 110-325 (S 3406)
September 25, 2008

An Act to restore the intent and protections of the Americans with Disabilities Act of 1990.

Be it enacted by the Senate and House of Representatives of the United States of America in Congress assembled,

SEC. 1. SHORT TITLE *[42 USCA § 12101 NOTE]*

This Act may be cited as the "ADA Amendments Act of 2008".

SEC. 2. FINDINGS AND PURPOSES *[42 USCA § 12101 NOTE]*

(a) FINDINGS. Congress finds that –

(1) in enacting the Americans with Disabilities Act of 1990 (ADA), Congress intended that the Act "provide a clear and comprehensive national mandate for the elimination of discrimination against individuals with disabilities" and provide broad coverage;

(2) in enacting the ADA, Congress recognized that physical and mental disabilities in no way diminish a person's right to fully participate in all aspects of society, but that people with physical or mental disabilities are frequently precluded from doing so because of prejudice, antiquated attitudes, or the failure to remove societal and institutional barriers;

(3) while Congress expected that the definition of disability under the ADA would be interpreted consistently with how courts had applied the definition of a handicapped individual under the Rehabilitation Act of 1973, that expectation has not been fulfilled;

(4) the holdings of the Supreme Court in Sutton v. United Air Lines, Inc., 527 U.S. 471 (1999) and its companion cases have narrowed the broad scope of protection intended to be afforded by the ADA, thus eliminating protection for many individuals whom Congress intended to protect;

(5) the holding of the Supreme Court in Toyota Motor Manufacturing, Kentucky, Inc. v. Williams, 534 U.S. 184 (2002) further narrowed the broad scope of protection intended to be afforded by the ADA;

(6) as a result of these Supreme Court cases, lower courts have incorrectly found in individual cases that people with a range of substantially limiting impairments are not people with disabilities;

(7) in particular, the Supreme Court, in the case of Toyota Motor Manufacturing, Kentucky, Inc. v. Williams, 534 U.S. 184 (2002),

interpreted the term "substantially limits" to require a greater degree of limitation than was intended by Congress; and

(8) Congress finds that the current Equal Employment Opportunity Commission ADA regulations defining the term "substantially limits" as "significantly restricted" are inconsistent with congressional intent, by expressing too high a standard.

(b) PURPOSES. The purposes of this Act are—

(1) to carry out the ADA's objectives of providing "a clear and comprehensive national mandate for the elimination of discrimination" and "clear, strong, consistent, enforceable standards addressing discrimination" by reinstating a broad scope of protection to be available under the ADA;

(2) to reject the requirement enunciated by the Supreme Court in Sutton v. United Air Lines, Inc., 527 U.S. 471 (1999) and its companion cases that whether an impairment substantially limits a major life activity is to be determined with reference to the ameliorative effects of mitigating measures;

(3) to reject the Supreme Court's reasoning in Sutton v. United Air Lines, Inc., 527 U.S. 471 (1999) with regard to coverage under the third prong of the definition of disability and to reinstate the reasoning of the Supreme Court in School Board of Nassau County v. Arline, 480 U.S. 273 (1987) which set forth a broad view of the third prong of the definition of handicap under the Rehabilitation Act of 1973;

(4) to reject the standards enunciated by the Supreme Court in Toyota Motor Manufacturing, Kentucky, Inc. v. Williams, 534 U.S. 184 (2002), that the terms "substantially" and "major" in the definition of disability under the ADA "need to be interpreted strictly to create a demanding standard for qualifying as disabled," and that to be substantially limited in performing a major life activity under the ADA "an individual must have an impairment that prevents or severely restricts the individual from doing activities that are of central importance to most people's daily lives";

(5) to convey congressional intent that the standard created by the Supreme Court in the case of Toyota Motor Manufacturing, Kentucky, Inc. v. Williams, 534 U.S. 184 (2002) for "substantially limits", and applied by lower courts in numerous decisions, has created an inappropriately high level of limitation necessary to obtain coverage under the ADA, to convey that it is the intent of Congress that the primary object of attention in cases brought under the ADA should be whether entities covered under the ADA have complied with their obligations, and to convey that the question of whether an individual's impairment is a disability under the ADA should not demand extensive analysis; and

(6) to express Congress' expectation that the Equal Employment Opportunity Commission will revise that portion of its current regulations that defines the term "substantially limits" as "significantly restricted" to be consistent with this Act, including the amendments made by this Act.

SEC. 3. CODIFIED FINDINGS.

Section 2(a) of the Americans with Disabilities Act of 1990 (42 U.S.C. 12101) is amended—

(1) by amending paragraph (1) to read as follows: "(1) physical or mental disabilities in no way diminish a person's right to fully participate in all aspects of society, yet many people with physical or mental disabilities have been precluded from doing so because of discrimination; others who have a record of a disability or are regarded as having a disability also have been subjected to discrimination;";

(2) by striking paragraph (7); and

(3) by redesignating paragraphs (8) and (9) as paragraphs (7) and (8), respectively.

SEC. 4. DISABILITY DEFINED AND RULES OF CONSTRUCTION.

(a) DEFINITION OF DISABILITY.—Section 3 of the Americans with Disabilities Act of 1990 (42 U.S.C. 12102) is amended to read as follows:
SEC. 3. DEFINITION OF DISABILITY.
As used in this Act:

(1) DISABILITY.—The term 'disability' means, with respect to an individual—

    (A)   a physical or mental impairment that substantially limits one or more major life activities of such individual;

    (B)   a record of such an impairment; or

    (C)   being regarded as having such an impairment (as described in paragraph (3)).

    (2) MAJOR LIFE ACTIVITIES.—

        (A)   IN GENERAL.—For purposes of paragraph (1), major life activities include, but are not limited to, caring for oneself, performing manual tasks, seeing, hearing, eating, sleeping, walking, standing, lifting, bending, speaking, breathing, learning, reading, concentrating, thinking, communicating, and working.

        (B)   MAJOR BODILY FUNCTIONS.—For purposes of paragraph (1), a major life activity also includes the operation of a major bodily function, including but not limited to, functions of the immune system, normal cell growth, digestive, bowel, bladder, neurological, brain, respiratory, circulatory, endocrine, and reproductive functions.

    (3) REGARDED AS HAVING SUCH AN IMPAIRMENT.—For purposes of paragraph (1)(C):

        (A)   An individual meets the requirement of 'being regarded as having such an impairment' if the individual establishes that he or she has been subjected to an action prohibited under this Act because of an actual or perceived physical or mental impairment whether or not the impairment limits or is perceived to limit a major life activity.

(B) Paragraph (1)(C) shall not apply to impairments that are transitory and minor. A transitory impairment is an impairment with an actual or expected duration of 6 months or less.

(4) RULES OF CONSTRUCTION REGARDING THE DEFINITION OF DISABILITY.—The definition of 'disability' in paragraph (1) shall be construed in accordance with the following:

(A) The definition of disability in this Act shall be construed in favor of broad coverage of individuals under this Act, to the maximum extent permitted by the terms of this Act.

(B) The term 'substantially limits' shall be interpreted consistently with the findings and purposes of the ADA Amendments Act of 2008.

(C) An impairment that substantially limits one major life activity need not limit other major life activities in order to be considered a disability.

(D) An impairment that is episodic or in remission is a disability if it would substantially limit a major life activity when active.

(E) (i) The determination of whether an impairment substantially limits a major life activity shall be made without regard to the ameliorative effects of mitigating measures such as—

(I) medication, medical supplies, equipment, or appliances, low-vision devices (which do not include ordinary eyeglasses or contact lenses), prosthetics including limbs and devices, hearing aids and cochlear implants or other implantable hearing devices, mobility devices, or oxygen therapy equipment and supplies;

(II) use of assistive technology;

(III) reasonable accommodations or auxiliary aids or services; or

(IV) learned behavioral or adaptive neurological modifications.

(ii) The ameliorative effects of the mitigating measures of ordinary eyeglasses or contact lenses shall be considered in determining whether an impairment substantially limits a major life activity.

(iii) As used in this subparagraph—

(I) the term 'ordinary eyeglasses or contact lenses' means lenses that are intended to fully correct visual acuity or eliminate refractive error; and

(II) the term 'low-vision devices' means devices that magnify, enhance, or otherwise augment a visual image.

(b) CONFORMING AMENDMENT.—The Americans with Disabilities Act of 1990 (42 U.S.C. 12101 et seq.) is further amended by adding after section 3 the following:

SEC. 4. ADDITIONAL DEFINITIONS.

As used in this Act:

(1) AUXILIARY AIDS AND SERVICES.—The term 'auxiliary aids and services' includes—

      (A)   qualified interpreters or other effective methods of making aurally delivered materials available to individuals with hearing impairments;

      (B)   qualified readers, taped texts, or other effective methods of making visually delivered materials available to individuals with visual impairments;

      (C)   acquisition or modification of equipment or devices; and

      (D)   other similar services and actions.

   (2)   STATE.—The term 'State' means each of the several States, the District of Columbia, the Commonwealth of Puerto Rico, Guam, American Samoa, the Virgin Islands of the United States, the Trust Territory of the Pacific Islands, and the Commonwealth of the Northern Mariana Islands.

(c) AMENDMENT TO THE TABLE OF CONTENTS.—The table of contents contained in section 1(b) of the Americans with Disabilities Act of 1990 is amended by striking the item relating to section 3 and inserting the following items:

Sec. 3. Definition of disability.

Sec. 4. Additional definitions.

SEC. 5. DISCRIMINATION ON THE BASIS OF DISABILITY.

(a) ON THE BASIS OF DISABILITY.—Section 102 of the Americans with Disabilities Act of 1990 (42 U.S.C. 12112) is amended—

   (1) in subsection (a), by striking "with a disability because of the disability of such individual" and inserting "on the basis of disability"; and

   (2) in subsection (b) in the matter preceding paragraph (1), by striking "discriminate" and inserting "discriminate against a qualified individual on the basis of disability".

(b) QUALIFICATION STANDARDS AND TESTS RELATED TO UNCORRECTED VISION.—Section 103 of the Americans with Disabilities Act of 1990 (42 U.S.C. 12113) is amended by redesignating subsections (c) and (d) as subsections (d) and (e), respectively, and inserting after subsection (b) the following new subsection:

  "(c) QUALIFICATION STANDARDS AND TESTS RELATED TO UNCORRECTED VISION.— Notwithstanding section 3(4)(E)(ii), a covered entity shall not use qualification standards, employment tests, or other selection criteria based on an individual's uncorrected vision unless the standard, test, or other selection criteria, as used by the covered entity, is shown to be job-related for the position in question and consistent with business necessity."

(c) CONFORMING AMENDMENTS.—

   (1) Section 101(8) of the Americans with Disabilities Act of 1990 (42 U.S.C. 12111(8)) is amended—

     (A) in the paragraph heading, by striking "WITH A DISABILITY"; and

     (B) by striking "with a disability" after "individual" both places it appears.

(2) Section 104(a) of the Americans with Disabilities Act of 1990 (42 U.S.C. 12114(a)) is amended by striking "the term 'qualified individual with a disability' shall" and inserting "a qualified individual with a disability shall".

SEC. 6. RULES OF CONSTRUCTION.

(a) Title V of the Americans with Disabilities Act of 1990 (42 U.S.C. 12201 et seq.) is amended—

   (1) by adding at the end of section 501 the following:

      (e) BENEFITS UNDER STATE WORKER'S COMPENSATION LAWS.—Nothing in this Act alters the standards for determining eligibility for benefits under State worker's compensation laws or under State and Federal disability benefit programs.

      (f) FUNDAMENTAL ALTERATION.—Nothing in this Act alters the provision of section 302(b)(2)(A)(ii), specifying that reasonable modifications in policies, practices, or procedures shall be required, unless an entity can demonstrate that making such modifications in policies, practices, or procedures, including academic requirements in postsecondary education, would fundamentally alter the nature of the goods, services, facilities, privileges, advantages, or accommodations involved.

      (g) CLAIMS OF NO DISABILITY.—Nothing in this Act shall provide the basis for a claim by an individual without a disability that the individual was subject to discrimination because of the individual's lack of disability.

      (h) REASONABLE ACCOMMODATIONS AND MODIFICATIONS. — A covered entity under title I, a public entity under title II, and any person who owns, leases (or leases to), or operates a place of public accommodation under title III, need not provide a reasonable accommodation or a reasonable modification to policies, practices, or procedures to an individual who meets the definition of disability in section 3(1) solely under subparagraph (C) of such section.;

   (2) by redesignating section 506 through 514 as sections 507 through 515, respectively, and adding after section 505 the following:

      SEC. 506. RULE OF CONSTRUCTION REGARDING REGULATORY AUTHORITY. "The authority to issue regulations granted to the Equal Employment Opportunity Commission, the Attorney General, and the Secretary of Transportation under this Act includes the authority to issue regulations implementing the definitions of disability in section 3 (including rules of construction) and the definitions in section 4, consistent with the ADA Amendments Act of 2008."; and

(3) in section 511 (as redesignated by paragraph (2)) (42 U.S.C. 12211), in subsection (c), by striking 511(b)(3) and inserting 512(b)(3).

(b) The table of contents contained in section 1(b) of the Americans with Disabilities Act of 1990 is amended by redesignating the items relating to sections 506 through 514 as the items relating to sections 507 through 515, respectively, and by inserting after the item relating to section 505 the following new item:

Sec. 506. Rule of construction regarding regulatory authority.

SEC. 7. CONFORMING AMENDMENTS.

Section 7 of the Rehabilitation Act of 1973 (29 U.S.C. 705) is amended—

(1) in paragraph (9)(B), by striking "a physical" and all that follows through "major life activities", and inserting "the meaning given it in section 3 of the Americans with Disabilities Act of 1990 (42 U.S.C. 12102)"; and

(2) in paragraph (20)(B), by striking "any person who" and all that follows through the period at the end, and inserting "any person who has a disability as defined in section 3 of the Americans with Disabilities Act of 1990 (42 U.S.C. 12102).".

SEC. 8. EFFECTIVE DATE *[29 USCA § 705 NOTE]*

This Act and the amendments made by this Act shall become effective on January 1, 2009.

Approved September 25, 2008.

The Equal Pay Act of 1963

MINIMUM WAGE

SEC. 206. [Section 6]

(d) Prohibition of sex discrimination

(1) No employer having employees subject to any provisions of this section shall discriminate, within any establishment in which such employees are employed, between employees on the basis of sex by paying wages to employees in such establishment at a rate less than the rate at which he pays wages to employees of the opposite sex in such establishment for equal work on jobs the performance of which requires equal skill, effort, and responsibility, and which are performed under similar working conditions, except where such payment is made pursuant to (i) a seniority system; (ii) a merit system; (iii) a system which measures earnings by quantity or quality of production; or (iv) a differential based on any other factor other than sex: *Provided*, That an employer who is paying a wage rate differential in violation of this subsection shall not, in order to comply with the provisions of this subsection, reduce the wage rate of any employee.

(2) No labor organization, or its agents, representing employees of an employer having employees subject to any provisions of this section shall cause or attempt to cause such an employer to discriminate against an employee in violation of paragraph (1) of this subsection.

(3) For purposes of administration and enforcement, any amounts owing to any employee which have been withheld in violation of this subsection shall be deemed to be unpaid minimum wages or unpaid overtime compensation under this chapter.

(4) As used in this subsection, the term "labor organization" means any organization of any kind, or any agency or employee representation committee or plan, in which employees participate and which exists for the purpose, in whole or in part, of dealing with employers concerning grievances, labor disputes, wages, rates of pay, hours of employment, or conditions of work.

ADDITIONAL PROVISIONS OF EQUAL PAY ACT OF 1963

An Act

To prohibit discrimination on account of sex in the payment of wages by employers engaged in commerce or in the production of goods for commerce.

Be it enacted by the Senate and House of Representatives of the United States of America in Congress assembled, that this Act may be cited as the "Equal Pay Act of 1963."

***

DECLARATION OF PURPOSE

Not Reprinted in U.S. Code *[Section 2]*

  (a) The Congress hereby finds that the existence in industries engaged in commerce or in the production of goods for commerce of wage differentials based on sex—

    (1) depresses wages and living standards for employees necessary for their health and efficiency;

    (2) prevents the maximum utilization of the available labor resources;

    (3) tends to cause labor disputes, thereby burdening, affecting, and obstructing commerce;

    (4) burdens commerce and the free flow of goods in commerce; and

    (5) constitutes an unfair method of competition.

  (b) It is hereby declared to be the policy of this Act, through exercise by Congress of its power to regulate commerce among the several States and with foreign nations, to correct the conditions above referred to in such industries.

[Section 3 of the Equal Pay Act of 1963 amends section 6 of the Fair Labor Standards Act by adding a new subsection (d). The amendment is incorporated in the revised text of the Fair Labor Standards Act.]

EFFECTIVE DATE

Not Reprinted in U.S. Code *[Section 4]*

The amendments made by this Act shall take effect upon the expiration of one year from the date of its enactment: Provided, That in the case of employees covered by a bona fide collective bargaining agreement in effect at least thirty days prior to the date of enactment of this Act entered into by a labor organization (as defined in section 6(d)(4) of the Fair Labor Standards Act of 1938, as amended) [subsection (d) (4) of this section], the amendments made by this Act shall take effect upon the termination of such collective bargaining agreement or upon the expiration of two years from the date of enactment of this Act, whichever shall first occur.

Approved June 10, 1963, 12 m.

[In the following excerpts from the Fair Labor Standards Act of 1938, as amended, authority given to the Secretary of Labor is exercised by the Equal Employment Opportunity Commission for purposes of enforcing the Equal Pay Act of 1963.]

ATTENDANCE OF WITNESSES

SEC. 209 [Section 9]

For the purpose of any hearing or investigation provided for in this chapter, the provisions of sections 49 and 50 of title 15 *[Federal Trade Commission Act of September 16, 1914, as amended (U.S.C., 1934 edition)]* (relating to the attendance of witnesses and the production of books, papers, and documents), are made applicable to the jurisdiction, powers, and duties of the Administrator, the Secretary of Labor, and the industry committees.

COLLECTION OF DATA

SEC. 211 [Section 11]

(a) Investigations and inspections

The Administrator or his designated representatives may investigate and gather data regarding the wages, hours, and other conditions and practices of employment in any industry subject to this chapter, and may enter and inspect such places and such records (and make such transcriptions thereof), question such employees, and investigate such facts, conditions, practices, or matters as he may deem necessary or appropriate to determine whether any person has violated any provision of this chapter, or which may aid in the enforcement of the provisions of this chapter. Except as provided in section 212 *[section 12]* of this title and in subsection (b) of this section, the Administrator shall utilize the bureaus and divisions of the Department of Labor for all the investigations and inspections necessary under this section. Except as provided in section 212 *[section 12] of this title*, the Administrator shall bring all actions under section 217 *[section 17]* of this title to restrain violations of this chapter.

(b) State and local agencies and employees

With the consent and cooperation of State agencies charged with the administration of State labor laws, the Administrator and the Secretary of Labor may, for the purpose of carrying out their respective functions and duties under this chapter, utilize the services of State and local agencies and their employees and, notwithstanding any other provision of law, may reimburse such State and local agencies and their employees for services rendered for such purposes.

(c) Records

Every employer subject to any provision of this chapter or of any order issued under this chapter shall make, keep, and preserve such records of the persons employed by him and of the wages, hours, and other conditions and practices of employment maintained by him, and shall preserve such records for such periods of time, and shall make such reports therefrom to the Administrator as he shall prescribe by regulation or order as necessary or appropriate for the enforcement of the provisions of this chapter or the regulations or orders thereunder. The employer of an employee who performs substitute work described in section 207(p) (3) *[section 7(p)(3)]* of this title may not be required under this subsection to keep a record of the hours of the substitute work.

(d) Homework regulations

The Administrator is authorized to make such regulations and orders regulating, restricting, or prohibiting industrial homework as are necessary or appropriate to prevent the circumvention or evasion of and to safeguard the minimum wage rate prescribed in this chapter, and all existing regulations or orders of the Administrator relating to industrial homework are continued in full force and effect.

EXEMPTIONS

SEC. 213 [Section 13]

(a) Minimum wage and maximum hour requirements

The provisions of sections 206 *[section 6]* (except subsection (d) in the case of paragraph (1) of this subsection) and section 207 *[section 7]* of this title shall not apply with respect to—

(1) any employee employed in a bona fide executive, administrative, or professional capacity (including any employee employed in the capacity of academic administrative personnel or teacher in elementary or secondary schools), or in the capacity of outside salesman (as such terms are defined and delimited from time to time by regulations of the Secretary, subject to the provisions of subchapter II of chapter 5 of Title 5 *[the Administrative Procedure Act]*, except that an employee of a retail or service establishment shall not be excluded from the definition of employee employed in a bona fide executive or administrative capacity because of the number of hours in his workweek which he devotes to activities not directly or closely related to the performance of executive or administrative activities, if less than 40 per centum of his hours worked in the workweek are devoted to such activities); or

(2) *[Repealed]*

[Note: Section 13(a)(2) (relating to employees employed by a retail or service establishment) was repealed by Pub. L. 101-157, section 3(c)(1), November 17, 1989.]

(3) any employee employed by an establishment which is an amusement or recreational establishment, organized camp, or religious or non-profit educational conference center, if (A) it does not operate for more than seven months in any calendar year, or (B) during the preceding calendar year, its average receipts for any six months of such year were not more than 33 1/3 per centum of its average receipts for the other six months of such year, except that the exemption from sections 206 and 207 [sections 6 and 7] of this title provided by this paragraph does not apply with respect to any employee of a private entity engaged in providing services or facilities (other than, in the case of the exemption from section 206 [section 6] of this title, a private entity engaged in providing services and facilities directly related to skiing) in a national park or a national forest, or on land in the National Wildlife Refuge System, under a contract with the Secretary of the Interior or the Secretary of Agriculture; or

(4) *[Repealed]*

[Note: Section 13(a)(4) (relating to employees employed by an establishment which qualified as an exempt retail establishment) was repealed by Pub. L. 101-157, Section 3(c)(1), November 17, 1989.]

(5) any employee employed in the catching, taking, propagating, harvesting, cultivating, or farming of any kind of fish, shellfish, crustacea, sponges, seaweeds, or other aquatic forms of animal and vegetable life, or in the first processing, canning or packing such marine products at sea as an incident to, or in conjunction with, such fishing operations, including the going to and returning from work and loading and unloading when performed by any such employee; or

(6) any employee employed in agriculture (A) if such employee is employed by an employer who did not, during any calendar quarter during the preceding calendar year, use more than five hundred man-days of agricultural labor, (B) if such employee is the parent, spouse, child, or other member of his employer's immediate family, (C) if such employee (i) is employed as a hand harvest laborer and is paid on a piece rate basis in an operation which has been, and is customarily and generally recognized as having been, paid on a piece rate basis in the region of employment, (ii) commutes daily from his permanent residence to the farm on which he is so employed, and (iii) has been employed in agriculture less than thirteen weeks during the preceding calendar year, (D) if such employee (other than an employee described in clause (C) of this subsection) (i) is sixteen years of age or under and is employed as a hand harvest laborer, is paid on a piece rate basis in an operation which has been, and is customarily and generally recognized as having been, paid on a piece rate basis in the region of employment, (ii) is employed on the same farm as his parent or person standing in the place of his parent, and (iii) is paid at the same piece rate as employees over age sixteen are paid on the same farm, or (E) if such employee is principally engaged in the range production of livestock; or

(7) any employee to the extent that such employee is exempted by regulations, order, or certificate of the Secretary issued under section 214 [section 14] of this title; or

(8) any employee employed in connection with the publication of any weekly, semiweekly, or daily newspaper with a circulation of less than four thousand the major part of which circulation is within the county where published or counties contiguous thereto; or

(9) *[Repealed]*

[Note: Section 13(a)(9) (relating to motion picture theater employees) was repealed by section 23 of the Fair Labor Standards Amendments of 1974. The 1974 amendments created an exemption for such employees from the overtime provisions only in section 13(b)27.]

(10) any switchboard operator employed by an independently owned public telephone company which has not more than seven hundred and fifty stations; or

(11) *[Repealed]*

[Note: Section 13(a)(11) (relating to telegraph agency employees) was repealed by section 10 of the Fair Labor Standards Amendments of 1974. The 1974 amendments created an exemption from the overtime provisions only in section 13(b) (23), which was repealed effective May 1, 1976.]

(12) any employee employed as a seaman on a vessel other than an American vessel; or

(13) *[Repealed]*

[Note: Section 13(a)(13) (relating to small logging crews) was repealed by section 23 of the Fair Labor Standards Amendments of 1974. The 1974 amendments created an exemption for such employees from the overtime provisions only in section 13(b)(28).]

(14) *[Repealed]*

[Note: Section 13(a)(14) (relating to employees employed in growing and harvesting of shade grown tobacco) was repealed by section 9 of the Fair Labor Standards Amendments of 1974. The 1974 amendments created an exemption for certain tobacco producing employees from the overtime provisions only in section 13(b)(22). The section 13(b)(22) exemption was repealed, effective January 1, 1978, by section 5 of the Fair Labor Standards Amendments of 1977.]

(15) any employee employed on a casual basis in domestic service employment to provide babysitting services or any employee employed in domestic service employment to provide companionship services for individuals who (because of age or infirmity) are unable to care for themselves (as such terms are defined and delimited by regulations of the Secretary); or

(16) a criminal investigator who is paid availability pay under section 5545a of Title 5 [Law Enforcement Availability Pay Act of 1994]; or

(17) any employee who is a computer systems analyst, computer programmer, software engineer, or other similarly skilled worker, whose primary duty is—

(A) the application of systems analysis techniques and procedures, including consulting with users, to determine hardware, software, or system functional specifications;

(B) the design, development, documentation, analysis, creation, testing, or modification of computer systems or programs, including prototypes, based on and related to user or system design specifications;

(C) the design, documentation, testing, creation, or modification of computer programs related to machine operating systems; or

    (D)   a combination of duties described in subparagraphs (A), (B), and (C) the performance of which requires the same level of skills, and who, in the case of an employee who is compensated on an hourly basis, is compensated at a rate of not less than $27.63 an hour.

                    SEC. 213 [Section 13]

  (g) Certain employment in retail or service establishments, agriculture

The exemption from section 206 *[section 6]* of this title provided by paragraph (6) of subsection (a) of this section shall not apply with respect to any employee employed by an establishment (1) which controls, is controlled by, or is under common control with, another establishment the activities of which are not related for a common business purpose to, but materially support the activities of the establishment employing such employee; and (2) whose annual gross volume of sales made or business done, when combined with the annual gross volume of sales made or business done by each establishment which controls, is controlled by, or is under common control with, the establishment employing such employee, exceeds $10,000,000 (exclusive of excise taxes at the retail level which are separately stated).

PROHIBITED ACTS

SEC. 215 [Section 15]

  (a) After the expiration of one hundred and twenty days from June 25, 1938 *[the date of enactment of this Act]*, it shall be unlawful for any person—

    (1) to transport, offer for transportation, ship, deliver, or sell in commerce, or to ship, deliver, or sell with knowledge that shipment or delivery or sale thereof in commerce is intended, any goods in the production of which any employee was employed in violation of section 206 *[section 6]* or section 207 *[section 7]* of this title, or in violation of any regulation or order of the Secretary issued under section 214 *[section 14]* of this title, except that no provision of this chapter shall impose any liability upon any common carrier for the transportation in commerce in the regular course of its business of any goods not produced by such common carrier, and no provision of this chapter shall excuse any common carrier from its obligation to accept any goods for transportation; and except that any such transportation, offer, shipment, delivery, or sale of such goods by a purchaser who acquired them in good faith in reliance on written assurance from the producer that the goods were produced in compliance with the requirements of this chapter, and who acquired such goods for value without notice of any such violation, shall not be deemed unlawful;

    (2) to violate any of the provisions of section 206 *[section 6]* or section 207 *[section 7]* of this title, or any of the provisions of any regulation or order of the Secretary issued under section 214 *[section 14]* of this title;

    (3) to discharge or in any other manner discriminate against any employee because such employee has filed any complaint or instituted or caused

to be instituted any proceeding under or related to this chapter, or has testified or is about to testify in any such proceeding, or has served or is about to serve on an industry committee;

(4) to violate any of the provisions of section 212 *[section 12]* of this title;

(5) to violate any of the provisions of section 211(c) *[section 11(c)]* of this title, or any regulation or order made or continued in effect under the provisions of section 211(d) *[section 11(d)]* of this title, or to make any statement, report, or record filed or kept pursuant to the provisions of such section or of any regulation or order thereunder, knowing such statement, report, or record to be false in a material respect.

(b) For the purposes of subsection (a)(1) of this section proof that any employee was employed in any place of employment where goods shipped or sold in commerce were produced, within ninety days prior to the removal of the goods from such place of employment, shall be prima facie evidence that such employee was engaged in the production of such goods.

PENALTIES

SEC. 216 [Section 16]

(a) Fines and imprisonment

Any person who willfully violates any of the provisions of section 215 *[section 15]* of this title shall upon conviction thereof be subject to a fine of not more than $10,000, or to imprisonment for not more than six months, or both. No person shall be imprisoned under this subsection except for an offense committed after the conviction of such person for a prior offense under this subsection.

(b) Damages; right of action; attorney's fees and costs; termination of right of action

Any employer who violates the provisions of section 206 *[section 6]* or section 207 *[section 7]* of this title shall be liable to the employee or employees affected in the amount of their unpaid minimum wages, or their unpaid overtime compensation, as the case may be, and in an additional equal amount as liquidated damages. Any employer who violates the provisions of section 215(a)(3) *[section 15(a)(3)]* of this title shall be liable for such legal or equitable relief as may be appropriate to effectuate the purposes of section 215(a)(3) *[section 15(a)(3)]* of this title, including without limitation employment, reinstatement, promotion, and the payment of wages lost and an additional equal amount as liquidated damages. An action to recover the liability prescribed in either of the preceding sentences may be maintained against any employer (including a public agency) in any Federal or State court of competent jurisdiction by any one or more employees for and in behalf of himself or themselves and other employees similarly situated. No employee shall be a party plaintiff to any such action unless he gives his consent in writing to become such a party and such consent is filed in the court in which such action is brought. The court in such action shall, in addition to any judgment awarded to the plaintiff or plaintiffs, allow a reasonable attorney's fee to be paid by the defendant, and costs of the action. The right provided by this subsection to bring an action by or on behalf of any employee, and the right of any employee to become a party plaintiff to any such action, shall terminate upon the filing of a complaint by the Secretary of Labor in an action under section 217 *[section 17]* of this title in which (1) restraint is sought of any further delay in the payment of unpaid minimum wages, or the amount of unpaid overtime compensation, as the case may be, owing to such employee under section 206 *[section 6]* or section 207 *[section 7]* of this title by an employer liable therefor*[sic]* under the provisions of this subsection or (2) legal or equitable relief is sought as a result of alleged violations of section 215(a)(3) *[section 15(a)(3)]* of this title.

(c) Payment of wages and compensation; waiver of claims; actions by the Secretary; limitation of actions

The Secretary is authorized to supervise the payment of the unpaid minimum wages or the unpaid overtime compensation owing to any employee or employees under section 206 *[section 6]* or section 207 *[section 7]* of this title, and the agreement of any employee to accept such payment shall upon payment in full constitute a waiver by such employee of any right he may have under subsection (b) of this section to such unpaid minimum wages or unpaid overtime compensation and an additional equal amount as liquidated damages. The Secretary may bring an action in any court of competent jurisdiction to recover the amount of the unpaid minimum wages or overtime compensation and an equal amount as liquidated damages. The right provided by subsection (b) of this section to bring an action by or on behalf of any employee to recover the liability specified in the first sentence of such subsection and of any employee to become a party plaintiff to any such action shall terminate upon the filing of a complaint by the Secretary in an action under this subsection in which a recovery is sought of unpaid minimum wages or unpaid overtime compensation under sections 206 and 207 *[sections 6 and 7]* of this title or liquidated or other damages provided by this subsection owing to such employee by an employer liable under the provisions of subsection (b) of this section, unless such action is dismissed without prejudice on motion of the Secretary. Any sums thus recovered by the Secretary of Labor on behalf of an employee pursuant to this subsection shall be held in a special deposit account and shall be paid, on order of the Secretary of Labor, directly to the employee or employees affected. Any such sums not paid to an employee because of inability to do so within a period of three years shall be covered into the Treasury of the United States as miscellaneous receipts. In determining when an action is commenced by the Secretary of Labor under this subsection for the purposes of the statutes of limitations provided in section 255(a) of this title *[section 6(a) of the Portal-to-Portal Act of 1947]*, it shall be considered to be commenced in the case of any individual claimant on the date when the complaint is filed if he is specifically named as a party plaintiff in the complaint, or if his name did not so appear, on the subsequent date on which his name is added as a party plaintiff in such action.

(d) Savings provisions

In any action or proceeding commenced prior to, on, or after August 8, 1956 *[the date of enactment of this subsection]*, no employer shall be subject to any liability or punishment under this chapter or the Portal-to-Portal Act of 1947 [29 U.S.C. 251 et seq.] on account of his failure to comply with any provision or provisions of this chapter or such Act (1) with respect to work heretofore or hereafter performed in a workplace to which the exemption in section 213(f) *[section 13(f)]* of this title is applicable, (2) with respect to work performed in Guam, the Canal Zone or Wake Island before the effective date of this amendment of subsection (d), or (3) with respect to work performed in a possession named in section 206(a)(3) *[section 6(a)(3)]* of this title at any time prior to the establishment by the Secretary, as provided therein, of a minimum wage rate applicable to such work.

(e)

(1)

(A) Any person who violates the provisions of sections 212 or 213(c) *[sections 12 or 13(c)]* of this title, relating to child labor, or any regulation issued pursuant to such sections, shall be subject to a civil penalty of not to exceed—

(i) $11,000 for each employee who was the subject of such a violation; or

(ii) $50,000 with regard to each such violation that causes the death or serious injury of any employee under the age of 18 years, which penalty may be doubled where the violation is a repeated or willful violation.

(B) For purposes of subparagraph (A), the term "serious injury" means—

(i) permanent loss or substantial impairment of one of the senses (sight, hearing, taste, smell, tactile sensation);

(ii) permanent loss or substantial impairment of the function of a bodily member, organ, or mental faculty, including the loss of all or part of an arm, leg, foot, hand or other body part; or

(iii) permanent paralysis or substantial impairment that causes loss of movement or mobility of an arm, leg, foot, hand or other body part.

(2) Any person who repeatedly or willfully violates section 206 or 207 *[section 6 or 7]*, relating to wages, shall be subject to a civil penalty not to exceed $1,100 for each such violation.

(3) In determining the amount of any penalty under this subsection, the appropriateness of such penalty to the size of the business of the person charged and the gravity of the violation shall be considered. The amount of any penalty under this subsection, when finally determined, may be—

(A) deducted from any sums owing by the United States to the person charged;

(B) recovered in a civil action brought by the Secretary in any court of competent jurisdiction, in which litigation the Secretary shall be represented by the Solicitor of Labor; or

(C) ordered by the court, in an action brought for a violation of section 215(a)(4) *[section 15(a)(4)]* of this title or a repeated or willful violation of section 215(a)(2) *[section 15(a)(2)]* of this title, to be paid to the Secretary.

(4) Any administrative determination by the Secretary of the amount of any penalty under this subsection shall be final, unless within 15 days after receipt of notice thereof by certified mail the person charged with the violation takes exception to the determination that the violations for which the penalty is imposed occurred, in which event final determination of the penalty shall be made in an administrative proceeding after opportunity for hearing in accordance with section 554 of Title 5 *[Administrative Procedure Act]*, and regulations to be promulgated by the Secretary.

(5) Except for civil penalties collected for violations of section 212 *[section 12]* of this title, sums collected as penalties pursuant to this section shall be applied toward reimbursement of the costs of determining the violations and assessing and collecting such penalties, in accordance with the provision of section 9a of Title 29 *[An Act to authorize the Department of Labor to make special statistical studies upon payment of the cost thereof and for other purposes]*. Civil penalties collected for violations of section 212 *[section 12]* of this title shall be deposited in the general fund of the Treasury.

INJUNCTION PROCEEDINGS

SEC. 217 [Section 17]

The districts courts, together with the United States District Court for the District of the Canal Zone, the District Court of the Virgin Islands, and the District Court of Guam shall have jurisdiction, for cause shown, to restrain violations of section 215 *[section 15]* of this title, including in the case of violations of section 215(a)(2) of this title the restraint of any withholding of payment of minimum wages or overtime compensation found by the court to be due to employees under this chapter (except sums which employees are barred from recovering, at the time of the commencement of the action to restrain the violations, by virtue of the provisions of section 255 of this title *[section 6 of the Portal-to-Portal Act of 1947]*.

RELATION TO OTHER LAWS

SEC. 218 [Section 18]

(a) No provision of this chapter or of any order thereunder shall excuse noncompliance with any Federal or State law or municipal ordinance establishing a minimum wage higher than the minimum wage established under this chapter or a maximum work week lower than the maximum workweek established under this chapter, and no provision of this chapter relating to the employment of child labor shall justify noncompliance with any Federal or State law or municipal ordinance establishing a higher standard than the standard established under this chapter. No provision of this chapter shall justify any employer in reducing a wage paid by him which is in excess of the applicable minimum wage under this chapter, or justify any employer in increasing hours of employment maintained by him which are shorter than the maximum hours applicable under this chapter.

SEPARABILITY OF PROVISIONS

SEC. 219 [Section 19]

If any provision of this chapter or the application of such provision to any person or circumstance is held invalid, the remainder of this chapter and the application of such provision to other persons or circumstances shall not be affected thereby.

Approved June 25, 1938.

[In the following excerpts from the Portal-to-Portal Act of 1947, the authority given to the Secretary of Labor is exercised by the Equal Employment Opportunity Commission for purposes of enforcing the Equal Pay Act of 1963.]

PART IV—MISCELLANEOUS

STATUTE OF LIMITATIONS

SEC. 255 [Section 6]

Any action commenced on or after May 14, 1947 *[the date of the enactment of this Act]*, to enforce any cause of action for unpaid minimum wages, unpaid overtime compensation, or liquidated damages, under the Fair Labor Standards Act of 1938, as amended, [29 U.S.C. 201 et seq.], the Walsh-Healey Act [41 U.S.C. 35 et seq.], or the Bacon-Davis Act [40 U.S.C. 276a et seq.]-

(a) if the cause of action accrues on or after May 14, 1947 *[the date of the enactment of this Act]*-may be commenced within two years after the cause of action accrued, and every such action shall be forever barred unless commenced within two years after the cause of action accrued, except that a cause of action arising out of a willful violation may be commenced within three years after the cause of action accrued;

DETERMINATION OF COMMENCEMENT OF FUTURE ACTIONS

SEC. 256 [Section 7]

In determining when an action is commenced for the purposes of section 255 *[section 6]* of this title, an action commenced on or after May 14, 1947 *[the date of the enactment of this Act]* under the Fair Labor Standards Act of 1938, as amended, [29 U.S.C. 201 et seq.], the Walsh-Healey Act [41 U.S.C. 35 et seq.], or the Bacon-Davis

Act [40 U.S.C. 276a et seq.], shall be considered to be commenced on the date when the complaint is filed; except that in the case of a collective or class action instituted under the Fair Labor Standards Act of 1938, as amended, or the Bacon-Davis Act, it shall be considered to be commenced in the case of any individual claimant—

   (a) on the date when the complaint is filed, if he is specifically named as a party plaintiff in the complaint and his written consent to become a party plaintiff is filed on such date in the court in which the action is brought; or
   (b) if such written consent was not so filed or if his name did not so appear—on the subsequent date on which such written consent is filed in the court in which the action was commenced.

RELIANCE IN FUTURE ON ADMINISTRATIVE RULINGS, ETC.

SEC. 259 [Section 10]

   (a) In any action or proceeding based on any act or omission on or after May 14, 1947 *[the date of the enactment of this Act]*, no employer shall be subject to any liability or punishment for or on account of the failure of the employer to pay minimum wages or overtime compensation under the Fair Labor Standards Act of 1938, as amended, [29 U.S.C. 201 et seq.], the Walsh-Healey Act [41 U.S.C. 35 et seq.], or the Bacon-Davis Act [40 U.S.C. 276a et seq.], if he pleads and proves that the act or omission complained of was in good faith in conformity with and in reliance on any written administrative regulation, order, ruling, approval, or interpretation, of the agency of the United States specified in subsection (b) of this section, or any administrative practice or enforcement policy of such agency with respect to the class of employers to which he belonged. Such a defense, if established, shall be a bar to the action or proceeding, notwithstanding that after such act or omission, such administrative regulation, order, ruling, approval, interpretation, practice, or enforcement policy is modified or rescinded or is determined by judicial authority to be invalid or of no legal effect.
   (b) The agency referred to in subsection (a) shall be—
       (1) in the case of the Fair Labor Standards Act of 1938, as amended [29 U.S.C. 201 et seq.]- the Administrator of the Wage and Hour Division of the Department of Labor;

LIQUIDATED DAMAGES

SEC. 260 *[Section 11]*
   In any action commenced prior to or on or after May 14, 1947 *[the date of the enactment of this Act]* to recover unpaid minimum wages, unpaid overtime compensation, or liquidated damages, under the Fair Labor Standards Act of 1938, as amended [29 U.S.C. 201 et seq.], if the employer shows to the satisfaction of the court that the act or omission giving rise to such action was in good faith and that he had reasonable grounds for believing that his act or omission was not a

violation of the Fair Labor Standards Act of 1938, as amended *[29 U.S.C. 201 et seq.]*,the court may, in its sound discretion, award no liquidated damages or award any amount thereof not to exceed the amount specified in section 216 *[section 16]* of this title.

DEFINITIONS

SEC. 262 [Section 13]

(a) When the terms "employer", "employee", and "wage" are used in this chapter in relation to the Fair Labor Standards Act of 1938, as amended [29 U.S.C. 201 et seq.], they shall have the same meaning as when used in such Act of 1938.

SEPARABILITY

Not Reprinted in U.S. Code *[Section 14]*

If any provision of this Act or the application of such provision to any person or circumstance is held invalid, the remainder of this Act and the application of such provision to other persons or circumstances shall not be affected thereby.

SHORT TITLE

Not Reprinted in U.S. Code *[Section 15]*

This Act may be cited as the 'Portal-to-Portal Act of 1947.'
Approved May 14, 1947.

The Age Discrimination in Employment Act of 1967

An Act,

To prohibit age discrimination in employment.

Be it enacted by the Senate and House of Representatives of the United States of America in Congress assembled, that this Act may be cited as the "Age Discrimination in Employment Act of 1967."

* * *

CONGRESSIONAL STATEMENT OF FINDINGS AND PURPOSE

SEC. 621. [Section 2]

(a) The Congress hereby finds and declares that—
   (1) in the face of rising productivity and affluence, older workers find themselves disadvantaged in their efforts to retain employment, and especially to regain employment when displaced from jobs;
   (2) the setting of arbitrary age limits regardless of potential for job performance has become a common practice, and certain otherwise desirable practices may work to the disadvantage of older persons;
   (3) the incidence of unemployment, especially long-term unemployment with resultant deterioration of skill, morale, and employer acceptability is, relative to the younger ages, high among older workers; their numbers are great and growing; and their employment problems grave;
   (4) the existence in industries affecting commerce, of arbitrary discrimination in employment because of age, burdens commerce and the free flow of goods in commerce.
(b) It is therefore the purpose of this chapter to promote employment of older persons based on their ability rather than age; to prohibit arbitrary age discrimination in employment; to help employers and workers find ways of meeting problems arising from the impact of age on employment.

EDUCATION AND RESEARCH PROGRAM; RECOMMENDATION TO CONGRESS

SEC. 622. [Section 3]

(a) The EEOC *[originally, the Secretary of Labor]* shall undertake studies and provide information to labor unions, management, and the general public

379

concerning the needs and abilities of older workers, and their potentials for continued employment and contribution to the economy. In order to achieve the purposes of this chapter, the EEOC *[originally, the Secretary of Labor]* shall carry on a continuing program of education and information, under which he may, among other measures—

(1) undertake research, and promote research, with a view to reducing barriers to the employment of older persons, and the promotion of measures for utilizing their skills;

(2) publish and otherwise make available to employers, professional societies, the various media of communication, and other interested persons the findings of studies and other materials for the promotion of employment;

(3) foster through the public employment service system and through cooperative effort the development of facilities of public and private agencies for expanding the opportunities and potentials of older persons;

(4) sponsor and assist State and community informational and educational programs.

(b) Not later than six months after the effective date of this chapter, the Secretary shall recommend to the Congress any measures he may deem desirable to change the lower or upper age limits set forth in section 631 of this title *[section 12]*.

PROHIBITION OF AGE DISCRIMINATION

SEC. 623. [Section 4]

(a) Employer practices

It shall be unlawful for an employer—

(1) to fail or refuse to hire or to discharge any individual or otherwise discriminate against any individual with respect to his compensation, terms, conditions, or privileges of employment, because of such individual's age;

(2) to limit, segregate, or classify his employees in any way which would deprive or tend to deprive any individual of employment opportunities or otherwise adversely affect his status as an employee, because of such individual's age; or

(3) to reduce the wage rate of any employee in order to comply with this chapter.

(b) It shall be unlawful for an employment agency to fail or refuse to refer for employment, or otherwise to discriminate against, any individual because of such individual's age, or to classify or refer for employment any individual on the basis of such individual's age.

(c) Labor organization practices

It shall be unlawful for a labor organization—

    (1) to exclude or to expel from its membership, or otherwise to discriminate against, any individual because of his age;

    (2) to limit, segregate, or classify its membership, or to classify or fail or refuse to refer for employment any individual, in any way which would deprive or tend to deprive any individual of employment opportunities, or would limit such employment opportunities or otherwise adversely affect his status as an employee or as an applicant for employment, because of such individual's age;

    (3) to cause or attempt to cause an employer to discriminate against an individual in violation of this section.

(d) Opposition to unlawful practices; participation in investigations, proceedings, or litigation

It shall be unlawful for an employer to discriminate against any of his employees or applicants for employment, for an employment agency to discriminate against any individual, or for a labor organization to discriminate against any member thereof or applicant for membership, because such individual, member or applicant for membership has opposed any practice made unlawful by this section, or because such individual, member or applicant for membership has made a charge, testified, assisted, or participated in any manner in an investigation, proceeding, or litigation under this chapter.

(e) Printing or publication of notice or advertisement indicating preference, limitation, etc.

It shall be unlawful for an employer, labor organization, or employment agency to print or publish, or cause to be printed or published, any notice or advertisement relating to employment by such an employer or membership in or any classification or referral for employment by such a labor organization, or relating to any classification or referral for employment by such an employment agency, indicating any preference, limitation, specification, or discrimination, based on age.

(f) Lawful practices; age an occupational qualification; other reasonable factors; laws of foreign workplace; seniority system; employee benefit plans; discharge or discipline for good cause

It shall not be unlawful for an employer, employment agency, or labor organization—

    (1) to take any action otherwise prohibited under subsections (a), (b), (c), or (e) of this section where age is a bona fide occupational qualification reasonably necessary to the normal operation of the particular business, or where the differentiation is based on reasonable factors other than age, or where such practices involve an employee in a workplace in a foreign country, and compliance with such subsections would cause such employer, or a corporation controlled by such employer, to violate the laws of the country in which such workplace is located;

(2) to take any action otherwise prohibited under subsection (a), (b), (c), or (e) of this section—

    (A) to observe the terms of a bona fide seniority system that is not intended to evade the purposes of this chapter, except that no such seniority system shall require or permit the involuntary retirement of any individual specified by section 631(a) of this title because of the age of such individual; or

    (B) to observe the terms of a bona fide employee benefit plan—

        (i) where, for each benefit or benefit package, the actual amount of payment made or cost incurred on behalf of an older worker is no less than that made or incurred on behalf of a younger worker, as permissible under section 1625.10, title 29, Code of Federal Regulations (as in effect on June 22, 1989); or

        (ii) that is a voluntary early retirement incentive plan consistent with the relevant purpose or purposes of this chapter.

        Notwithstanding clause (i) or (ii) of subparagraph (B), no such employee benefit plan or voluntary early retirement incentive plan shall excuse the failure to hire any individual, and no such employee benefit plan shall require or permit the involuntary retirement of any individual specified by section 631(a) of this title, because of the age of such individual. An employer, employment agency, or labor organization acting under subparagraph (A), or under clause (i) or (ii) of subparagraph (B), shall have the burden of proving that such actions are lawful in any civil enforcement proceeding brought under this chapter; or

(3) to discharge or otherwise discipline an individual for good cause.

(g) *[Repealed]*

(h) Practices of foreign corporations controlled by American employers; foreign employers not controlled by American employers; factors determining control

    (1) If an employer controls a corporation whose place of incorporation is in a foreign country, any practice by such corporation prohibited under this section shall be presumed to be such practice by such employer.

    (2) The prohibitions of this section shall not apply where the employer is a foreign person not controlled by an American employer.

    (3) For the purpose of this subsection the determination of whether an employer controls a corporation shall be based upon the—

        (A) interrelation of operations,

        (B) common management,

        (C) centralized control of labor relations, and

        (D) common ownership or financial control, of the employer and the corporation.

(i) Employee pension benefit plans; cessation or reduction of benefit accrual or of allocation to employee account; distribution of benefits after attainment of normal retirement age; compliance; highly compensated employees

    (1) Except as otherwise provided in this subsection, it shall be unlawful for an employer, an employment agency, a labor organization, or any

combination thereof to establish or maintain an employee pension benefit plan which requires or permits—

(A) in the case of a defined benefit plan, the cessation of an employee's benefit accrual, or the reduction of the rate of an employee's benefit accrual, because of age, or

(B) in the case of a defined contribution plan, the cessation of allocations to an employee's account, or the reduction of the rate at which amounts are allocated to an employee's account, because of age.

(2) Nothing in this section shall be construed to prohibit an employer, employment agency, or labor organization from observing any provision of an employee pension benefit plan to the extent that such provision imposes (without regard to age) a limitation on the amount of benefits that the plan provides or a limitation on the number of years of service or years of participation which are taken into account for purposes of determining benefit accrual under the plan.

(3) In the case of any employee who, as of the end of any plan year under a defined benefit plan, has attained normal retirement age under such plan—

(A) if distribution of benefits under such plan with respect to such employee has commenced as of the end of such plan year, then any requirement of this subsection for continued accrual of benefits under such plan with respect to such employee during such plan year shall be treated as satisfied to the extent of the actuarial equivalent of in-service distribution of benefits, and

(B) if distribution of benefits under such plan with respect to such employee has not commenced as of the end of such year in accordance with section 1056(a)(3) of this title *[section 206(a)(3) of the Employee Retirement Income Security Act of 1974]* and section 401(a)(14)(C) of Title 26 *[the Internal Revenue Code of 1986]*, and the payment of benefits under such plan with respect to such employee is not suspended during such plan year pursuant to section 1053(a)(3)(B) of this title or section 411(a)(3)(B) of Title 26 *[the Internal Revenue Code of 1986]*, then any requirement of this subsection for continued accrual of benefits under such plan with respect to such employee during such plan year shall be treated as satisfied to the extent of any adjustment in the benefit payable under the plan during such plan year attributable to the delay in the distribution of benefits after the attainment of normal retirement age.

The provisions of this paragraph shall apply in accordance with regulations of the Secretary of the Treasury. Such regulations shall provide for the application of the preceding provisions of this paragraph to all employee pension benefit plans subject to this subsection and may provide for the application of such provisions, in the case of any such employee, with respect to any period of time within a plan year.

(4) Compliance with the requirements of this subsection with respect to an employee pension benefit plan shall constitute compliance with the requirements of this section relating to benefit accrual under such plan.

(5) Paragraph (1) shall not apply with respect to any employee who is a highly compensated employee (within the meaning of section 414(q) of Title 26 *[the Internal Revenue Code of 1986]*) to the extent provided in regulations prescribed by the Secretary of the Treasury for purposes of precluding discrimination in favor of highly compensated employees within the meaning of subchapter D of chapter 1 of Title 26 *[the Internal Revenue Code of 1986]*.

(6) A plan shall not be treated as failing to meet the requirements of paragraph (1) solely because the subsidized portion of any early retirement benefit is disregarded in determining benefit accruals or it is a plan permitted by subsection (m) of this section.

(7) Any regulations prescribed by the Secretary of the Treasury pursuant to clause (v) of section 411(b)(1)(H) of Title 26 *[the Internal Revenue Code of 1986]* and subparagraphs (C) and (D), of section 411(b)(2) of Title 26 *[the Internal Revenue Code of 1986]* shall apply with respect to the requirements of this subsection in the same manner and to the same extent as such regulations apply with respect to the requirements of such sections 411(b)(1)(H) and 411(b)(2).

(8) A plan shall not be treated as failing to meet the requirements of this section solely because such plan provides a normal retirement age described in section 1002(24)(B) *[section 2(24)(B) of the Employee Retirement Income Security Act of 1974]* of this title and section 411(a) (8)(B) of Title 26 *[the Internal Revenue Code of 1986]*.

(9) For purposes of this subsection—

  (A) The terms "employee pension benefit plan", "defined benefit plan", "defined contribution plan", and "normal retirement age" have the meanings provided such terms in section 1002 of this title *[section 3 of the Employee Retirement Income Security Act of 1974]*.

  (B) The term "compensation" has the meaning provided by section 414(s) of Title 26 *[the Internal Revenue Code of 1986]*.

(10) Special rules relating to age

  (A) Comparison to similarly situated younger individual

    (i) In general—A plan shall not be treated as failing to meet the requirements of paragraph (1) if a participant's accrued benefit, as determined as of any date under the terms of the plan, would be equal to or greater than that of any similarly situated, younger individual who is or could be a participant.

    (ii) Similarly situated—For purposes of this subparagraph, a participant is similarly situated to any other individual if such participant is identical to such other individual in every respect (including period of service, compensation, position, date of hire, work history, and any other respect) except for age.

(iii)   Disregard of subsidized early retirement benefits—In determining the accrued benefit as of any date for purposes of this
clause, the subsidized portion of any early retirement benefit or
retirement-type subsidy shall be disregarded.

(iv)   Accrued benefit—For purposes of this subparagraph, the
accrued benefit may, under the terms of the plan, be expressed
as an annuity payable at normal retirement age, the balance of
a hypothetical account, or the current value of the accumulated
percentage of the employee's final average compensation.

(B)   Applicable defined benefit plans

(i)   Interest credits

(I)   In general—An applicable defined benefit plan shall be treated
as failing to meet the requirements of paragraph (1) unless the
terms of the plan provide that any interest credit (or an equivalent amount) for any plan year shall be at a rate which is not
greater than a market rate of return. A plan shall not be treated
as failing to meet the requirements of this subclause merely
because the plan provides for a reasonable minimum guaranteed rate of return or for a rate of return that is equal to the
greater of a fixed or variable rate of return

(II)   Preservation of capital—An interest credit (or an equivalent
amount) of less than zero shall in no event result in the account
balance or similar amount being less than the aggregate
amount of contributions credited to the account.

(III)   Market rate of return—The Secretary of the Treasury may
provide by regulation for rules governing the calculation of a
market rate of return for purposes of subclause (I) and for permissible methods of crediting interest to the account (including fixed or variable interest rates) resulting in effective rates
of return meeting the requirements of subclause (I). In the
case of a governmental plan (as defined in the first sentence
of section 414(d) of Title 26 [the Internal Revenue Code of
1986], a rate of return or a method of crediting interest established pursuant to any provision of Federal, State, or local
law (including any administrative rule or policy adopted in
accordance with any such law) shall be treated as a market
rate of return for purposes of subclause (I) and a permissible method of crediting interest for purposes of meeting the
requirements of subclause (I), except that this sentence shall
only apply to a rate of return or method of crediting interest if
such rate or method does not violate any other requirement of
this chapter.

(ii)   Special rule for plan conversions—If, after June 29, 2005,
an applicable plan amendment is adopted, the plan shall be
treated as failing to meet the requirements of paragraph (1)(H)
unless the requirements of clause (iii) are met with respect to

each individual who was a participant in the plan immediately before the adoption of the amendment.

(iii) Rate of benefit accrual—Subject to clause (iv), the requirements of this clause are met with respect to any participant if the accrued benefit of the participant under the terms of the plan as in effect after the amendment is not less than the sum of—

(I) the participant's accrued benefit for years of service before the effective date of the amendment, determined under the terms of the plan as in effect before the amendment, plus

(II) the participant's accrued benefit for years of service after the effective date of the amendment, determined under the terms of the plan as in effect after the amendment.

(iv) Special rules for early retirement subsidies—For purposes of clause (iii)(I), the plan shall credit the accumulation account or similar amount with the amount of any early retirement benefit or retirement-type subsidy for the plan year in which the participant retires if, as of such time, the participant has met the age, years of service, and other requirements under the plan for entitlement to such benefit or subsidy.

(v) Applicable plan amendment—For purposes of this subparagraph—

(I) In general—The term "applicable plan amendment" means an amendment to a defined benefit plan which has the effect of converting the plan to an applicable defined benefit plan.

(II) Special rule for coordinated benefits—If the benefits of 2 or more defined benefit plans established or maintained by an employer are coordinated in such a manner as to have the effect of the adoption of an amendment described in subclause (I), the sponsor of the defined benefit plan or plans providing for such coordination shall be treated as having adopted such a plan amendment as of the date such coordination begins.

(III) Multiple amendments—The Secretary of the Treasury shall issue regulations to prevent the avoidance of the purposes of this subparagraph through the use of 2 or more plan amendments rather than a single amendment.

(IV) Applicable defined benefit plan—For purposes of this subparagraph, the term "applicable defined benefit plan" has the meaning given such term by section 1053(f)(3) of this title *[section 203(f)(3) of the Employee Retirement Income Security Act of 1974]*.

(vi) Termination requirements—An applicable defined benefit plan shall not be treated as meeting the requirements of clause (i) unless the plan provides that, upon the termination of the plan—

(I) if the interest credit rate (or an equivalent amount) under the plan is a variable rate, the rate of interest used to determine accrued benefits under the plan shall be equal to the average of the rates of interest used under the plan during the 5-year period ending on the termination date, and

(II)   the interest rate and mortality table used to determine the amount of any benefit under the plan payable in the form of an annuity payable at normal retirement age shall be the rate and table specified under the plan for such purpose as of the termination date, except that if such interest rate is a variable rate, the interest rate shall be determined under the rules of subclause (I).

(C)   Certain offsets permitted—A plan shall not be treated as failing to meet the requirements of paragraph (1) solely because the plan provides offsets against benefits under the plan to the extent such offsets are allowable in applying the requirements of section 401(a) of Title 26 *[the Internal Revenue Code of 1986]*.

(D)   Permitted disparities in plan contributions or benefits—A plan shall not be treated as failing to meet the requirements of paragraph (1) solely because the plan provides a disparity in contributions or benefits with respect to which the requirements of section 401(l) of Title 26 *[the Internal Revenue Code of 1986]* are met.

(E)   Indexing permitted—

(i)   In general—A plan shall not be treated as failing to meet the requirements of paragraph (1) solely because the plan provides for indexing of accrued benefits under the plan.

(ii)   Protection against loss—Except in the case of any benefit provided in the form of a variable annuity, clause (i) shall not apply with respect to any indexing which results in an accrued benefit less than the accrued benefit determined without regard to such indexing.

(iii)   Indexing—For purposes of this subparagraph, the term "indexing" means, in connection with an accrued benefit, the periodic adjustment of the accrued benefit by means of the application of a recognized investment index or methodology.

(F)   Early retirement benefit or retirement-type subsidy—For purposes of this paragraph, the terms "early retirement benefit" and "retirement-type subsidy" have the meaning given such terms in section 1053(g)(2)(A) of this title *[section 203(g)(2)(A) of the Employee Retirement Income Security Act of 1974]*.

(G)   Benefit accrued to date—For purposes of this paragraph, any reference to the accrued benefit shall be a reference to such benefit accrued to date.

(j) Employment as firefighter or law enforcement officer

It shall not be unlawful for an employer which is a State, a political subdivision of a State, an agency or instrumentality of a State or a political subdivision of a State, or an interstate agency to fail or refuse to hire or to discharge any individual because of such individual's age if such action is taken—

(1) with respect to the employment of an individual as a firefighter or as a law enforcement officer, the employer has complied with section 3(d)(2) of the Age Discrimination in Employment Amendments of 1996 if the individual was discharged after the date described in such section, and the individual has attained—

    (A) the age of hiring or retirement, respectively, in effect under applicable State or local law on March 3, 1983; or

    (B)

        (i) if the individual was not hired, the age of hiring in effect on the date of such failure or refusal to hire under applicable State or local law enacted after September 30, 1996; or

        (ii) if applicable State or local law was enacted after September 30, 1996, and the individual was discharged, the higher of—

        (I) the age of retirement in effect on the date of such discharge under such law; and

        (II) age 55; and

(2) pursuant to a bona fide hiring or retirement plan that is not a subterfuge to evade the purposes of this chapter.

(k) Seniority system or employee benefit plan; compliance

A seniority system or employee benefit plan shall comply with this chapter regardless of the date of adoption of such system or plan.

(l) Lawful practices; minimum age as condition of eligibility for retirement benefits; deductions from severance pay; reduction of long-term disability benefits

Notwithstanding clause (i) or (ii) of subsection (f)(2)(B) of this section—

(1)

    (A) It shall not be a violation of subsection (a), (b), (c), or (e) of this section solely because—

        (i) an employee pension benefit plan (as defined in section 1002(2) of this title *[section 2(2) of the Employee Retirement Income Security Act of 1974]*) provides for the attainment of a minimum age as a condition of eligibility for normal or early retirement benefits; or

        (ii) a defined benefit plan (as defined in section 1002(35) of this title *[section 2(35) of the Employee Retirement Income Security Act]*) provides for—

        (I) payments that constitute the subsidized portion of an early retirement benefit; or

        (II) social security supplements for plan participants that commence before the age and terminate at the age (specified by the plan) when participants are eligible to receive reduced or unreduced old-age insurance benefits under title II of the Social Security Act (42 U.S.C. 401 et seq.), and that do not exceed such old-age insurance benefits.

    (B) A voluntary early retirement incentive plan that—

        (i) is maintained by—

      (I)   a local educational agency (as defined in section 7801 of Title 20 *[the Elementary and Secondary Education Act of 1965]*, or

     (II)   an education association which principally represents employees of 1 or more agencies described in subclause (I) and which is described in section 501(c) (5) or (6) of Title 26 *[the Internal Revenue Code of 1986]* and exempt from taxation under section 501(a) of Title 26 *[the Internal Revenue Code of 1986]*, and

    (ii)   makes payments or supplements described in subclauses (I) and (II) of subparagraph (A)(ii) in coordination with a defined benefit plan (as so defined) maintained by an eligible employer described in section 457(e)(1) (A) of Title 26 *[the Internal Revenue Code of 1986]* or by an education association described in clause (i)(II), shall be treated solely for purposes of subparagraph (A)(ii) as if it were a part of the defined benefit plan with respect to such payments or supplements. Payments or supplements under such a voluntary early retirement incentive plan shall not constitute severance pay for purposes of paragraph (2).

(2)

  (A)   It shall not be a violation of subsection (a), (b), (c), or (e) of this section solely because following a contingent event unrelated to age—

     (i)   the value of any retiree health benefits received by an individual eligible for an immediate pension;

    (ii)   the value of any additional pension benefits that are made available solely as a result of the contingent event unrelated to age and following which the individual is eligible for not less than an immediate and unreduced pension; or

   (iii)   the values described in both clauses (i) and (ii); are deducted from severance pay made available as a result of the contingent event unrelated to age.

  (B)   For an individual who receives immediate pension benefits that are actuarially reduced under subparagraph (A)(i), the amount of the deduction available pursuant to subparagraph (A)(i) shall be reduced by the same percentage as the reduction in the pension benefits.

  (C)   For purposes of this paragraph, severance pay shall include that portion of supplemental unemployment compensation benefits (as described in section 501(c)(17) of Title 26 *[the Internal Revenue Code of 1986]*) that—

     (i)   constitutes additional benefits of up to 52 weeks;

    (ii)   has the primary purpose and effect of continuing benefits until an individual becomes eligible for an immediate and unreduced pension; and

   (iii)   is discontinued once the individual becomes eligible for an immediate and unreduced pension.

(D)   For purposes of this paragraph and solely in order to make the deduction authorized under this paragraph, the term "retiree health benefits" means benefits provided pursuant to a group health plan covering retirees, for which (determined as of the contingent event unrelated to age)—

(i)   the package of benefits provided by the employer for the retirees who are below age 65 is at least comparable to benefits provided under title XVIII of the Social Security Act (42 U.S.C. 1395 et seq.);

(ii)   the package of benefits provided by the employer for the retirees who are age 65 and above is at least comparable to that offered under a plan that provides a benefit package with one-fourth the value of benefits provided under title XVIII of such Act; or

(iii)   the package of benefits provided by the employer is as described in clauses (i) and (ii).

(E)

(i)   If the obligation of the employer to provide retiree health benefits is of limited duration, the value for each individual shall be calculated at a rate of $3,000 per year for benefit years before age 65, and $750 per year for benefit years beginning at age 65 and above.

(ii)   If the obligation of the employer to provide retiree health benefits is of unlimited duration, the value for each individual shall be calculated at a rate of $48,000 for individuals below age 65, and $24,000 for individuals age 65 and above.

(iii)   The values described in clauses (i) and (ii) shall be calculated based on the age of the individual as of the date of the contingent event unrelated to age. The values are effective on October 16, 1990, and shall be adjusted on an annual basis, with respect to a contingent event that occurs subsequent to the first year after October 16, 1990, based on the medical component of the Consumer Price Index for all-urban consumers published by the Department of Labor.

(iv)   If an individual is required to pay a premium for retiree health benefits, the value calculated pursuant to this subparagraph shall be reduced by whatever percentage of the overall premium the individual is required to pay.

(F)   If an employer that has implemented a deduction pursuant to subparagraph (A) fails to fulfill the obligation described in subparagraph (E), any aggrieved individual may bring an action for specific performance of the obligation described in subparagraph (E). The relief shall be in addition to any other remedies provided under Federal or State law.

(3)   It shall not be a violation of subsection (a), (b), (c), or (e) of this section solely because an employer provides a bona fide employee benefit plan

or plans under which long-term disability benefits received by an individual are reduced by any pension benefits (other than those attributable to employee contributions)—

(A) paid to the individual that the individual voluntarily elects to receive; or

(B) for which an individual who has attained the later of age 62 or normal retirement age is eligible.

(m) Voluntary retirement incentive plans

Notwithstanding subsection (f)(2)(b) of this section, it shall not be a violation of subsection (a), (b), (c), or (e) of this section solely because a plan of an institution of higher education (as defined in section 1001 of Title 20 [the Higher Education Act of 1965]) offers employees who are serving under a contract of unlimited tenure (or similar arrangement providing for unlimited tenure) supplemental benefits upon voluntary retirement that are reduced or eliminated on the basis of age, if—

(1) such institution does not implement with respect to such employees any age-based reduction or cessation of benefits that are not such supplemental benefits, except as permitted by other provisions of this chapter;

(2) such supplemental benefits are in addition to any retirement or severance benefits which have been offered generally to employees serving under a contract of unlimited tenure (or similar arrangement providing for unlimited tenure), independent of any early retirement or exit-incentive plan, within the preceding 365 days; and

(3) any employee who attains the minimum age and satisfies all non-age-based conditions for receiving a benefit under the plan has an opportunity lasting not less than 180 days to elect to retire and to receive the maximum benefit that could then be elected by a younger but otherwise similarly situated employee, and the plan does not require retirement to occur sooner than 180 days after such election.

STUDY BY SECRETARY OF LABOR; REPORTS TO PRESIDENT AND CONGRESS; SCOPE OF STUDY; IMPLEMENTATION OF STUDY; TRANSMITTAL DATE OF REPORTS

SEC. 624. [Section 5]

(a)

(1) The EEOC [originally, the Secretary of Labor] is directed to undertake an appropriate study of institutional and other arrangements giving rise to involuntary retirement, and report his findings and any appropriate legislative recommendations to the President and to the Congress. Such study shall include—

(A) an examination of the effect of the amendment made by section 3(a) of the Age Discrimination in Employment Act Amendments of 1978 in raising the upper age limitation established by section 631(a) of this title *[section 1(a)]* to 70 years of age;

(B) a determination of the feasibility of eliminating such limitation;

(C) a determination of the feasibility of raising such limitation above 70 years of age; and

(D) an examination of the effect of the exemption contained in section 631(c) of this title *[section 1(c)]*, relating to certain executive employees, and the exemption contained in section 631(d) of this title *[section 1(d)]*, relating to tenured teaching personnel.

(2) The EEOC *[originally, the Secretary of Labor]* may undertake the study required by paragraph (1) of this subsection directly or by contract or other arrangement.

(b) The report required by subsection (a) of this section shall be transmitted to the President and to the Congress as an interim report not later than January 1, 1981, and in final form not later than January 1, 1982.

Transfer of Functions *[All functions relating to age discrimination administration and enforcement vested by Section 6 in the Secretary of Labor or the Civil Service Commission were transferred to the Equal Employment Opportunity Commission effective January 1, 1979 under the President's Reorganization Plan No. 1.]*

ADMINISTRATION

SEC. 625. [Section 6]

The EEOC *[originally, the Secretary of Labor]* shall have the power—

(a) Delegation of functions; appointment of personnel; technical assistance to make delegations, to appoint such agents and employees, and to pay for technical assistance on a fee for service basis, as he deems necessary to assist him in the performance of his functions under this chapter;

(b) Cooperation with other agencies, employers, labor organizations, and employment agencies to cooperate with regional, State, local, and other agencies, and to cooperate with and furnish technical assistance to employers, labor organizations, and employment agencies to aid in effectuating the purposes of this chapter.

RECORDKEEPING, INVESTIGATION, AND ENFORCEMENT

SEC. 626. [Section 7]

(a) Attendance of witnesses; investigations, inspections, records, and homework regulations

The Equal Employment Opportunity Commission shall have the power to make investigations and require the keeping of records necessary or appropriate for the administration of this chapter in accordance with the powers and procedures provided in sections 209 and 211 of this title *[sections 9 and 11 of the Fair Labor Standards Act of 1938, as amended].*

(b) Enforcement; prohibition of age discrimination under fair labor standards; unpaid minimum wages and unpaid overtime compensation; liquidated damages; judicial relief; conciliation, conference, and persuasion

The provisions of this chapter shall be enforced in accordance with the powers, remedies, and procedures provided in sections 211(b), 216 (except for subsection (a) thereof), and 217 of this title *[sections 11(b), 16 (except for subsection (a) thereof), and 17 of the Fair Labor Standards Act of 1938, as amended],* and subsection (c) of this section. Any act prohibited under section 623 of this title *[section 4]* shall be deemed to be a prohibited act under section 215 of this title *[section 15 of the Fair Labor Standards Act of 1938, as amended].* Amounts owing to a person as a result of a violation of this chapter shall be deemed to be unpaid minimum wages or unpaid overtime compensation for purposes of sections 216 and 217 of this title *[sections 16 and 17 of the Fair Labor Standards Act of 1938, as amended]: Provided,* That liquidated damages shall be payable only in cases of willful violations of this chapter. In any action brought to enforce this chapter the court shall have jurisdiction to grant such legal or equitable relief as may be appropriate to effectuate the purposes of this chapter, including without limitation judgments compelling employment, reinstatement or promotion, or enforcing the liability for amounts deemed to be unpaid minimum wages or unpaid overtime compensation under this section. Before instituting any action under this section, the Equal Employment Opportunity Commission shall attempt to eliminate the discriminatory practice or practices alleged, and to effect voluntary compliance with the requirements of this chapter through informal methods of conciliation, conference, and persuasion.

(c) Civil actions; persons aggrieved; jurisdiction; judicial relief; termination of individual action upon commencement of action by Commission; jury trial

(1) Any person aggrieved may bring a civil action in any court of competent jurisdiction for such legal or equitable relief as will effectuate the purposes of this chapter: *Provided,* That the right of any person to bring such action shall terminate upon the commencement of an action by the Equal Employment Opportunity Commission to enforce the right of such employee under this chapter.

(2) In an action brought under paragraph (1), a person shall be entitled to a trial by jury of any issue of fact in any such action for recovery of amounts owing as a result of a violation of this chapter, regardless of whether equitable relief is sought by any party in such action.

(d)

(1) Filing of charge with Commission; timeliness; conciliation, conference, and persuasion

No civil action may be commenced by an individual under this section until 60 days after a charge alleging unlawful discrimination has been filed with the Equal Employment Opportunity Commission. Such a charge shall be filed—

(A) within 180 days after the alleged unlawful practice occurred; or

    (B) in a case to which section 633(b) of this title applies, within 300 days after the alleged unlawful practice occurred, or within 30 days after receipt by the individual of notice of termination of proceedings under State law, whichever is earlier.

  (2) Upon receiving such a charge, the Commission shall promptly notify all persons named in such charge as prospective defendants in the action and shall promptly seek to eliminate any alleged unlawful practice by informal methods of conciliation, conference, and persuasion.

  (3) For purposes of this section, an unlawful practice occurs, with respect to discrimination in compensation in violation of this Act, when a discriminatory compensation decision or other practice is adopted, when a person becomes subject to a discriminatory compensation decision or other practice, or when a person is affected by application of a discriminatory compensation decision or other practice, including each time wages, benefits, or other compensation is paid, resulting in whole or in part from such a decision or other practice.

(e) Reliance on administrative rulings; notice of dismissal or termination; civil action after receipt of notice

Section 259 of this title [section 10 of the Portal to Portal Act of 1947] shall apply to actions under this chapter. If a charge filed with the Commission under this chapter is dismissed or the proceedings of the Commission are otherwise terminated by the Commission, the Commission shall notify the person aggrieved. A civil action may be brought under this section by a person defined in section 630(a) of this title [section 11(a)] against the respondent named in the charge within 90 days after the date of the receipt of such notice.—

(f) Waiver

  (1) An individual may not waive any right or claim under this chapter unless the waiver is knowing and voluntary. Except as provided in paragraph (2), a waiver may not be considered knowing and voluntary unless at a minimum—

    (A) the waiver is part of an agreement between the individual and the employer that is written in a manner calculated to be understood by such individual, or by the average individual eligible to participate;

    (B) the waiver specifically refers to rights or claims arising under this chapter;

    (C) the individual does not waive rights or claims that may arise after the date the waiver is executed;

    (D) the individual waives rights or claims only in exchange for consideration in addition to anything of value to which the individual already is entitled;

(E)　the individual is advised in writing to consult with an attorney prior to executing the agreement;

(F)
   (i)　the individual is given a period of at least 21 days within which to consider the agreement; or
   (ii)　if a waiver is requested in connection with an exit incentive or other employment termination program offered to a group or class of employees, the individual is given a period of at least 45 days within which to consider the agreement;

(G)　the agreement provides that for a period of at least 7 days following the execution of such agreement, the individual may revoke the agreement, and the agreement shall not become effective or enforceable until the revocation period has expired;

(H)　if a waiver is requested in connection with an exit incentive or other employment termination program offered to a group or class of employees, the employer (at the commencement of the period specified in subparagraph (F)) informs the individual in writing in a manner calculated to be understood by the average individual eligible to participate, as to—
   (i)　any class, unit, or group of individuals covered by such program, any eligibility factors for such program, and any time limits applicable to such program; and
   (ii)　the job titles and ages of all individuals eligible or selected for the program, and the ages of all individuals in the same job classification or organizational unit who are not eligible or selected for the program.

(2)　A waiver in settlement of a charge filed with the Equal Employment Opportunity Commission, or an action filed in court by the individual or the individual's representative, alleging age discrimination of a kind prohibited under section 623 or 633a of this title [section 4 or 15] may not be considered knowing and voluntary unless at a minimum—
(A)　subparagraphs (A) through (E) of paragraph (1) have been met; and
(B)　the individual is given a reasonable period of time within which to consider the settlement agreement.

(3)　In any dispute that may arise over whether any of the requirements, conditions, and circumstances set forth in subparagraph (A), (B), (C), (D), (E), (F), (G), or (H) of paragraph (1), or subparagraph (A) or (B) of paragraph (2), have been met, the party asserting the validity of a waiver shall have the burden of proving in a court of competent jurisdiction that a waiver was knowing and voluntary pursuant to paragraph (1) or (2).

(4) No waiver agreement may affect the Commission's rights and responsibilities to enforce this chapter. No waiver may be used to justify interfering with the protected right of an employee to file a charge or participate in an investigation or proceeding conducted by the Commission.

NOTICES TO BE POSTED

SEC. 627. [Section 8]

Every employer, employment agency, and labor organization shall post and keep posted in conspicuous places upon its premises a notice to be prepared or approved by the Equal Employment Opportunity Commission setting forth information as the Commission deems appropriate to effectuate the purposes of this chapter.

RULES AND REGULATIONS

SEC. 628. [Section 9]

In accordance with the provisions of subchapter II of chapter 5 of title 5 *[Administrative Procedures Act, 5 U.S.C. § 551 et seq.]*, the Equal Employment Opportunity Commission may issue such rules and regulations as it may consider necessary or appropriate for carrying out this chapter, and may establish such reasonable exemptions to and from any or all provisions of this chapter as it may find necessary and proper in the public interest.

CRIMINAL PENALTIES

SEC. 629. *[Section 10]*

Whoever shall forcibly resist, oppose, impede, intimidate or interfere with a duly authorized representative of the Equal Employment Opportunity Commission while it is engaged in the performance of duties under this chapter shall be punished by a fine of not more than $500 or by imprisonment for not more than one year, or by both: *Provided, however,* That no person shall be imprisoned under this section except when there has been a prior conviction hereunder.

DEFINITIONS

SEC. 630. *[Section 11]*

For the purposes of this chapter—

(a) The term "person" means one or more individuals, partnerships, associations, labor organizations, corporations, business trusts, legal representatives, or any organized groups of persons.
(b) The term "employer" means a person engaged in an industry affecting commerce who has twenty or more employees for each working day in each of twenty or more calendar weeks in the current or preceding calendar year: *Provided,* That prior to June 30, 1968, employers having fewer than fifty employees shall not be considered employers. The term also means (1) any agent of such a person, and (2) a State or political subdivision of a State and any agency or instrumentality of a State or a political subdivision of a State, and any interstate agency, but such term does not include the United States, or a corporation wholly owned by the Government of the United States.

(c) The term "employment agency" means any person regularly undertaking with or without compensation to procure employees for an employer and includes an agent of such a person; but shall not include an agency of the United States.

(d) The term "labor organization" means a labor organization engaged in an industry affecting commerce, and any agent of such an organization, and includes any organization of any kind, any agency, or employee representation committee, group, association, or plan so engaged in which employees participate and which exists for the purpose, in whole or in part, of dealing with employers concerning grievances, labor disputes, wages, rates of pay, hours, or other terms or conditions of employment, and any conference, general committee, joint or system board, or joint council so engaged which is subordinate to a national or international labor organization.

(e) A labor organization shall be deemed to be engaged in an industry affecting commerce if (1) it maintains or operates a hiring hall or hiring office which procures employees for an employer or procures for employees opportunities to work for an employer, or (2) the number of its members (or, where it is a labor organization composed of other labor organizations or their representatives, if the aggregate number of the members of such other labor organization) is fifty or more prior to July 1, 1968, or twenty-five or more on or after July 1, 1968, and such labor organization—

　(1)　is the certified representative of employees under the provisions of the National Labor Relations Act, as amended *[29 U.S.C. 151 et seq.]*, or the Railway Labor Act, as amended *[45 U.S.C. 151 et seq.]*; or

　(2)　although not certified, is a national or international labor organization or a local labor organization recognized or acting as the representative of employees of an employer or employers engaged in an industry affecting commerce; or

　(3)　has chartered a local labor organization or subsidiary body which is representing or actively seeking to represent employees of employers within the meaning of paragraph (1) or (2); or

　(4)　has been chartered by a labor organization representing or actively seeking to represent employees within the meaning of paragraph (1) or (2) as the local or subordinate body through which such employees may enjoy membership or become affiliated with such labor organization; or

　(5)　is a conference, general committee, joint or system board, or joint council subordinate to a national or international labor organization, which includes a labor organization engaged in an industry affecting commerce within the meaning of any of the preceding paragraphs of this subsection.

(f) The term "employee" means an individual employed by any employer except that the term "employee" shall not include any person elected to public office in any State or political subdivision of any State by the qualified voters thereof, or any person chosen by such officer to be on such officer's personal staff, or an appointee on the policymaking level or an immediate adviser with respect to the exercise of the constitutional or legal powers of the office. The exemption set forth in the preceding sentence shall not include employees subject to the civil service laws of a State government, governmental agency, or political subdivision. The term "employee" includes any individual who is a citizen of the United States employed by an employer in a workplace in a foreign country.

[The exclusion from the term "employee" of any person chosen by an elected official "to be on such official's personal staff, or an appointee on the policymaking level or an immediate advisor with respect to the exercise of the constitutional or legal powers of the office," remains in section 11(f). However, the Civil Rights Act of 1991 now provides special procedures for such persons who feel they are victims of age and other types of discrimination prohibited by EEOC enforced statutes. See section 321 of the Civil Rights Act of 1991.]

(g) The term "commerce" means trade, traffic, commerce, transportation, transmission, or communication among the several States; or between a State and any place outside thereof; or within the District of Columbia, or a possession of the United States; or between points in the same State but through a point outside thereof.

(h) The term "industry affecting commerce" means any activity, business, or industry in commerce or in which a labor dispute would hinder or obstruct commerce or the free flow of commerce and includes any activity or industry "affecting commerce" within the meaning of the Labor-Management Reporting and Disclosure Act of 1959 *[29 U.S.C. 401 et seq.]*.

(i) The term "State" includes a State of the United States, the District of Columbia, Puerto Rico, the Virgin Islands, American Samoa, Guam, Wake Island, the Canal Zone, and Outer Continental Shelf lands defined in the Outer Continental Shelf Lands Act *[43 U.S.C. 1331 et seq.]*.

(j) The term "firefighter" means an employee, the duties of whose position are primarily to perform work directly connected with the control and extinguishment of fires or the maintenance and use of firefighting apparatus and equipment, including an employee engaged in this activity who is transferred to a supervisory or administrative position.

(k) The term "law enforcement officer" means an employee, the duties of whose position are primarily the investigation, apprehension, or detention of individuals suspected or convicted of offenses against the criminal laws of a State, including an employee engaged in this activity who is transferred to a supervisory or administrative position. For the purpose of this subsection, "detention" includes the duties of employees assigned to guard individuals incarcerated in any penal institution.

(l) The term "compensation, terms, conditions, or privileges of employment" encompasses all employee benefits, including such benefits provided pursuant to a bona fide employee benefit plan.

AGE LIMITS

SEC. 631. [Section 12]

(a) Individuals of at least 40 years of age
   The prohibitions in this chapter shall be limited to individuals who are at least 40 years of age.
(b) Employees or applicants for employment in Federal Government
   In the case of any personnel action affecting employees or applicants for employment which is subject to the provisions of section 633a of this title *[section 15]*, the prohibitions established in section 633a of this title *[section 15]* shall be limited to individuals who are at least 40 years of age.
(c) Bona fide executives or high policymakers
   (1) Nothing in this chapter shall be construed to prohibit compulsory retirement of any employee who has attained 65 years of age and who, for the 2-year period immediately before retirement, is employed in a bona fide executive or a high policymaking position, if such employee is entitled to an immediate nonforfeitable annual retirement benefit from a pension, profit-sharing, savings, or deferred compensation plan, or any combination of such plans, of the employer of such employee, which equals, in the aggregate, at least $44,000.
   (2) In applying the retirement benefit test of paragraph (1) of this subsection, if any such retirement benefit is in a form other than a straight life annuity (with no ancillary benefits), or if employees contribute to any such plan or make rollover contributions, such benefit shall be adjusted in accordance with regulations prescribed by the Equal Employment Opportunity Commission, after consultation with the Secretary of the Treasury, so that the benefit is the equivalent of a straight life annuity (with no ancillary benefits) under a plan to which employees do not contribute and under which no rollover contributions are made.

ANNUAL REPORT

SEC. 632. *[Section 13]*
   [Repealed]

FEDERAL–STATE RELATIONSHIP

SEC. 633. *[Section 14]*

(a) Federal action superseding State action
   Nothing in this chapter shall affect the jurisdiction of any agency of any State performing like functions with regard to discriminatory employment practices on account of age except that upon commencement of action under this chapter such action shall supersede any State action.

(b) Limitation of Federal action upon commencement of State proceedings
In the case of an alleged unlawful practice occurring in a State which has
a law prohibiting discrimination in employment because of age and
establishing or authorizing a State authority to grant or seek relief from
such discriminatory practice, no suit may be brought under section 626
of this title *[section 7]* before the expiration of sixty days after proceedings have been commenced under the State law, unless such proceedings have been earlier terminated: *Provided*, That such sixty-day period
shall be extended to one hundred and twenty days during the first year
after the effective date of such State law. If any requirement for the
commencement of such proceedings is imposed by a State authority
other than a requirement of the filing of a written and signed statement
of the facts upon which the proceeding is based, the proceeding shall
be deemed to have been commenced for the purposes of this subsection
at the time such statement is sent by registered mail to the appropriate
State authority.

NONDISCRIMINATION ON ACCOUNT OF AGE IN FEDERAL GOVERNMENT EMPLOYMENT

SEC. 633a. [Section 15]

(a) Federal agencies affected
All personnel actions affecting employees or applicants for employment who
are at least 40 years of age (except personnel actions with regard to aliens
employed outside the limits of the United States) in military departments
as defined in section 102 of Title 5 *[5 U.S.C. § 102]*, in executive agencies
as defined in section 105 of Title 5 *[5 U.S.C. § 105]* (including employees and applicants for employment who are paid from nonappropriated
funds), in the United States Postal Service and the Postal Regulatory
Commission, in those units in the government of the District of Columbia
having positions in the competitive service, and in those units of the
judicial branch of the Federal Government having positions in the competitive service, in the Smithsonian Institution, and in the Government
Printing Office, the Government Accountability Office, and the Library
of Congress shall be made free from any discrimination based on age.

(b) Enforcement by Equal Employment Opportunity Commission and by
Librarian of Congress in the Library of Congress; remedies; rules, regulations,
orders, and instructions of Commission: compliance by Federal agencies;
powers and duties of Commission; notification of final action on complaint
of discrimination; exemptions: bona fide occupational qualification
Except as otherwise provided in this subsection, the Equal Employment
Opportunity Commission is authorized to enforce the provisions of
subsection (a) of this section through appropriate remedies, including reinstatement or hiring of employees with or without backpay, as

will effectuate the policies of this section. The Equal Employment Opportunity Commission shall issue such rules, regulations, orders, and instructions as it deems necessary and appropriate to carry out its responsibilities under this section. The Equal Employment Opportunity Commission shall—

(1) be responsible for the review and evaluation of the operation of all agency programs designed to carry out the policy of this section, periodically obtaining and publishing (on at least a semiannual basis) progress reports from each department, agency, or unit referred to in subsection (a) of this section;

(2) consult with and solicit the recommendations of interested individuals, groups, and organizations relating to nondiscrimination in employment on account of age; and

(3) provide for the acceptance and processing of complaints of discrimination in Federal employment on account of age.

The head of each such department, agency, or unit shall comply with such rules, regulations, orders, and instructions of the Equal Employment Opportunity Commission which shall include a provision that an employee or applicant for employment shall be notified of any final action taken on any complaint of discrimination filed by him thereunder. Reasonable exemptions to the provisions of this section may be established by the Commission but only when the Commission has established a maximum age requirement on the basis of a determination that age is a bona fide occupational qualification necessary to the performance of the duties of the position. With respect to employment in the Library of Congress, authorities granted in this subsection to the Equal Employment Opportunity Commission shall be exercised by the Librarian of Congress.

(c) Civil actions; jurisdiction; relief

Any person aggrieved may bring a civil action in any Federal district court of competent jurisdiction for such legal or equitable relief as will effectuate the purposes of this chapter.

(d) Notice to Commission; time of notice; Commission notification of prospective defendants; Commission elimination of unlawful practices

When the individual has not filed a complaint concerning age discrimination with the Commission, no civil action may be commenced by any individual under this section until the individual has given the Commission not less than thirty days' notice of an intent to file such action. Such notice shall be filed within one hundred and eighty days after the alleged unlawful practice occurred. Upon receiving a notice of intent to sue, the Commission shall promptly notify all persons named therein as prospective defendants in the action and take any appropriate action to assure the elimination of any unlawful practice.

(e) Duty of Government agency or official

Nothing contained in this section shall relieve any Government agency or official of the responsibility to assure nondiscrimination on account of age in employment as required under any provision of Federal law.

(f) Applicability of statutory provisions to personnel action of Federal departments, etc.

Any personnel action of any department, agency, or other entity referred to in subsection (a) of this section shall not be subject to, or affected by, any provision of this chapter, other than the provisions of sections 7(d)(3) and 631(b) of this title *[section 12(b)]* and the provisions of this section.

(g) Study and report to President and Congress by Equal Employment Opportunity Commission; scope

(1) The Equal Employment Opportunity Commission shall undertake a study relating to the effects of the amendments made to this section by the Age Discrimination in Employment Act Amendments of 1978, and the effects of section 631(b) of this title *[section 12(b)]*.

(2) The Equal Employment Opportunity Commission shall transmit a report to the President and to the Congress containing the findings of the Commission resulting from the study of the Commission under paragraph (1) of this subsection. Such report shall be transmitted no later than January 1, 1980.

EFFECTIVE DATE (SECTION 16 OF THE ADEA—NOT REPRODUCED IN THE U.S. CODE)

This Act shall become effective one hundred and eighty days after enactment, except (a) that the Secretary of Labor may extend the delay in effective date of any provision of this Act up to an additional ninety days thereafter if he finds that such time is necessary in permitting adjustments to the provisions hereof, and (b) that on or after the date of enactment the EEOC [originally, the Secretary of Labor] is authorized to issue such rules and regulations as may be necessary to carry out its provisions.

AUTHORIZATION OF APPROPRIATIONS

SEC. 634. (Section 17)

There are hereby authorized to be appropriated such sums as may be necessary to carry out this chapter

Index